Wild Plant Species as Potential Horticultural Crops: An Opportunity for Farmers and Consumers—2nd Edition

Wild Plant Species as Potential Horticultural Crops: An Opportunity for Farmers and Consumers—2nd Edition

Guest Editors

Roberta Bulgari
Ada Baldi
Antonios Chrysargyris

MDPI

Basel • Beijing • Wuhan • Barcelona • Belgrade • Novi Sad • Cluj • Manchester

Guest Editors
Roberta Bulgari
Department of Agricultural and Environmental Sciences
University of Milan
Milan
Italy

Ada Baldi
Dipartimento di Scienze e Tecnologie Agrarie, Alimentari Ambientali e Forestali
Università degli Studi di Firenze
Firenze
Italy

Antonios Chrysargyris
Department of Agricultural Sciences, Biotechnology and Food Science
Cyprus University of Technology
Limassol
Cyprus

Editorial Office
MDPI AG
Grosspeteranlage 5
4052 Basel, Switzerland

This is a reprint of the Special Issue, published open access by the journal *Horticulturae* (ISSN 2311-7524), freely accessible at: https://www.mdpi.com/journal/horticulturae/special_issues/QN1ZTT7M16.

For citation purposes, cite each article independently as indicated on the article page online and as indicated below:

Lastname, A.A.; Lastname, B.B. Article Title. *Journal Name* **Year**, *Volume Number*, Page Range.

ISBN 978-3-7258-8193-2 (Hbk)
ISBN 978-3-7258-8194-9 (PDF)
https://doi.org/10.3390/books978-3-7258-8194-9

Cover image courtesy of Antonios Chrysargyris

Contents

About the Editors

Roberta Bulgari

Roberta Bulgari is a researcher at the Department of Agricultural and Environmental Sciences (DiSAA), University of Milan, working on vegetable production with a focus on abiotic stress, biostimulants, soilless cultivation, vegetable quality, microgreens, ornamentals, and postharvest quality. Her research also explores sustainable strategies to improve crop performance and resilience.

Ada Baldi

Ada Baldi is a researcher at the Department of Agriculture, Food, Environment and Forestry (DAGRI), University of Florence, working on wild and underutilized plant species, sustainable horticultural production systems, crop quality, soilless cultivation, vertical farming, and urban agriculture. Her research also explores innovative strategies for the domestication, cultivation, and valorization of emerging crops.

Antonios Chrysargyris

Antonios Chrysargyris is a Special Teaching Staff at Cyprus University of Technology, at the Department of Agricultural Sciences, Biotechnology and Food Science. His main research work is focused on the introduction of wild and unexplored plant species (ornamental, medicinal/aromatic, edible species) to cultivation schemes (extensive and intensive-hydroponics), identification of their mineral needs and nutritional value, and evaluation of the biological activities of their extracts and essential oils.

Preface

This Reprint presents contributions on wild and underutilized plant species as emerging horticultural crops. It aims to highlight their potential to support more sustainable production systems in the face of environmental and climate challenges. It is addressed to researchers, students, and professionals in horticulture, agronomy, plant science and related fields.

Roberta Bulgari, Ada Baldi, and Antonios Chrysargyris
Guest Editors

horticulturae

MDPI

Editorial

Wild Plant Species as Potential Horticultural Crops: An Opportunity for Farmers and Consumers, 2nd Edition

Ada Baldi [1], Antonios Chrysargyris [2,*] and Roberta Bulgari [3]

[1] Department of Agriculture, Food, Environment and Forestry (DAGRI), University of Florence, Piazzale delle Cascine 18, 50144 Firenze, Italy; ada.baldi@unifi.it
[2] Department of Agricultural Sciences, Biotechnology and Food Science, Cyprus University of Technology, Limassol 3603, Cyprus
[3] Department of Agricultural and Environmental Sciences-Production, Landscape, Agroenergy (DiSAA), University of Milan, Via Celoria 2, 20133 Milan, Italy; roberta.bulgari@unimi.it
* Correspondence: a.chrysargyris@cut.ac.cy

Increasing pressure on global food systems, driven by climate change, biodiversity loss, and rising demand for nutritious foods, has intensified interest in alternative, climate-resilient, and underutilized crop resources. In this context, wild plant species are the subject of renewed scientific and agronomic attention for their potential to diversify horticultural production, enhance sustainability, and provide novel economic opportunities for farmers while satisfying evolving consumer preferences. Traditionally gathered from natural ecosystems, many wild edible and medicinal plants possess valuable nutritional and phytochemical properties, as well as adaptive traits that make them promising candidates for sustainable horticulture. Their domestication and integration into modern horticultural systems create new opportunities for farmers to diversify production and income sources while enabling consumers to access novel, health-promoting products aligned with local food traditions and environmentally responsible diets. Moreover, their cultivation may reduce harvesting pressure on wild populations and support biodiversity conservation efforts. In line with this perspective, recent research has progressively moved beyond descriptive ethnobotanical studies toward tackling the practical challenges of domestication and commercialization, with increasing consideration of cultivation feasibility, agronomic management, product quality, and the functional traits of wild species.

Building on the framework proposed by Chrysargyris et al. [1], the contributions to this second edition demonstrate how wild and underutilized species can move from conceptual opportunities to applied horticultural innovations while preserving their functional and cultural value.

Several contributions in this Special Issue highlight the potential of wild leafy vegetables as alternative crops for diversified and climate-smart horticulture, with promising applications in fresh markets and functional food development. Botella et al. [2] demonstrated the cultivation potential and nutritional profile of *Sanguisorba verrucosa* (Link ex G.Don) Ces., *Eruca vesicaria* (L.) Cav., and *Scorzonera laciniata* L. Sensory acceptance was good for all three species, and all exhibited high concentrations of bioactive compounds, including phenolics, organic acids, vitamin C, and β-carotene, often exceeding levels found in many commonly consumed vegetables. Although domestication resulted in a partial reduction in certain phytochemicals, the overall nutritional quality remained significant, supporting their introduction into horticultural production systems. Complementarily, Paschoalinotto et al. [3] investigated the effects of different water regimes (rain-fed, deficit, and full supply) and legume-based crop rotation on the growth performance and biochemical profile of *Cichorium spinosum* L., a species strongly linked to Mediterranean culinary

 https://doi.org/10.3390/horticulturae12030346

heritage. The results demonstrated that deficit irrigation combined with crop rotation enhances fatty-acid quality, phenolic content, and antioxidant activity while maintaining yield performance. These findings highlight the potential of integrating sustainable agronomic practices to valorize wild edible species within a climate-resilient and resource-efficient horticultural system. The integration of wild taxa into intensive horticultural systems was further explored by Ntanasi et al. [4], who evaluated the performance of *Portulaca oleracea* L., *Mesembryanthemum crystallinum* L., and *Valerianella locusta* L. under hydroponic cultivation. By balancing the NH_4-N/total-N ratio and applying protein hydrolysates and seaweed extracts, the authors demonstrated species-specific responses in yield, nitrogen assimilation, micronutrient uptake, and nitrate accumulation. Notably, tailored nitrogen management and biostimulant application improved product quality and reduced leaf nitrate levels without compromising yield. These findings are particularly relevant for commercial soilless systems, for which food safety and production efficiency are essential requirements.

Several contributions in this Special Issue also focus on aromatic and medicinal plants, emphasizing their importance for both horticultural production and high-value markets. Ložiene [5], by analyzing 124 *Thymus pulegioides* L. plants belonging to distinct chemotypes and sexual morphs sampled from 23 natural populations, identified peltate glandular trichome density and diameter in the upper leaf epidermis as reliable anatomical markers for predicting essential oil content. The author concluded that these traits may serve as cost-effective screening tools to support the selection and domestication of essential oil-rich genotypes, although further investigations are required before extending this strategy to other species. The study by Tzortzakis et al. [6] investigated the effects of organic and conventional fertilizers, combined with foliar applications of iron (Fe) and zinc (Zn), on the development and quality of two Mediterranean aromatic species of emerging interest. The findings clearly demonstrate species-specific nutritional responses. *Sideritis cypria* Post maintained comparable yields under both fertilization systems, highlighting its suitability for low-input cultivation systems, whereas *Origanum dubium* L. showed a significantly higher biomass under conventional fertilization. Foliar Fe and Zn applications differentially affected phenolic content, antioxidant activity, mineral uptake, and stress-related parameters, with responses strongly dependent on the fertilization regime. Zinc in particular enhanced water use efficiency under certain conditions, while conventional fertilization improved nutrient use efficiency. Overall, their study demonstrates that the integration of fertilization regimes with targeted micronutrient applications can substantially influence both productivity and phytochemical quality in aromatic crops. These findings underscore the importance of species-specific nutrient management, especially within sustainable and organic cultivation systems.

In addition, woody and perennial wild species are emerging as promising candidates for crop diversification, contributing to improved food security and enhanced resilience of food systems under environmental change. Bai et al. [7] evaluated the physical and chemical characteristics of *Terminalia catappa* L., a traditional Pacific Forest nut, across the Solomon Islands, Vanuatu, and Fiji to assess their potential for commercialization. Significant geographic variation in kernel size and kernel-to-fruit ratios, key attributes for economic viability, has been observed. Across all regions, kernels contained higher proportions of unsaturated than saturated fatty acids, particularly oleic and linoleic acids, indicating potential health benefits. However, an accelerated aging test showed rapid oil oxidation, suggesting potential shelf-life limitations under high-temperature storage. Overall, the results highlight the Solomon Islands populations as promising candidates for selection and breeding and indicate *T. catappa* as a nutritionally valuable indigenous tree species with potential to enhance food and nutrition security and rural livelihoods. Similarly, Shahbazian et al. observed morphological and biochemical intraspecific diversity among eight wild black myrtle (*Myrtus communis* L.) populations in southwestern Iran [8].

 https://doi.org/10.3390/horticulturae12030346

Fruit size and seed traits varied significantly among populations. All populations showed high phenolic content, elevated antioxidant capacity, and strong radical scavenging activity, with these patterns strongly influenced by altitude and soil characteristics. Together, these studies demonstrate that natural genetic diversity in forest and wild fruit species represents a strategic resource for horticulture.

Beyond primary production, this collection also addresses value-chain development and product differentiation. Ma et al. [9] compared the volatile and non-volatile profiles of Kujingcha, a traditional Chinese substitute tea obtained from the fresh leaves of *Ilex dabieshanensis* K. Yao and M. B. Deng, with those of green tea (*Camellia sinensis* L.). Although processed using green tea techniques, Kujingcha differs substantially in both flavor profile and bioactive composition. Using an integrated metabolomic approach combining HPLC, UPLC-QQQ-MS, and HS-SPME-GC-MS, the study identified high flavonoid and water-soluble sugar contents, together with low levels of amino-acids, as key determinants of its intense bitter taste and sweet aftertaste, while heterocyclic compounds contribute to its characteristic bean-like aroma. This research not only strengthens knowledge of the distinctive biochemical profile and health-related potential of Kujingcha but also supports its further valorization as a cultivated alternative tea product with a distinct geographical and cultural identity.

In addition to these applied studies, two comprehensive reviews further expand the perspective on Mediterranean wild edible Asteraceae. Mandriota and Renna [10] critically synthesized current knowledge on *Urospermum dalechampii* (L.) Scop. ex F.W. Schmidt, highlighting its nutritional composition, phytochemical profile, biological activities, and agronomic potential and identifying research gaps due to the lack of standardized cultivation and post-harvest protocols and limited understanding of consumer acceptance. Similarly, Ruiz-Rocamora et al. [11] provided an extensive overview of *Sonchus* species. By integrating taxonomy, geographical distribution, traditional knowledge, and bioactive composition, their review underscores the agronomic feasibility and nutraceutical relevance of these taxa while emphasizing the need for coordinated conservation and development strategies.

Collectively, the contributions included in this Special Issue demonstrate that the transition from wild harvesting to structured cultivation is technically feasible when supported by genotype selection, adaptive agronomic practices, and quality-oriented management. However, important knowledge gaps persist, including limited long-term assessments of domestication effects, insufficient understanding of genotype × environment interactions, and a lack of standardized cultivation protocols. Furthermore, post-harvest performance, shelf life, food safety, and consumer acceptance also remain underexplored for many of these emerging crops. Addressing these gaps will be essential to ensuring that wild plant species can be reliably integrated into commercial horticulture. Future research should adopt integrated and multidisciplinary approaches that combine agronomy, breeding, food science, and sustainability assessment. Particular emphasis should be placed on low-input and organic cultivation systems, conservation of genetic diversity, and the development of resilient value chains that align ecological sustainability with cultural heritage. In this way, wild and underutilized plant resources can move beyond their marginal status to serve as key components of a diversified, resilient, and quality-oriented production system.

Conflicts of Interest: The authors declare no conflicts of interest.

References

1. Chrysargyris, A.; Baldi, A.; Lenzi, A.; Bulgari, R. Wild plant species as potential horticultural crops: An opportunity for farmers and consumers. *Horticulturae* **2023**, *9*, 1193. [CrossRef]
2. Botella, M.Á.; Hellín, P.; Hernández, V.; Dabauza, M.; Robledo, A.; Sánchez, A.; Fenoll, J.; Flores, P. Phytochemical composition, sensory acceptance, and cultivation potential of *Sanguisorba verrucosa*, *Eruca vesicaria*, and *Scorzonera laciniata*. *Horticulturae* **2025**, *11*, 1021. [CrossRef]

 https://doi.org/10.3390/horticulturae12030346

3. Paschoalinotto, B.H.; Polyzos, N.; Liava, V.; Mandim, F.; Pires, T.C.S.P.; Añibarro-Ortega, M.; Ferreira, I.C.F.R.; Dias, M.I.; Barros, L.; Petropoulos, S.A. The effect of cropping system and irrigation regime on the plant growth and biochemical profile of *Cichorium spinosum*. *Horticulturae* **2025**, *11*, 306. [CrossRef]
4. Ntanasi, T.; Karavidas, I.; Giannothanasis, E.; Spyrou, G.P.; Karaviti, T.; Napoli, S.; Neocleous, D.; Ntatsi, G. enhancing soilless production of *Portulaca oleracea*, *Mesembryanthemum crystallinum* and *Valerianella locusta* through nitrogen form ratio optimization and biostimulant application. *Horticulturae* **2025**, *11*, 1076. [CrossRef]
5. Ložienė, K. Peltate glandular trichomes in relation to their parameters, essential oil amount, chemotype, plant sex and habitat characteristics in *Thymus pulegioides*. *Horticulturae* **2025**, *11*, 871. [CrossRef]
6. Tzortzakis, N.; Hajisolomou, E.; Zavarelli, N.; Chrysargyris, A. Foliar Application of iron and zinc affected aromatic plants grown under conventional and organic agriculture differently. *Horticulturae* **2025**, *11*, 967. [CrossRef]
7. Bai, S.B.; Randall, B.; Gama, R.; Gua, B.; Keli, D.; Jones, K.; Elliott, B.; Wallace, H. Variations in physical and chemical characteristics of *Terminalia catappa* nuts. *Horticulturae* **2025**, *11*, 540. [CrossRef]
8. Shahbazian, D.; Karami, A.; Gruda, N.S. Biochemical and morphological traits of wild myrtle populations for horticultural use. *Horticulturae* **2025**, *11*, 233. [CrossRef]
9. Mal, L.; Peng, Y.; Cao, D.; Fan, P.; Wang, L.; Feng, G.; Lei, A.; Hu, B.; Liu, Y.; Jin, X. Difference analysis of non-volatile and volatile components in Kujingcha (*Ilex dabieshanensis*) compared with green tea (*Camellia sinensis*). *Horticulturae* **2025**, *11*, 804. [CrossRef]
10. Mandriota, L.; Renna, M. *Urospermum dalechampii* (Asteraceae): A neglected mediterranean wild edible plant with nutritional, phytochemical and agronomic potential. *Horticulturae* **2026**, *12*, 186. [CrossRef]
11. Ruiz-Rocamora, A.; Obón, C.; Ríos, S.; Alcara, F.; Rivera, D. *Sonchus* Species of the Mediterranean Region: From wild food to horticultural innovation—Exploring taxonomy, cultivation, and health benefits. *Horticulturae* **2025**, *11*, 893. [CrossRef]

 https://doi.org/10.3390/horticulturae12030346

horticulturae

MDPI

Review

Urospermum dalechampii (Asteraceae): A Neglected Mediterranean Wild Edible Plant with Nutritional, Phytochemical and Agronomic Potential

Luigia Mandriota and Massimiliano Renna *

Department of Soil, Plant and Food Sciences, University of Bari Aldo Moro, Via Amendola 165/A, 70126 Bari, Italy; luigia.mandriota@uniba.it

* Correspondence: massimiliano.renna@uniba.it

Abstract

Wild edible plants represent an underexploited resource for enhancing agrobiodiversity and promoting more sustainable food systems. *Urospermum dalechampii* (Asteraceae) is a traditional Mediterranean wild edible species widely used in local cuisines, yet it remains marginal in modern horticulture. This review provides an overview of the current knowledge on the nutritional composition, phytochemical profile, biological activities, and agronomic potential of *U. dalechampii*. Available evidence indicates that its edible tissues are characterized by a favorable nutritional profile, including appreciable levels of dietary fiber, minerals, and beneficial fatty acids. Moreover, the species is a rich source of bioactive compounds—particularly phenolic acids, flavonoids, and terpenoids—which have been associated with antioxidant, anti-inflammatory, and antimicrobial activities. Recent studies have demonstrated the feasibility of domesticating *U. dalechampii* through both soil-based and soilless cultivation systems, including baby-leaf and microgreen production, thereby enabling controlled yields and reducing harvesting pressure on wild populations. Nevertheless, research on this species remains fragmented, and several knowledge gaps persist, notably the absence of standardized agronomic protocols, limited information on post-harvest management and safety, and potential market constraints related to bitterness and variability in product quality. Overall, *U. dalechampii* emerges as a promising candidate for sustainable horticulture and the development of functional foods.

Keywords: agrobiodiversity conservation; climate-smart agriculture; edible plant biodiversity; domestication; ethnobotanical knowledge; food system resilience; functional food resource; smooth golden fleece; traditional use; underutilized crop species

1. Introduction

The concept of biodiversity, developed by entomologist Edward O. Wilson (1986), refers to the overall variety of living forms. Agrobiodiversity includes the stock of plant, animal and microbial genetic resources, specific to agricultural systems (agroecosystems). The conservation of and enhancement in agrobiodiversity are necessary for sustainability and food security, as well as to ensure the resilience of agroecosystems and adaptation to climate change [1,2]. Unfortunately, the increase in the cultivation and consumption of high-yield commercial varieties lead to a progressive genetic erosion of crops and a limitation of the plant diversity available for human consumption [3]. Therefore, spontaneous edible species, or wild edible plants (WEPs), can be understood as resources to ensure healthy and eco-sustainable nutrition, as well as the cultural identity of local communities; furthermore, from an ecological and nutritional point of view, WEPs take on a crucial role in the diversification of food sources and in the conservation of agricultural genetic biodiversity [4]. Spontaneous plants, particularly those typical of the Mediterranean region, have for centuries been an integral part of the traditional diet and medicine [5]. For example, in southern Italy, specifically in the Apulia region, about 2500 wild herbaceous species have been identified, and, among these, more than 500 are assimilated into local gastronomic dishes [6–8]. The Mediterranean basin, with its climate, geography and soil quality, has allowed for the development of abundant flora, including multiple wild horticultural species, which have adapted to hostile environmental conditions, such as the scarcity of water resources [9]. In periods of war or famine, Mediterranean WEPs represented a primary food resource [10]. However, they take on a peculiar role in the Mediterranean diet, owing to the healthy compounds they contain [6,9]. For example, many wild edible greens are not sufficiently rich in fat to be considered a good source of fatty acids. However, they have a healthy fat profile due to their proportions of n-6 and n-3 fatty acids [1].

In this regard, the scientific literature demonstrates that species belonging to the Asteraceae (Bercht. & J.Presl), in particular, contain significant quantities of vitamins, fiber, minerals, polyunsaturated fatty acids and phenolic compounds [11,12]. The Asteraceae, in fact, constitutes one of the most diverse and significant botanical families in the nutritional and medicinal fields; it includes over 1600 genera and approximately 25,000 species [4]. Many species belonging to this family, such as *Sìlybum marianum* (L.) Gaertn., *Cynara cardunculus* L. and *Sonchus oleraceus* L. are traditionally adopted for their therapeutic properties and in territorial gastronomy [13,14].

Among the species belonging to the Asteraceae, the genus *Urospermum* Scop. stands out for its evolutionary and applied value [15]. It includes two closely related species: *Urospermum dalechampii* (L.) Scop. ex F.W. Schmidt ("smooth golden fleece") and *Urospermum picroides* (L.) Scop. ex F.W. Schmidt ("prickly golden fleece") [16]. Both are widespread in the Mediterranean and Macaronesia, and exhibit morphological and genomic differences associated with environmental adaptation phenomena and episodes of homoploid hybridization [17–20].

Urospermum dalechampii is a species known for its traditional uses as food, a tonic, and a purifying and diuretic plant [21]. It is frequently collected in rural landscapes of southern Italy and Tunisia, where its leaves and inflorescences are consumed raw or cooked, often in association with other spontaneous horticultural species [22]. Its nutritional qualities have recently been confirmed: the aerial and root parts contain high percentages of carbohydrates (46.6–78.0%), fiber (10.4–24.6%), proteins (5.4–13.5%) and a mineral profile dominated by potassium, calcium and iron. The low oil content (1.0–1.5%), is characterized by a high percentage of unsaturated fatty acids, with a favorable UFA/SFA profile and excellent nutritional indices [21,23].

Regarding the phytochemical characteristics, *U. dalechampii* contains significant concentrations of phenolic and flavonoid compounds (quercetin, luteolin, naringenin) [24], as well as sesquiterpene lactones, which contribute to antioxidant and anti-inflammatory activity. Its use in traditional Mediterranean medicine is supported by the remarkable analgesic, anti-inflammatory and antioxidant effects found in the aerial parts of the plant and in related methanolic extracts [25,26].

Likewise, *U. picroides* has aroused scientific curiosity for its capabilities as a functional food. In fact, the most recent studies have recognized 77 metabolites in vitro, including the 12 new sesquiterpene-amino acid conjugates, which give the species anti-inflammatory activity [24]. As an affirmation of the function of WEPs in Mediterranean gastronomic

https://doi.org/10.3390/horticulturae12020186

culture, especially in Egypt and Southern Italy, *U. picroides* and *U. dalechampii* are used in traditional dishes such as "Chòrta Vramena" and "mixed leaves" [26].

In addition to its nutritional and pharmacological qualities, *U. dalechampii* exhibits significant allelopathic activity, acting as a natural bioherbicide: the germination of *Lactuca sativa* can be inhibited by leaf and root extracts. This property enhances its relevance in sustainable agriculture, which is founded on ecological weed management strategies and the valorization of spontaneous plant resources [12].

Urospermum dalechampii, in all its peculiarities, represents the convergence of agricultural biodiversity, ethnobotanical value and nutraceutical potential. Its recovery and scientific study contribute to the conservation of the Mediterranean WEPs and the diversification of horticultural species, providing new production possibilities for agri-food supply chains, based on the use of local varieties adapted to climate change [27].

In recent years, Renna et al. have focused extensively on the valorization and domestication of various spontaneous Mediterranean species, primarily used in healthy diet rich in nutritious elements, such as *Heracleum sphondylium* L. and *Crithmum maritimum* L. [28–31].

In the age of climate and food crises, integrating spontaneous species, such as *U. dalechampii* and *U. picroides,* into current agricultural systems and dietary regimes becomes a strong strategy to combine sustainability, innovation and protection of the Mediterranean ethnobotanical heritage [19,23]. However, this species remains the least studied among the Asteraceae family.

In this context, the available knowledge on *U. dalechampii* remains fragmented across botanical descriptions, phytochemical investigations, ethnobotanical reports and occasional nutritional studies. What is currently lacking is an integrated synthesis that combines these perspectives and provides a comprehensive overview of the agronomic, gastronomic and potential nutraceutical value of the species. Therefore, this manuscript presents a narrative review aimed at collecting, organizing and critically discussing the existing literature on *U. dalechampii*. Specifically, this review intends to (i) describe the morphological, biological and ecological features of the species within Mediterranean agroecosystems; (ii) summarize traditional harvesting practices and culinary uses, with particular attention to their cultural significance in local communities; (iii) analyze the available evidence concerning the nutritional profile and bioactive compounds, highlighting the species' potential nutraceutical relevance; and (iv) explore possible agronomic applications, including prospects for crop valorization and integration into sustainable farming systems.

Taken together, these elements are intended to provide an integrated and up-to-date overview of the current state of knowledge while emphasizing both the opportunities and the research gaps that emerge in view of a future valorization of *U. dalechampii* as an agricultural, gastronomic and cultural resource.

2. Materials and Methods

This work intends to provide an interdisciplinary and international overview of *Urospermum dalechampii* (L.) F.W. Schmidt, as no previous reviews exclusively dedicated to this species have been identified in the literature.

Initially, bibliographic databases and main websites were selected, and target research questions and relevant search terms were defined. Subsequently, operational criteria were adopted for the selection of sources considered relevant. In a third step, precise methodological criteria for this review were implemented; finally, the information and data retrieved were examined and organized into the various chapters of this review.

 https://doi.org/10.3390/horticulturae12020186

2.1. Step 1: Selecting Research Questions, Databases, Websites, and Appropriate Search Terms

The research questions have taken on a rather broad approach, as no comprehensive scientific contributions relating to *U dalechampii* (L.) F.W. Schmidt have been found. Therefore, questions such as "How is it used?", "What are the salient features?", and "In which territory is it mostly found?" were asked.

A bibliometric analysis was conducted using the Scopus database to assess the evolution and development of research on *Urospermum dalechampii* to date, with the aim of evaluating the level of scientific interest in this species. This study concludes with an examination of the results of more than 30 recent research papers, highlighting the current knowledge on this still-understudied plant with significant phytochemical and biological properties. Scopus analytical tools and Boolean operators were used to review and analyze the literature from 1970 to 2026. The search, based on the descriptor "Urospermum AND dalechampii" within the title, abstract, and keyword fields, retrieved a total of 12 articles (Figure 1).

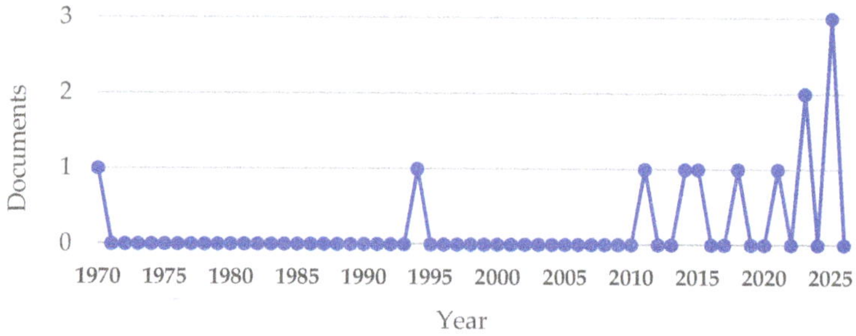

Figure 1. Number of articles on *U. dalechampii* published from 1970 to 2025 (data retrieved from Scopus® database).

Considering the limited number of scientific articles retrieved using these criteria in the Scopus database (Figure 1), the literature search was extended to additional sources. This is in accordance with the recommendations of Tranfield et al. [32], who suggested that bibliographic surveys should not be limited exclusively to academic databases. In fact, this review was conducted through a systematic analysis of the scientific literature and grey literature published from the 1970s to the present. The literature search was carried out using the scientific name *Urospermum dalechampii* (L.) F.W. Schmidt, and querying major international databases, including Scopus, Web of Science, and Google Scholar. Contributions of various types (scientific articles, book chapters, conference proceedings, theses, reports, unpublished studies, doctoral theses, internal company reports, and university handouts) were included, without limitations related to the language of publication, geographic area, author affiliation, or subject area. The initial survey produced approximately 89 contributions, which were subsequently subjected to a critical content evaluation. Works in which the species was cited only in the bibliography or mentioned without providing relevant information from a botanical, ethnobotanical, ecological, nutritional, or pharmacological perspective were excluded. To ensure more up-to-date and reliable data, priority was given to publications after the 2000s. The application of the selection criteria led to the final inclusion of 41 sources. To complement the indexed scientific literature, 13 specialized and institutional websites were consulted, used to delve deeper into the food uses, ethnobotanical knowledge, and traditional uses of Urospermum dalechampii, as well as to find floristic and popular information relating to the distribution and territorial development of the species. To this end, non-strictly academic bibliographic research has been conducted using, in addition to the scientific name, several common and vulgar names attributed to the

 https://doi.org/10.3390/horticulturae12020186

species, including Big Boccione, Daléchamp's Lettuce, Dalechamp's Urospermo, Bunommo, Amarago, Bitter Ciconia, Dairy, and Scorzanera trifoliata [33]. Particular attention was paid to avoiding information overlapping with related species belonging to the Asteraceae Bercht family.

2.2. Step 2: Applying Practical Screening Criteria

The selection of materials involved scientific articles, monographs, book chapters, research reports, and conference proceedings, integrating both empirical contributions and works of a conceptual or theoretical nature. The present review aims to provide a comprehensive overview of *U. dalechampii*, and therefore any formal quality criteria, such as rankings or journal scores, were excluded.

2.3. Step 3: Applying Methodological Screening Criteria

The review protocol, developed for the systematic analysis of the content of the selected documents, was divided into three main sections. The first summarizes the essential bibliographic information of each source, i.e., the authors, title, year of publication, affiliation of the authors, type of contribution, and name of the journal, in the case of articles. The second section illustrates the methodology used, distinguishing between theoretical-conceptual, empirical, and applied studies. The last section examines the texts, i.e., the works relating to "biology and morphology", which were used for drafting the section describing the plant, while those concerning "use" were identified as relevant for the preparation of the chapter on food uses.

Finally, the stored data and information were structured and extensively discussed in the following sections, also incorporating the authors' thoughts on the prospects of the species.

3. Results and Discussion

3.1. Biology and Morphology

The main morphological, cytogenetic and phylogenetic traits are summarized in Table 1. *U. dalechampii* is a perennial species of the Asteraceae, widespread in the Mediterranean, with an erect stem, basal rosette of leaves and large yellow flower heads that flower in spring [23].

It is a perennial species widespread in Central Europe, North Africa and Western Asia, typical of uncultivated lands, meadows and fields: erect pubescent stems (10–60 cm), panduriform/oblong-lanceolate basal leaves, large flower heads with yellow flowers and external purple shades, tomentose lanceolate bract shells (Figure 2), and black achenes with rostrum and pappus [23,34].

In Apulia, it is documented as a traditional alimurgic species: perennial plant with pale yellow flower heads and large basal leaves with a bitter taste, an integral part of the regional ethnobotanical heritage [22]. As a Mediterranean WEP (smooth golden fleece), it has been historically consumed: leaves, buds and roots have been used, respectively, as substitutes for spinach, capers and potatoes. From a morphological perspective, it is characterized by large basal leaves, showy yellow-pale flower heads and a generally smoother habitus than that of *U. picroides* [35].

Morphologically, it has a robust taproot and an erect, hollow, velvety floral scape, sometimes exceeding 80 cm. The basal leaves have an opaque green color, reaching 20 cm in length, and vary from lanceolate to deeply runcinate or pinnatifid with wavy and toothed margins. The lower cauline leaves are progressively larger and less divided and the upper cauline leaves are lanceolate, subentire, auriculate and amplexicaul, sometimes in whorls of three. A dense pubescence (tomentosity) covers the leaves, stems and calyx,

 https://doi.org/10.3390/horticulturae12020186

giving the plants a velvety–bristly appearance. The upper cauline leaves, smaller and amplexicaul, have mostly entire margins. The solitary flower heads consist exclusively of hermaphroditic ligulate flowers and are borne on a long, hollow, and enlarged peduncle. Flowering occurs from March to August; in deep soils and during humid years, the peduncle may reach up to 6 cm in diameter. Each capitulum bears 7–8 uniseriate involucral bracts with reddish margins and contains more than 200 florets. The fruit is a rostrate black achene (1000 achenes weigh approximately 1.6 g) with a long beak surmounted by a dark, feathery pappus arranged in two series (dandelion type), which enables anemophilous dispersal over long distances. Biologically, the species is a scapose hemicryptophyte, i.e., a perennial plant with overwintering buds at ground level and an elongated floral axis. It as an erect, weakly branched stem (20–40 cm), pinnately divided petiolate basal leaves (up to 6 cm), opposite upper lanceolate leaves, and sulfur-yellow ligules (approximately 20 mm long) with central red streaks [36–42].

Figure 2. Plants of *U. dalechampii* on a plot of uncultivated land as typical example of natural habitat for this species. The image is original and was created by the authors.

U. dalechampii is distinguished by its morphological and genomic characters: it has large flower heads with soft, tomentose involucral bracts, in contrast to the spinulose bracts of *U. picroides*. Cytogenetically, the species is diploid, with a chromosome number of $2n = 14$ ($x = 7$), and possesses a relatively large genome (2.26 pg/2C) compared to U. picroides (1.93 pg/2C), of which approximately 74% consists of repetitive DNA, dominated by LTR Ty1/Copia retrotransposons. These traits allow for a clear distinction from *U. picroides* and have enabled the species to be identified as a maternal progenitor in homoploid hybridization events. The integration of morphological, cytogenetic, and genomic data confirms its autonomous evolutionary identity and its key role in the diversification of the genus *Urospermum*, providing new insights into adaptation and evolution within the Asteraceae [23,34].

In the study by Fernández-Mazuecos et al. [16], *U. dalechampii* was analyzed within a phylogenetic framework of the tribe Cichorieae (Asteraceae), based on nuclear ITS

https://doi.org/10.3390/horticulturae12020186

(Internal Transcribed Spacer) sequences and the plastid matK (maturase K) gene. The results showed that *U. dalechampii* and *U. picroides* form a monophyletic clade, closely related to *Avellara fistulosa* and *Prenanthes purpurea*, together constituting the so-called APU clade (Avellara–Prenanthes–Urospermum). A BEAST-dated analysis indicated a divergence between *Avellara* and *Urospermum* in the middle–late Miocene (15.58–8.62 Ma), followed by a subsequent divergence between *U. dalechampii* and *U. picroides* during the Late Miocene–Quaternary (7.00–2.34 Ma), highlighting a long evolutionary history and prolonged genetic isolation within the Mediterranean region.

Table 1. *U. dalechampii* main morphological, cytogenetic and phylogenetic traits.

References	Parameter	Observation
[23]	Life form	Perennial herb of Asteraceae with erect stem and basal rosette
[36,42]	Vegetative traits	Erect pubescent stems (10–60 cm), basal leaves 10–20 cm, tomentose and runcinate
[36,42]	Reproductive traits	Large yellow capitula (≈6 cm), ligules ≈ 20 mm; flowering March–August
[36,42]	Fruit	Black rostrate achenes (~1.6 g/1000) with feathery pappus for wind dispersal
[23,34]	Cytogenetics	Diploid (2n = 14); nuclear DNA = 2.26 pg/2C; ≈74% repetitive DNA (LTR Ty1/Copia)
[23,34]	Comparative genomics	Larger genome than *U. picroides* (1.93 pg); maternal progenitor in homoploid hybridization
[16]	Phylogeny	Forms monophyletic APU clade (Avellara–Prenanthes–Urospermum); Miocene divergence (15.6–2.3 Ma)

Note. Abbreviations and units—cm: centimeter; mm: millimeter; g: gram; pg: picogram; Ma: million years ago; 2C: diploid nuclear DNA content; LTR Ty1/Copia: Long Terminal Repeat Ty1/Copia retrotransposons.

3.2. Ethnobotanical Knowledge and Habitat

The genus *Urospermum* derives from the Greek ourá ("tail") and spérma ("seed"), in reference to the dark achenes with a long caudate appendage; the epithet "Dalechampii" honors Jacques Daléchamps (1513–1588), author of *Historia Generalis Plantarum* [36]. In addition to its botanical interest, *U. dalechampii* has ornamental value for its yellow flowers and emits white latex when cut. Ethnobotanical uses are summarized in Table 2.

In Apulia (Southern Italy), the species is rooted in popular and culinary tradition, especially in Manduria town, where it is called *"cequaire all'ammérse"* for its curved leaves (Figure 3); it also inspired photographic works by Stephen Sharnoff [36]. Since 2021, it has been included among the Traditional Apulian Agri-food Products (TAFP) [43]. It is visited by small Apoidea and, to a lesser extent, by bees, and serves as a natural reservoir of AILV in the Apulian artichoke. Basal leaves and buds are collected in late winter and spring. There are numerous regional vernacular names (e.g., *Daléchamps lettuce, Bitter Ciconia, Dairy, Amaragous, Scorzanera trifoliata, Beliommi/Bunommi*) [33,36,43–46]. Known as smooth golden fleece, it belongs to the Euro-Western Mediterranean and Steno-Mediterranean corotypes and is widespread in the Mediterranean basin, temperate Europe, North Africa, and Asia Minor [17,47]. It is common in Central–Southern Italy but absent in Alpine regions [34,38,48].

https://doi.org/10.3390/horticulturae12020186

Figure 3. Appearance of a plant of *U. dalechampii* just harvested, with its recurved leaves, from which the dialectal term "cequaire all'ammérse" derives, meaning an edible wild plant with its leaves turned upside down. The image is original and was created by the authors.

A strictly Mediterranean species, it tolerates calcareous and poor soils and contributes to xerophilous ecosystems. It prefers sunny, well-drained environments, from 0 to 1200 m [33]. Flowering runs from March to August [38,48].

In ethnobotany, it is prized for its diuretic, depurative, and digestive properties; infusions are used as calming and hepatoprotective agents. Agronomically, it is rustic, undemanding, easily propagated by seed and sometimes confused with *T. officinale* or *Sonchus* spp., but distinguishable by larger flower heads and robust habit. It contributes to Mediterranean biodiversity and resilience [49,50].

Systematically, it belongs to the tribe *Hypochaeridinae (Asteraceae)*, as *Hypochaeris radicata* L. Traditionally thought to be galactogenic, it is now recognized mainly for its diuretic, depurative and hepatoprotective effects [37,39,51]. Ethnobotanical knowledge persists marginally, indicating little-explored pharmacological potential. Its high adaptability makes it a model for ecological and evolutionary studies [52]. Homoploid hybridization is documented in Catalonia and southern France [34]. In Tunisia, aqueous extracts inhibit lettuce germination by up to 100%, suggesting use as a natural bioherbicide [12]. Culinary resilience practices integrate it with interest as a "functional food", rich in sesquiterpenes, flavonoids, terpenoids, mucilages, and phenols with antioxidant, analgesic, and antimicrobial activities; no acute toxicity emerges [23–25].

Ethnobotanical use reflects both historical scarcity and the revaluation of the Mediterranean diet [27]. It is among the most common WEPs in Italy: raw or cooked tender leaves; shoots and roots as substitutes for spinach, capers and potatoes [35,44–46]. Ecologically, it is important for pollinators: In Corsica, it represents 41% of spring visits to *Asteraceae* [53,54]. In Alta Murgia, it has limited coverage (0.2%) but is relevant for functional biodiversity [55]. Mallorca is home to numerous Cetoniinae [56]. In Sicily, it constitutes 10.1% of the pastures, influencing the forage quality and traceability of Ragusano DOP cheese [57].

In folk medicine, galactogenic attribution was based on morphological analogy. Bitter compounds, mucilages, mineral salts, and vitamins support antibacterial, anti-inflammatory, antioxidant, hepatoprotective, and digestive activities [36,48]. In Sar-

https://doi.org/10.3390/horticulturae12020186

dinia, the raw root is used for gastric disorders; the extracts show antimicrobial action [37,50,58,59].

Table 2. Research regarding the ethnobotanical uses of *U. dalechampii*.

References	Geographical Area	Ethnobotanical Use	Plant Part(s) Used
[33,35,36,43,46]	Central–Southern Italy (Apulia, Tuscany, Liguria); calcareous, dry soils, 0–1200 m a.s.l.	Edible wild vegetable (leaves, shoots); used in mixed greens; ornamental relevance	Leaves, young shoots
[17,34,38,47,48]	Mediterranean Basin; widespread in Central–Northern Italy, absent from Alpine regions	Hardy species enhancing agroecosystem stability and biodiversity	Whole plant
[37,39,49–52]	Mediterranean drylands, fallows, olive groves	Diuretic, depurative, hepatoprotective, digestive; used in herbal preparations	Aerial parts, roots
[23–25,27,35,44–46]	Central–Southern Italy; traditional Mediterranean diet contexts	Functional food with antioxidant, analgesic, antimicrobial properties (infusions)	Aerial parts, Leaves
[12,34,53,56]	Spain, Southern France, Tunisia, Mediterranean islands	Pollinator resource; allelopathic potential; spontaneous hybridization zones	Flowers
[55,57]	Apulia (Murgia), Sicily (Iblean Plateau); rocky semi-natural pastures	Pasture quality indicator; linked to Ragusano PDO cheese traceability	Whole plant
[33]	Mediterranean lowlands and rural slopes (vineyards, olive groves, roadsides)	Fodder and minor folk uses; contributes to soil biodiversity and ecological resilience	Aerial parts
[37]	General European folk tradition	Medicinal purposes with hepatoprotective effects	Leaves, roots
[36,48]	Not region-specific (laboratory studies)	Antimicrobial activity from acid–rich extracts and essential oils	Roots, aerial parts
[37,50,58,59]	Sardinia (Italy)	Roots for gastric issues; decoctions digestive/depurative; infusions for bile disorders	Roots, aerial parts

3.3. Food Uses

Among the main bitter herbs of Italian cuisine, the seasonal culinary use of *U. dalechampii* confirms its role in the traditional Mediterranean diet as both food and a digestive remedy [58,60]. The plant, morphologically distinguished by large, soft flower heads, is still appreciated today in Mediterranean markets and gastronomy [22]. Before flowering, the basal leaves are harvested (Figure 4); although bitter, they are edible and commonly used in soups, omelets, or sautéed with oil and garlic. The green, streaked buds are preserved in brine or pickle as caper substitutes and feature in traditional recipes [30,33], confirming their local alimurgic (famine-food) value [61]. The versatility of the species makes it a characteristic element of local cuisines. In central Italy, *U. dalechampii* is employed in a traditional wild herb mixture known in peasant dialect as "mad grass" (*Ciociarian erua pazza*). This dish consists of a salad prepared from a mix of wild herbs and flowers, assembled according to seasonal availability and historically associated with periods of food scarcity [38,59]. In Sicily (Italy), *U. dalechampii* is consumed as a spontaneous vegetable: boiled or fried leaves and shoots enter mixed dishes and soups [58].

 https://doi.org/10.3390/horticulturae12020186

Figure 4. Leaves of *U. dalechampii* cleaned and ready to be used in various gastronomic preparations. The image is original and was created by the authors.

In Umbria (Italy), the buds are used as caper substitutes, while the young leaves, characterized by a bitter taste, are boiled and simply seasoned. Harvesting before flowering reduces bitterness, and the species remains a defining element of rural cuisine and of the local health tradition [50,59].

The Sardinian soup S'Erbuzu combines about seventeen wild herbs (including *Foeniculum vulgare, Reichardia picroides, Sonchus oleraceus, Chondrilla juncea, Andryala integrifolia*), cooked with lard, beans, and pickled curd to a creamy consistency. The young leaves of *U. dalechampii*, with a delicate but slightly bitter taste, are sometimes eaten raw but more often cooked, and are commonly combined with species having a sweeter flavor to balance bitterness, in keeping with Mediterranean culinary traditions based on mixed herbs [49,50].

In Liguria, following periods of intense rains, wild herbs rapidly recolonize fields. When a leaf of *U. dalechampii* is cut, an artichoke-like aroma can be perceived, attributed to its bitter sesquiterpene lactones. In the Ligurian Apennines, a portion of preboggiòn—a traditional mixture of wild herbs briefly boiled—also including *U. dalechampii*, is used as a filling for a stuffed meat dish, cooked in a flavored broth and served sliced [37].

As part of the Ligurian prebuggiun, the species provides the bitter component of the mixture. When young, it can also be eaten raw; however, due to the intensity of its flavor, it is rarely consumed alone and is more commonly cooked. It is often combined with other wild herbs, such as *Sonchus oleraceus, Tragopogon pratensis, Apium nodiflorum, Hypochaeris radicata*, and *Papaver rhoeas*, which are boiled and seasoned with oil or sautéed with garlic, oil, and chili pepper. Harvesting is carried out by cutting at the base of the rosette in order to preserve the leaves and roots. The plant is used in soups and as a filling for piadina with watercress, and these preparations are appreciated for their yield and nutritional value [38,62].

The young rosettes, harvested at the end of winter, are boiled and used in omelets, savory pies, soups, or as a filling for *ravioli* and *pansotti* [37]. In Apulia, the dialect expression "foje creste" refers to blanched and sautéed mixtures of wild herbs used as a

 https://doi.org/10.3390/horticulturae12020186

condiment [44–46]. In Abbruzzo, similar preparations are employed in the traditional dish "pizza e foje", consisting of a soft corn-based flatbread baked in the oven and served with wild vegetables sautéed with garlic and chili pepper, accompanied by browned sardines and roasted sweet peppers [63]. In addition, in Abruzzo (Italy), a sauce obtained by macerating the aerial parts (excluding the flowers) in cooking water, then puréeing and acidulating it with lemon juice, is prepared to season meat dishes with a characteristic bitter note [36,48].

In Apulia, *U. dalechampii* is used in typical dishes such as "Callarredda", a lamb-based preparation with wild herbs from the Murgia area, where it serves as an alternative to cultivated chicory (*Cichorium intybus*) [22,43–46,61]. Moreover, the term "Fògghj ammìske" ("mixed leaves") refers to a collection of different wild edible species—including *U. dalechampii*—gathered from uncultivated fields. This mixture of wild greens is cooked and used as an ingredient in various traditional dishes, such as *"Fave bianche e cicorie"* (Figure 5).

Figure 5. *Fave bianche e cicorie*: purée of husked broad beans presented with boiled "Fògghj ammìske" (mixed wild leaves) and Cipolla rossa di Acquaviva (an Apulian local variety of onion). The image is original and was created by the authors.

As a Mediterranean alimurgic species, the green, dark-streaked buds of *U. dalechampii* are preserved in brine as caper substitutes or prepared in a sweet-and-sour style. After washing and drying, the buds are mixed daily with coarse salt for about one week until dehydrated, and then placed in glass jars and stored in a cool, dry environment. The resulting product, characterized by an intense flavor, is used as a condiment [37,59,61,64,65]. The roots are also cooked and used as a substitute for potatoes [63].

In the western Mediterranean basin, particularly in Spain and Morocco, the species belongs to the group of wild vegetables employed in mixed preparations ("mixed recipes of cooked vegetables"), in which the basal leaves are traditionally boiled together with other species and seasoned simply with salt and oil [66,67]. This consumption pattern, based on the use of the basal rosette and boiling, represents a characteristic feature of rural Ibero-Maghreb cuisine [67]. The core distribution of the species lies in the Iberian area, where it is especially exploited for food purposes [68]. The species shows remarkable

 https://doi.org/10.3390/horticulturae12020186

versatility in Iberian and North African culinary traditions: in addition to the leaves, the flower buds and roots are also used, generally cooked and mixed with other herbs. The buds may be preserved in brine and used as such as capers substitute, while the root, resembling a small potato, is consumed as a cooked vegetable. Preparations include boiling, raw consumption in salads, incorporation into savory wild herb cakes, and use in mixed cooked vegetable dishes [50,69].

The tradition of wild edible greens is shared with Greek, Cypriot and Levantine cuisines, where *U. dalechampii* is commonly boiled and seasoned with oil and lemon, or sautéed with onion, garlic, and spices, as in the Greek recipe *Tsigarelli* (*horta tsigari*) [70,71]. In Cyprus, *U. dalechampii* is boiled, sautéed with onion and blended with bitter citrus fruits, while many preparations flow into hortopita, savory pies filled with wild vegetables [70,72]. In Greece and in Crete, the congener *Urospermum* is consumed as a raw wild edible green, sautéed, mixed with other herbs or boiled and seasoned with oil [25,52].

Overall, *U. dalechampii* is embedded in the Mediterranean tradition of spring wild vegetables, prepared through boiling, stewing, raw consumption, inclusion in mixed cooked vegetable dishes, preservation of the buds as caper substitutes, and use of the roots as small tubers. These practices collectively define the species as a characteristic gastronomic element of the western Mediterranean.

3.4. Genomic, Nutritional and Bioactive Profile

U. dalechampii has been the subject of an integrated cytogenetics and genomic phytocomposition analysis, with a focus on the repetitive structure of DNA and its role in genome evolution. Approximately 74.16% of the genome is made up of repetitive sequences, predominantly LTR retrotransposons of the Ty1/Copia (24%) and Ty3/Gypsy (11%) families, with the SIRE and Tekay lines predominating. The homoploid hybrid (*U. dalechampii* × *U. picroides*) shows an intermediate genome (1.89 pg/2 C), reflecting a balanced combination of the parental genomes. Retrotransposon activity and partial loss of ribosomal DNA indicate post-hybridative reorganization processes, with implications for adaptation and genomic stability of the species [34].

From a nutritional point of view, Hammami et al. state that *U. dalechampii* presents a high-quality profile, with carbohydrates (46.6–78.0%), proteins (5.4–13.5%), raw fibers (10.4–24.6%), ash (5.2–13.8%), and a modest but valuable lipid fraction (1.0–1.5%) [21]. The plant appears rich in macroelements such as potassium (404 mg/100 g), calcium (91 mg/100 g), and magnesium (20 mg/100 g), and shows a secondary metabolism oriented towards the production of phenols and flavonoids, positively correlated with antiradical activity [35]. It should be noted that quantitative differences reported among studies may partly reflect variability in analytical methods and sample preparation procedures. Furthermore, the trace elements (Fe, Zn, Cu) contribute to remineralizing and antioxidant properties. The lipid profile, dominated by oleic (33.3%), palmitic (19.5%), stearic (18.4%), α-linolenic (12.9%), and linoleic acids (8.4%), results in a UFA/SFA ratio > 1.3 and low atherogenic (AI = 0.51) and thrombogenic (TI = 0.62) indices, supporting the nutraceutical and cardioprotective potential of the species [21].

Anaclerio et al. [35] further support the relevance of the phytochemical and nutraceutical profile of *U. dalechampii*. According to their study, the species is characterized by total polyphenols (90.17 mg CAE/100 g fw) and antioxidant activity (0.09 g Trolox/100 g fw) [35]. However, these values should be interpreted in light of the plant matrix analyzed and the extraction conditions adopted (Table 3).

In particular, Hammami et al. [12] point out that *U. dalechampii* shows a complex phenolic profile, with polyphenols (9.67 mg GAE/g DW), flavonoids (4.24 mg RE/g DW) and tannins (0.04 mg CE/g DW), accompanied by a high antioxidant capacity (50.99 mg AAE/g DW) and

https://doi.org/10.3390/horticulturae12020186

scavenger activity (IC_{50} = 0.032 g/mL). LC–MS analysis identified 17 phenolic compounds, including chlorogenic, protocatechuic, 1,3-di-O-capheoylquinic, salviolic and flavonoid acids such as luteolin (7.98 mg/g DW), quercetin, rutin and apigenin [12] Differences in compound abundance reported in the literature may also depend on whether aerial parts, roots, or whole plants were used (Table 3).

According to Bouzid et al., the methanolic extract of *U. dalechampii* (UDME) shows a high concentration of phenols (154.91 μg GAE/mg), flavonoids (29.96 μg QE/mg) and tannins (9.65 μg CE/mg), together with alkaloids, terpenoids, saponins, coumarins and steroids [24].

Phytochemical investigations on the essential oil of *U. dalechampii*, conducted by Ramdani et al. [23], reveal a composition dominated by fatty acids and oxygenated aromatic compounds. GC/MS analysis identified 26 major constituents (97.82% of the total), including palmitic acid (24.3%), henicosan (10.1%), 2-methyl-Z, Z-3,13-octadecadienol (7.1%), dill apiole (5.25%), myristic and lauric acids, myristicin and elemycin. These compounds, belonging to terpenoids, phenylpropanoids and saturated fatty acids, are associated with marked antimicrobial activity against Gram-negative bacteria (*A. baumannii, S. typhimurium*) and moderate against Gram-positive ones (*S. aureus*) [23]. Comparability with other studies is influenced by differences in distillation methods and analytical sensitivity.

U. dalechampii Methanolic Extract (UDME) also shows significant antioxidant activity (IC_{50} = 45.42 μg/mL) and iron chelation capacity (IC_{50} = 0.54 mg/mL). In vivo models indicate dose-dependent anti-inflammatory and analgesic effects, with inhibition of up to 80.6% of acetic acid-induced contortions, without antipyretic activity. The absence of acute toxicity (LD_{50} > 2000 mg/kg) supports the safety of use and potential pharmacological applications [24]. However, toxicological assessment and biosafety evaluation are crucial aspects that should be considered for the potential use of products derived by this species.

Aqueous extracts of roots and aerial parts exhibit marked allelopathic activity, with total inhibition of lettuce (*Lactuca sativa* L.) germination at combined concentrations of 5%, attributable to the synergistic action of phenols and flavonoids [12]. This functional variability further emphasizes the influence of plant part selection on bioactivity outcomes.

Finally, *U. dalechampii* has been explored for innovative applications in metallurgy. Remache et al. [73] demonstrated that its extracts act as effective natural corrosion inhibitors (up to 83%) for carbon steel in acidic environments, functioning as environmentally friendly green inhibitors through combined physical and chemical adsorption mechanisms [73].

Table 3. Research regarding phytochemicals of *U. dalechampii*.

Reference	Phytochemical(s)	Chemical Class	Plant Part(s) Used	Analytical Method(s)
[12]	Chlorogenic acid, protocatechuic acid, salvianolic acid, 1,3-di-O-caffeoylquinic acid	Phenolic acids	Aerial parts (leaves and stems)	Aqueous extraction; spectrophotometric assays; LC–MS
[24]	Total flavonoids	Flavonoids	Aerial parts	Methanolic extraction; colorimetric assay
[12]	Luteolin, quercetin, rutin, apigenin	Flavonoids (flavones and flavonols)	Aerial parts	Aqueous extraction; LC–MS; colorimetric assay
[24]	Total polyphenols, tannins	Polyphenols	Aerial parts	Methanolic extraction; Folin–Ciocalteu assay
[24]	Sesquiterpene lactones	Terpenoids	Aerial parts	Qualitative phytochemical screening

 https://doi.org/10.3390/horticulturae12020186

Table 3. *Cont.*

Reference	Phytochemical(s)	Chemical Class	Plant Part(s) Used	Analytical Method(s)
[24]	Alkaloids, saponins, coumarins, steroids	Alkaloids and terpenoid-derived compounds	Aerial parts	Qualitative phytochemical screening
[23]	Palmitic, myristic and lauric acids	Saturated fatty acids	Whole plant	Hydrodistillation; GC–MS
[23]	Dill apiole, myristicin, elemicin	Phenylpropanoids	Whole plant	Hydrodistillation; GC–MS
[12]	Total phenolics, total flavonoids (aqueous extracts)	Polyphenols and flavonoids	Leaves and roots	Aqueous extraction; spectrophotometric assays
[35]	Total phenols, antioxidant compounds	Polyphenols	Leaves (wild and cultivated plants)	Methanolic extraction; Folin–Ciocalteu assay; ABTS test

Note. Spectrophotometric assays refer to colorimetric methods commonly used for the determination of total phenolics (Folin–Ciocalteu method) and total flavonoids ($AlCl_3$-based assay). LC–MS: liquid chromatography–mass spectrometry. GC–MS: gas chromatography–mass spectrometry. Qualitative phytochemical screening indicates preliminary tests aimed at detecting major classes of secondary metabolites.

3.5. Domestication

Both ethnobotanical knowledge and food uses suggest good potential for the domestication of *U. dalechampii*.

U. dalechampii is a hardy and xerophilous species. It prefers full-sun exposure but tolerates partial shade. The species withstands temperatures down to 5 °C. It requires protection from frost during winter. *U. dalechampii* grows on loose and well-drained soils. It requires little organic fertilization. It needs irrigation only during prolonged drought [45,46]. The species grows in soils with medium-to-low fertility. It tolerates pH values from slightly acidic to basic. The species flowers from March to August. Flowering is favored by full-sun conditions. Pollination is entomogamous. Seed dispersal occurs through a feathery pappus or, as in related species, through epizoochory [51]. Sowing takes place between winter and spring, with germination in 6–12 weeks. The leaves, bitter but tender, are harvested in March–April, cutting the apical parts to encourage regrowth [46]. The cut leaves tend to curve, resembling octopus tentacles, hence the dialect name "cequaire all'ammérse". The plant, which is also valued for ornamental purposes, flowers from March to July/August and is harvested from late winter to early spring, before flowering, when the basal leaves are more tender. In deep soils and rainy years, the flowers can exceed 6 cm in diameter. The greater snake is harvested from fields and uncultivated land by wild herb collectors [43].

It is resistant to parasites but can suffer powdery mildew or rot from humidity. In Italy, it coexists with *U. picroides* (lesser bush), and with annual, bristlier and light flowers. *U. dalechampii* is appreciated as an ornamental, melliferous and useful plant for vegetable gardens because it attracts entomophagous insects. The leaves are eaten raw or cooked in minestrone soups, fillings and rustic cakes, while the buds can be preserved in brine or used in sweet-and-sour applications [39].

U. dalechampii is a perennial vascular plant typical of temperate–Mediterranean climates (zone 8), capable of surviving prolonged droughts owing to its xerophytic physiology. This perennial plant is well adapted to Mediterranean climates and grows easily owing to its robust root system and photophilic physiology. It performs best under bright light at around 20 °C, with average humidity of about 50%, and requires at least six hours of full sunlight per day. The species prefers moderate weekly watering and sandy or calcareous soils with a pH between 6.0 and 7.5. It tolerates a wide range of temperatures, from −20 °C to 38 °C (zones 5–9), and does not require specific climate protection. To prevent root

https://doi.org/10.3390/horticulturae12020186

rot, water stagnation should be avoided, and balanced fertilizers may be applied monthly during the growing season. Pruning is recommended in spring and summer to remove dry tissues and improve air circulation, while the removal of withered flower heads helps to prolong flowering. Fertilization requirements are generally low, and repotting is usually carried out in spring. The species propagates naturally by seed after the last frost, but it can also be propagated by cuttings or spring transplantation. Plants typically reach 10–50 cm in height, produce characteristic bright yellow flowers up to 5 cm in diameter, and develop a crown up to 60 cm wide. *U. dalechampii* is not toxic to humans or animals, contains latex, and is considered a hardy medicinal herb with moderate maintenance requirements [51,74].

In the paper by Chenot-Lescure et al. [75], *U. dalechampii* is reported as one of the characteristic species of the Mediterranean arid meadows of the Crau plain (southern France). After five years of experimentation on different artificial substrates in former quarries, the species appeared only in unseeded lots, indicating spontaneous colonization favored by natural substrates. It is therefore considered an indicator of taxon of ecological success in the reconstitution of typical Mediterranean dry grasslands.

Spontaneous edible species (SEP), including *U. dalechampii*, represent an agrobiodiversity resource with cultural and genetic value, useful for the domestication of new crops. The increase in demand for fresh-cut products has led to studies of their cultivation above ground, in pots with sub-irrigation and in floating systems [76]. In support of these activities, some studies demonstrate that the cultivation and domestication of edible spontaneous species, in high-tech soilless systems, can be ecologically sustainable if integrated into a circular economy model. Buttaro et al. show that innovative photovoltaic greenhouses maintain acceptable yields and quality by reducing energy consumption and emissions, thus promoting energy self-sufficiency [77]. Gonnella and Renna [78] highlight that the evolved soilless systems allow for the efficient use of water and nutrients, the recycling of substrates and the employment of organic by-products, reducing the environmental impact. These crops, if managed with renewable materials and closed circuits, represent a concrete way for sustainable food production, even in urban contexts [78].

Therefore, Anaclerio et al. [76] studied three spontaneous edible species, *U. dalechampii* and *U. picroides*, in soilless culture and these showed good domestication potential, with average yields of approximately 26 t/ha in two months. The cultivated plants showed morphological differences compared to the spontaneous ones and a high mineral content (especially K^+, Ca^{2+} and Mg^{2+}). These species represent promising resources for expanding and diversifying the production of fresh-cut vegetables.

The same research group tested the domestication of *U. dalechampii* and *U. picroides* in floating hydroponic cultivation to obtain baby leaves. The plants, grown at two densities (412 and 824 plants/m^2), were harvested after 44–54 days for *U. dalechampii* and 35 days for *U. picroides*, showing good precocity. *U. dalechampii* showed more leaves and higher concentrations of sodium, chlorine and nitrates; *U. picroides* produced greater leaf weight, dry matter, potassium, calcium and phenols (+49%), with superior antioxidant activity (+61%). Lower densities favored phenols and antioxidant capacity, while higher densities increased yield and mineral content. Spontaneous plants, compared to cultivated ones, had more phenols and antioxidants, but lower water and nitrate contents. Edible microgreens were also obtained, demonstrating the feasibility of controlled large-scale production. Hydroponic cultivation confirmed the potential suitability of both species for range IV while requiring control of the nitrate content [35].

U. dalechampii, as a rustic Mediterranean species tolerant to drought and poor substrates (Table 4), is indicated as a plant useful for urban biodiversity and for use in green roofs and peri-urban agriculture owing to its ability to attract pollinators and resist extreme environmental conditions [79]. In parallel, the species is recognized for its ethnobotanical

 https://doi.org/10.3390/horticulturae12020186

and nutraceutical value, traditionally consumed as a wild herb and today re-evaluated for the production of microgreens and baby leaves in sustainable hydroponic systems, for the high content of antioxidants, vitamins and minerals, as well as for the bitter and aromatic flavor typical of Mediterranean alimurgic plants [80,81]. Overall, it represents a bridge between traditional use and agroecological innovation, with a potential role in the valorization of spontaneous plant resources.

Table 4. *U. dalechampii* pedoclimatic characteristics.

References	Parameter	Observation
[45,46,51,74]	Climate zones	5–9 (USDA hardiness zones)
[45,46,51,74]	Climate tolerance	−20 to 38 °C (optimal: 20 °C); 50% RH
[45,46]	Water and soil	Irrigation only under drought; avoid waterlogging; pH 6.0–7.5
[46]	Growth cycle	Sowing winter–spring; germination 6–12 weeks; harvest after 44–54 days
[35,46]	Harvest period	March–April

4. Prospect

Based on the integrated review of botanical, nutritional, phytochemical, ethnobotanical, and agronomic data, *U. dalechampii* emerges as a Mediterranean wild plant with significant multifunctional potential, offering concrete benefits in terms of biodiversity, food security, sustainability, and the valorization of ethnobotanical heritage. Therefore, in order to evaluate whether this species can be regarded as a new horticultural crop with a concrete chance to succeed, a SWOT analysis was performed (Table 5).

Table 5. SWOT analysis related to the exploitation of *U. dalechampii* as a new cash crop.

STRENGTHS	**WEAKNESSES**

Environmental adaptability Nutritional and nutraceutical value Multiple uses and functions Cultural and ethnobotanical value	Lack of standardized agronomic protocols Fragmented scientific knowledge Organoleptic and market limitations
OPPORTUNITIES	**THREATS**
	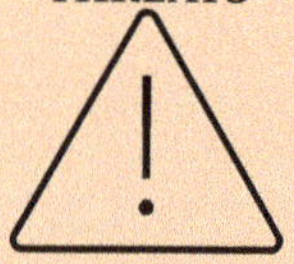
Domestication and sustainable cultivation Integration into contemporary Mediterranean diet Biodiversity conservation and agroecological resilience Development of functional and nutraceutical products Research and innovation	Genetic erosion and pressure on wild populations Regulatory and legal barriers Competition with established crops Qualitative and agronomic variability

https://doi.org/10.3390/horticulturae12020186

4.1. Strengths

U. dalechampii presents notable environmental adaptability, being hardy, xerophytic, and tolerant of poor, arid, or calcareous soils, water stress, Mediterranean conditions, and wide thermal ranges. These traits make it suitable for cultivation in marginal, hilly, or semi-arid areas, including regions impacted by climate change. The species also has high nutritional and nutraceutical value, with carbohydrates, fibers, proteins, minerals (K, Ca, Mg, trace elements), unsaturated fatty acids, and bioactive compounds (phenols, flavonoids, sesquiterpene lactones), supporting its use in functional foods, microgreens, fresh-cut products, and health-oriented dishes. Beyond dietary applications, *U. dalechampii* provides ornamental value, supports pollinators, enhances biodiversity in xerophilic ecosystems, and has potential nutraceutical and phytopharmaceutical applications. Its traditional use in Mediterranean cuisine further connects cultural heritage with agricultural innovation.

4.2. Weaknesses

Key limitations for wider adoption include:

- Lack of standardized agronomic protocols: optimal cultivation, post-harvest handling, variety selection, and bitterness management remain undefined.
- Fragmented scientific knowledge: data on agronomic yield, genetic stability, and trait variability under controlled conditions are limited.
- Organoleptic and market limitations: natural bitterness and variable sensory traits may reduce consumer acceptance, while regulatory barriers could constrain commercialization.

4.3. Opportunities and Future Research Directions

Addressing these weaknesses can guide research with measurable horticultural outcomes:

- Standardized cultivation protocols: optimize sowing density, substrates (soil, soilless, hydroponics), irrigation, nutrition, harvesting, and post-harvest handling to improve yield, quality, uniformity, and bitterness mitigation.
- Varietal selection and improvement: identify ecotypes with superior agronomic and sensory traits; assess genetic and metabolic stability to ensure consistent performance.
- Phytochemical and nutraceutical evaluation: quantify bioactive compounds and evaluate antioxidant, anti-inflammatory, and hepatoprotective properties, including preliminary safety and efficacy studies.
- Market and socio-economic assessment: investigate consumer perception, regulatory feasibility, and short-supply-chain potential to facilitate adoption.
- Agroecological integration: evaluate performance in crop associations, urban/peri-urban systems, and regenerative agriculture to enhance biodiversity, pollinator support, and ecosystem resilience.

4.4. Threats

Potential risks include:

- Genetic erosion and pressure on wild populations: overharvesting could threaten biodiversity and local germplasm; research should include conservation programs and sustainable domestication strategies.
- Regulatory and legal barriers: Compliance with novel food regulations, safety assessments, and toxicological requirements is essential; studies should generate data to support authorizations.
- Competition with established crops: new "traditional vegetables" may struggle to gain market share; breeding for desirable traits and marketing research can mitigate this risk.

 https://doi.org/10.3390/horticulturae12020186

- Qualitative and agronomic variability: inconsistent nutritional and phytochemical quality could affect consumer acceptance; future research should focus on standardization, varietal improvement, and controlled cultivation techniques.

By linking weaknesses and threats to specific research priorities, this approach provides a coherent roadmap for sustainable domestication, measurable horticultural outcomes, and valorization of *U. dalechampii*.

5. Conclusions

Urospermum dalechampii is a multifunctional Mediterranean wild species that combines notable ecological resilience, a favorable nutritional profile and a rich phytochemical composition. The available evidence indicates that *U. dalechampii* provides significant amounts of carbohydrates, fibers and essential minerals, a beneficial fatty acid profile, and a wide array of phenolic and terpenoid compounds associated with antioxidant, anti-inflammatory and antimicrobial activities. Cultivation trials and soilless production trials demonstrate that domestication and controlled production (baby leaves, microgreens and baby-leaf ready-to-eat chains) are feasible and can reduce pressure on wild populations. Despite these strengths, the body of literature remains fragmented and several research gaps limit immediate commercialization. Addressing these gaps will require coordinated agronomic, phytochemical and socio-economic studies, combined with germplasm conservation strategies to avoid genetic erosion. Future research activities should include standardized cultivation trials (soil and soilless), selection of low-bitterness genotypes, pilot-scale post-harvest assessments, and socio-economic analyses of supply chain feasibility. Furthermore, toxicological assessment and biosafety evaluation are crucial aspects that should be considered for the potential use of products derived by this species. When integrated with conservation measures and circular production systems, *U. dalechampii* could become a locally adapted crop that enhances Mediterranean agrobiodiversity while offering novel, health-promoting food products.

Author Contributions: All authors contributed equally to the drafting of all parts of this manuscript. All authors have read and agreed to the published version of the manuscript.

Funding: This study was carried out within the Agritech National Research Center (Spoke 7) and received funding from the European Union Next-GenerationEU (PIANO NAZIONALE DI RIPRESA E RESILIENZA (PNRR)—MISSIONE 4 COMPONENTE 2, INVESTIMENTO 1.4—D.D. 1032 17/06/2022, CN00000022). This manuscript reflects only the authors' views and opinions; neither the European Union nor the European Commission can be considered responsible for them. This research was also supported by the Regione Puglia Administration under Rural Development Program 2014–2020, Project 'Biodiversity of Apulian vegetable species (BiodiverSO Veg)', Measure 10, Sub measure 10.2, Operation 1 "Program for the conservation and the valorization of the genetic resources in agriculture" (DDS No. 04250182807, CUP: B97H22003760009)—No. 22.

Data Availability Statement: No new data were created or analyzed in this study.

Acknowledgments: The authors gratefully acknowledge Pietro Santamaria for his valuable assistance in the selection of photographic images.

Conflicts of Interest: The authors declare no conflicts of interest.

References

1. Elia, A.; Santamaria, P. Biodiversity in Vegetable Crops, a Heritage to Save: The Case of Puglia Region. *Ital. J. Agron.* **2013**, *8*, e4. [CrossRef]
2. Geraci, A.; Amato, F.; Di Noto, G.; Bazan, G.; Schicchi, R. The wild taxa utilized as vegetables in Sicily (Italy): A traditional component of the Mediterranean diet. *J. Ethnobiol. Ethnomed.* **2018**, *14*, 14. [CrossRef] [PubMed]

 https://doi.org/10.3390/horticulturae12020186

3. Mattirolo, O.; Gallino, B.; Pallavicin, G. *Phytoalimurgia Pedemontana. Come Alimentarsi con le Piante Selvatiche*; Blu Edizioni: Pevegnano, Italy, 2011.
4. Le Floc'h, E.; Boulos, L.; Vela, E. *Catalogue Synonymique Commenté de la Flore de Tunisie*; République Tunisienne Ministère de l'environnement et du Développement Durable Banque Nationale de Gènes: Tunis, Tunisia, 2010.
5. Cornara, L.; La Rocca, A.; Marsili, S.; Mariotti, M.G. Traditional uses of plants in the Eastern Riviera (Liguria, Italy). *J. Ethnopharmacol.* **2009**, *125*, 16–30. [CrossRef] [PubMed]
6. Renna, M.; Cocozza, C.; Gonnella, M.; Abdelrahman, H.; Santamaria, P. Elemental characterization of wild edible plants from countryside and urban areas. *Food Chem.* **2015**, *177*, 29–36. [CrossRef]
7. Bianco, V.V.; Renna, M.; Santamaria, P. *Ortaggi Liberati: L'e-Book-Dieci Prodotti Straordinari Della Biodiversità Pugliese e Qualcosa in Più*; Università degli Studi di Bari Aldo Moro: Bari, Italy, 2018.
8. Renna, M.; Montesano, F.F.; Signore, A.; Gonnella, M.; Santamaria, P. BiodiverSO: A Case Study of Integrated Project to Preserve the Biodiversity of Vegetable Crops in Puglia (Southern Italy). *Agriculture* **2018**, *8*, 128. [CrossRef]
9. Petropoulos, S.A.; Karkanis, A.; Martins, N.; Ferreira, I.C.F.R. Edible halophytes of the Mediterranean basin: Potential candidates for novel food products. *Trends Food Sci. Technol.* **2018**, *74*, 69–84. [CrossRef]
10. Redžić, S.; Ferrier, J. *The Use of Wild Plants for Human Nutrition During a War: Eastern Bosnia (Western Balkans)*; Springer: New York, NY, USA, 2014; pp. 149–182. [CrossRef]
11. Ogwu, M.C.; Osawaru, M.E. Traditional Methods of Plant Conservation for Sustainable Utilization and Development. In *Biodiversity in Africa: Potentials, Threats and Conservation*; Chibueze Izah, S., Ed.; Springer Nature: Singapore, 2022; pp. 451–472.
12. Hammami, N.; Benabderrahim, M.A.; Hannachi, H. Modeling approach of allelopathy effect of *Urospermum dalechampii* and *Andryala integrifolia* on lettuce-seed germination and seedling growth. *Euro-Mediterr. J. Environ. Integr.* **2025**, *10*, 1889–1899. [CrossRef]
13. Nouidha, S.; Selmi, S.; Guigonis, J.M.; Pourcher, T.; Chekir-Ghedira, L.; Kilani-Jaziri, S. Metabolomics Profiling of Tunisian *Sonchus oleraceus* L. Extracts and Their Antioxidant Activities. *Chem. Biodivers.* **2023**, *20*, e202300290. [CrossRef]
14. Maaloul, S.; Mahmoudi, M.; Mighri, H.; Ghzaiel, I.; Bouhamda, T.; Boughalleb, F.; El Midaoui, A.; Vejux, A.; Lizard, G.; Abdellaoui, R. Tunisian Silybum Species: Important Sources of Polyphenols, Organic Acids, Minerals, and Proteins across Various Plant Organs. *Plants* **2024**, *13*, 989. [CrossRef]
15. Talavera, S.; Talavera, M.; Berjano, R. *Urospermum* Scop. [nom. cons.]. In *Flora iberica. Plantas Vasculares de la Península Ibérica e Islas Baleares*; Talavera, S., Buira, A., Quintanar, A., García, M.Á., Talavera, M., Fernández Piedra, P., Aedo, C., Eds.; Real Jardín Botánico de Madrid: Madrid, Spain, 2017; Volume XVI, pp. 1063–1068.
16. Fernández-Mazuecos, M.; Jiménez-Mejías, P.; Martín-Bravo, S.; Buide, M.L.; Álvarez, I.; Vargas, P. Narrow endemics on coastal plains: Miocene divergence of the critically endangered genus Avellara (Compositae). *Plant Biol.* **2016**, *18*, 729–738. [CrossRef]
17. Berjano, R.; Talavera, M.; Talavera, S. El género *Urospermum* en el oeste de la región mediterránea. The genus *Urospermum* in Western Mediterranean. *Acta Bot. Malacit.* **2014**, *39*, 117–128. [CrossRef]
18. Abbott, R.; Albach, D.; Ansell, S.; Arntzen, J.W.; Baird, S.J.; Bierne, N.; Boughman, J.; Brelsford, A.; Buerkle, C.A.; Buggs, R.; et al. Hybridization and speciation. *J. Evol. Biol.* **2013**, *26*, 229–246. [CrossRef] [PubMed]
19. Roux, C.; Fraïsse, C.; Romiguier, J.; Anciaux, Y.; Galtier, N.; Bierne, N. Shedding Light on the Grey Zone of Speciation along a Continuum of Genomic Divergence. *PLoS Biol.* **2016**, *14*, e2000234. [CrossRef] [PubMed]
20. Nieto Feliner, G.; Álvarez, I.; Fuertes-Aguilar, J.; Heuertz, M.; Marques, I.; Moharrek, F.; Piñeiro, R.; Riina, R.; Rosselló, J.A.; Soltis, P.S.; et al. Is homoploid hybrid speciation that rare? An empiricist's view. *Heredity* **2017**, *118*, 513–516. [CrossRef]
21. Hammami, N.; Rodrigues, M.J.; M'rabet, Y.; Saadellaoui, W.; Hosni, K.; Stambouli-Essassi, S.; Hannachi, H. Nutritional composition of the unexplored Mediterranean plant *Urospermum dalechampii* (L.) Scop. ex F.W.Schmidt from Tunisia. *J. Food Compos. Anal.* **2025**, *148*, 108206. [CrossRef]
22. Biscotti, N.; Bonsanto, D.; Del Viscio, G. The traditional food use of wild vegetables in Apulia (Italy) in the light of Italian ethnobotanical literature. *Ital. Bot.* **2018**, *5*, 1–24.
23. Ramdani, M.; Lograda, T.; Chalard, P.; Figueredo, G.; Laidoudi, H.; Ounoughi, A. Chemical composition, Antimicrobial activity and chromosome number of *Urospermum dalechampii* from Algeria. *Sch. Acad. J. Pharm.* **2014**, *3*, 477–482.
24. Bouzid, W.; Yahia, M.; Benkiki, N.; Hachemi, M.; Meziti, A.; Haba, H. Antioxidant, anti-inflammatory, analgesic, and antipyretic effects of *Urospermum dalechampii* (L.) scop. ex F.W.Schmidt. *Current Bioactive Compounds. Pharmacol. Toxicol. Et Pharm.* **2018**, *14*, 264–272. [CrossRef]
25. Saber, F.R.; Elosaily, A.H.; Mahrous, E.A.; Pecio, Ł.; Pecio, S.; El-Amier, Y.A.; Korczak, M.; Piwowarski, J.P.; Świątek, Ł.; Skalicka-Woźniak, K. Detailed metabolite profiling and in vitro studies of *Urospermum picroides* as a potential functional food. *Food Chem.* **2023**, *427*, 136677. [CrossRef]

 https://doi.org/10.3390/horticulturae12020186

26. Nebel, S.; Pieroni, A.; Heinrich, M. Ta chòrta: Wild edible greens used in the Graecanic area in Calabria, Southern Italy. *Appetite* **2006**, *47*, 333–342. [CrossRef]
27. Guarrera, P.M.; Savo, V. Wild food plants used in traditional vegetable mixtures in Italy. *J. Ethnopharmacol.* **2016**, *185*, 202–234. [CrossRef]
28. Renna, M.; Gonnella, M. The use of the sea fennel as a new spice-colorant in culinary preparations. *Int. J. Gastron. Food Sci.* **2012**, *1*, 111–115. [CrossRef]
29. Renna, M.; Gonnella, M.; Caretto, S.; Mita, G.; Serio, F. Sea fennel (*Crithmum maritimum* L.): From underutilized crop to new dried product for food use. *Genet. Resour. CROP Evol.* **2017**, *64*, 205–216. [CrossRef]
30. Renna, M. Reviewing the Prospects of Sea Fennel (*Crithmum maritimum* L.) as Emerging Vegetable Crop. *Plants* **2018**, *7*, 92. [CrossRef] [PubMed]
31. Matarrese, E.; Renna, M. Prospects of Hogweed (*Heracleum sphondylium* L.) as a New Horticultural Crop for Food and Non-Food Uses: A Review. *Horticulturae* **2023**, *9*, 246. [CrossRef]
32. Tranfield, D.; Denyer, D.; Smart, P. Towards a Methodology for Developing Evidence-Informed Management Knowledge by Means of Systematic Review. *Br. J. Manag.* **2003**, *14*, 207–222. [CrossRef]
33. Trieste, U.O.; Sciences, D.o.L. Dryades-Flora of Rome. Available online: https://dryades.units.it/Roma/ (accessed on 2 November 2025).
34. Pellicer, J.; Balant, M.; Fernández, P.; Rodríguez González, R.; Hidalgo, O. Morphological and Genome-Wide Evidence of Homoploid Hybridisation in *Urospermum* (Asteraceae). *Plants* **2022**, *11*, 182. [CrossRef]
35. Anaclerio, M.; Renna, M.; Di Venere, D.; Sergio, L.; Santamaria, P. Smooth Golden Fleece and Prickly Golden Fleece as Potential New Vegetables for the Ready-to-Eat Production Chain. *Agriculture* **2021**, *11*, 74. [CrossRef]
36. Bianco, V.V.; Mariani, R. *Asteracee Spontanee Commestibili di Puglia. Saperi, Sapori e Salute*; E-book edition; Università degli studi di Bari "Aldo Moro": Bari, Italy, 2025.
37. Brattel, L. Nel prato di Laura-No. 16. Available online: https://www.quilianonline.it/ (accessed on 2 November 2025).
38. Rauzino, V. Boccione maggiore. Copyright © 2013 suryanamaskara.altervista.org. Available online: https://suryanamaskara.altervista.org/piantealimurgiche/journal1.php?id=23 (accessed on 2 November 2025).
39. Bennet, L. Boccione maggiore-*Urospermum dalechampii*. Apiweb s.r.l. © 2025—Casa e Giardino. Available online: https://www.casaegiardino.it/lifestyle/boccione-maggiore-urospermum-dalechampii.php (accessed on 2 November 2025).
40. Jana, B.K.; Mukherjee, S.K. Exomorphic and histological characters of fruits in some taxa of the tribe-*Lactuceae* (Asteraceae). *J. Econ. Taxon. Botany. India* **2013**, *37*, 372–378.
41. Thompson, I.R. A taxonomic treatment of tribe *Lactuceae* (Asteraceae) in Australia. *Muelleria* **2007**, *25*, 59–100. [CrossRef]
42. Park, S.J.; Korompai, E.J.; Francisco-Ortega, J.; Santos-Guerra, A.; Jansen, R.K. Phylogenetic relationships of Tolpis (Asteraceae: Lactuceae) based on ndhF sequence data. *Plant Syst. Evol.* **2001**, *226*, 23–33. [CrossRef]
43. Atlante dei PAT. Boccione Maggiore. Available online: https://www.patpuglia.it/it/ (accessed on 2 November 2025).
44. Bianco, V.V.; Mariani, R.; Santamaria, P. *Piante Spontanee Nella Cucina Tradizionale Molese*; Levante Editori: Bari, Italy, 2009.
45. Accogli, R.; Medagli, P. *Erbe Spontanee Salentine. Guida Al Riconoscimento e All'Uso Delle Piante Alimentari Tradizionali*; Ezio Grifo Editore: Lecce, Italy, 2014.
46. Serio, F.; Florio, G.; Aarente, P.; Santamaria, P. Specie erbacee della flora infestante pugliese utilizzabili come ortaggi e piante da condimento. *Accad. Pugl. Delle Sci.* **2005**, *46*, 11–27.
47. Jiménez López, J.; Talavera Solís, M.; Sánchez Casimiro-Soriguer, C.; Terrab Benjelloun, A.; Talavera Lozano, S. Tipificación de *Crepis bursifolia* y su distribución en Marruecos. Typification of *Crepis bursifolia* and chorology in Morocco. *Acta Bot. Malacit.* **2013**, *38*, 222–224. [CrossRef]
48. Pica, A.; Lupoletti, J. Il Boccione Maggiore. Available online: https://www.floradabruzzo.it/ (accessed on 2 November 2025).
49. Boccione Maggiore o Lattugaccio, *Urospermum Dalechampii* L. Available online: https://www.vivigreen.eu/ (accessed on 2 November 2025).
50. Ranfa, A.; Orlandi, F.; Maurizi, A.; Bodesmo, M. Ethnobotanical knowledge and nutritional properties of two edible wild plants from Central Italy: *Tordylium apulum* L. and *Urospermum dalechampii* (L.) F.W. Schmid. *J. Appl. Bot. Food Qual.* **2015**, *88*, 49–254. [CrossRef]
51. Come Coltivare e Curare Boccione Maggiore. Available online: https://www.picturethisai.com/it/ (accessed on 2 November 2025).
52. Christoforidi, I.; Daliakopoulos, I.N.; Papadimitriou, D.M.; Tzanakakis, V.A.; Louloudakis, I.; Papadopoulos, I.; Pnevmatikakis, M.; Tsitsiridou, M.; Kafkala, T.; Kontekas, K.; et al. Effect of foliar fertilization on *Urospermum picroides*, an indigenous plant of the traditional cretan-mediterranean diet. *Sci. Hortic.* **2024**, *332*, 113202. [CrossRef]
53. Maestracci, P.Y.; Plume, L.; Gibernau, M. Anthophilous insects' seasonal variation in Corsican thermo-Mediterranean shrubland maquis. *Biodivers. Data J.* **2025**, *13*, e144560. [CrossRef] [PubMed]

 https://doi.org/10.3390/horticulturae12020186

54. Maestracci, P.Y.; Plume, L.; Gibernau, M. Insect floral visitors of thermo-Mediterranean shrubland maquis (Ajaccio, Corsica, France). *Biodivers. Data J.* **2024**, *12*, e118614. [CrossRef]
55. Ciani, E.; Tedone, L.; Terzi, M.; De Cillis, F.M.; Castellana, E.; Fracchiolla, M. Characterization of plant diversity of pastures and volatile organic compound analysis in ewe's milk from a typical farm system in the Alta Murgia national Park (southern Italy). *Oppor. A Sustain. Land Use* **2012**, *7*, e19. [CrossRef]
56. Shields, J.M.; Singer-Pennell, L.M. Springtime Habitat Survey of Oxythyrea sp. and Tropinota sp. in Northern Mallorca, Plant Preferences, and Ecological and Conservation Concerns of Roadside Proximity. *Res. Sq.* **2024**, *preprint*. [CrossRef]
57. Copani, V.; Guarnaccia, P.; Biondi, L.; Lesina, S.C.; Longo, S.; Testa, G.; Cosentino, S.L. Pasture quality and cheese traceability index of Ragusano PDO cheese. *Ital. J. Agron.* **2015**, *10*, 667. [CrossRef]
58. Lentini, F.; Venza, F. Wild food plants of popular use in Sicily. *J. Ethnobiol. Ethnomed.* **2007**, *3*, 15. [CrossRef]
59. Ranfa, A.; Bodesmo, M. An Ethnobotanical investigation of traditional knowledge and uses of edible wild plants in the Umbria Region, Central Italy. *J. Appl. Bot. Food Qual.* **2017**, *90*, 246–258.
60. Renna, M. Wild edible plants as a source of mineral elements in the daily diet. *Prog. Nutr.* **2017**, *19*, 219–222. [CrossRef]
61. Monzani, P.; Lambardi, S. Boccione Maggiore Sott'Aceto. Available online: https://campiditeo-buonipropositi.blogspot.com/ (accessed on 2 November 2025).
62. Canepa, L. Bunommo…O Bell'Ommo? Available online: https://www.lellacanepa.com/ (accessed on 2 November 2025).
63. Pizza e Foje, Il Comfort Food della Tradizione. Available online: https://www.ilpescara.it/cucina/pizza-foje-ricetta-originale.html (accessed on 2 November 2025).
64. Boccione Maggiore Sottosale. Available online: https://cookpad.com/it/ricette/17018425 (accessed on 2 November 2025).
65. Plantarum, A. *Urospermum dalechampii* (L.) Scop. ex F.W.Schmidt-IPFI page, Acta Plantarum. Available online: https://www.actaplantarum.org/flora/flora_info.php?id=506219 (accessed on 2 November 2025).
66. Łuczaj, Ł.; Pieroni, A. Nutritional ethnobotany in Europe: From emergency foods to healthy folk cuisines and contemporary foraging trends. In *Mediterranean Wild Edible Plants: Ethnobotany and Food Composition Tables*; Springer: Cham, Switzerland, 2016; pp. 33–56.
67. Kozuharova, E.; Malfa, G.A.; Acquaviva, R.; Valdés, B.; Batovska, D.; Stoycheva, C.; Rejdali, M.; Marino, P.; Spadaro, V. Wild Species from the Asteraceae Family, Traditionally Consumed in Some Mediterranean Countries. *Plants* **2025**, *14*, 2006. [CrossRef] [PubMed]
68. Savo, V. Ethnobotanical Analysis in the Amalfi Coast and Evaluation of Results on A Scientific and Economic Point of View. Ph.D. Thesis, Environmental Biology Department, Università degli Studi Roma Tre, Rome, Italy, 2008.
69. Dop, M.C.; Kefi, F.; Karous, O.; Verger, E.O.; Bahrini, A.; Ghrabi, Z.; El Ati, J.; Kennedy, G.; Termote, C. Identification and frequency of consumption of wild edible plants over a year in central Tunisia: A mixed-methods approach. *Public Health Nutr.* **2020**, *23*, 782–794. [CrossRef]
70. Adams, A. Gathering Wild Greens and a Recipe for Hortopita, or Wild Greens Pie. 2011. Available online: https://theshepherdandtheolivetree.wordpress.com/2011/11/25/gathering-wild-greens-and-a-recipe-for-wild-greens-pie-2/ (accessed on 2 November 2025).
71. Bergo, A. Foraged Greens with Garlic and Paprika (Tsigarelli). Forager | Chef LLC 2024, Leafy Greens. 2024. Available online: https://foragerchef.com/tsigarelli-wild-greens-with-paprika/ (accessed on 2 November 2025).
72. Wild Greens with Bitter Orange. cyprusfoodmuseum. 2025. Available online: https://www.cyprusfoodmuseum.com/en/syllog es/paradosiakes-syntages/wild-greens-bitter-orange (accessed on 2 November 2025).
73. Remache, S.; Khamaysa, O.M.A.; Benahmed, M.; Lgaz, H.; Lee, H.S.; Sehnoune, D.; Aldalbahi, A.; Laouar, H. Mechanistic understanding and performance assessment of *Urospermum dalechampii*-based green inhibitors for enhanced corrosion resistance of carbon steel in acidic media. *Colloids Surf. A Physicochem. Eng. Asp.* **2025**, *727*, 138076. [CrossRef]
74. *Urospermum dalechampii* . Available online: https://earthone.io/plant/ (accessed on 2 November 2025).
75. Chenot-Lescure, J.; Jaunatre, R.; Buisson, E.; Ramone, H.; Dutoit, T. Using various artificial soil mixtures to restore dry grasslands in quarries. *Restor. Ecol.* **2022**, *30*, e13620. [CrossRef]
76. Anaclerio, M.; Buttaro, D.; Leoni, B.; Renna, M.; Serio, F.; Santamaria, P. Boccioni e grespini: Da piante spontanee a nuovi ortaggi da foglia per la IV gamma. In Proceedings of the Atti di Convegno-11 Convegno Nazionale Sulla Biodiversità, Matera, Italy, 9–10 June 2016.
77. Buttaro, D.; Renna, M.; Gerardi, C.; Blando, F.; Santamaria, P.; Serio, F. Soilless production of wild rocket as affected by greenhouse coverage with photovoltaic modules. *Acta Sci. Pol. Hortorum Cultus* **2016**, *15*, 129–142.
78. Gonnella, M.; Renna, M.J.A. The evolution of soilless systems towards ecological sustainability in the perspective of a circular economy. Is it really the opposite of organic agriculture? *Agronomy* **2021**, *11*, 950. [CrossRef]
79. Bretzel, F.; Vannucchi, F.; Benvenuti, S.; Rumble, H. Biodiversity of flora and fauna. In *Rooftop Urban Agriculture*; Springer: Cham, Switzerland, 2017; pp. 235–252.

 https://doi.org/10.3390/horticulturae12020186

80. Di Gioia, F.; Renna, M.; Santamaria, P. Sprouts, microgreens and "baby leaf" vegetables. In *Minimally Processed Refrigerated Fruits and Vegetables*; Springer: Cham, Switzerland, 2017; pp. 403–432.
81. Paradiso, V.M.; Castellino, M.; Renna, M.; Leoni, B.; Caponio, F.; Santamaria, P. Simple tools for monitoring chlorophyll in *broccoli raab* and *radish* microgreens on their growing medium during cold storage. *Prog. Nutr.* **2018**, *20*, 415–422.

horticulturae

MDPI

Review

Sonchus Species of the Mediterranean Region: From Wild Food to Horticultural Innovation—Exploring Taxonomy, Cultivation, and Health Benefits

Adrián Ruiz-Rocamora [1,*], Concepción Obón [2], Segundo Ríos [3], Francisco Alcaraz [1] and Diego Rivera [1,*]

[1] Departamento de Biología Vegetal, Facultad de Biología, Campus de Espinardo, Universidad de Murcia, 30100 Murcia, Murcia, Spain; falcaraz@um.es
[2] Instituto de Investigación e Innovación Agroalimentario y Agroambiental (CIAGRO), Escuela Politécnica Superior de Orihuela, Universidad Miguel Hernández, Ctra. Beniel, Km 3,2, 03312 Orihuela, Alicante, Spain; cobon@umh.es
[3] I.U. de Investigación—CIBIO, Universidad de Alicante, Carretera San Vicente del Raspeig s/n, 03690 San Vicente del Raspeig, Alicante, Spain; s.rios@ua.es
* Correspondence: ruiz.rocamora@yahoo.es (A.R.-R.); drivera@um.es (D.R.); Tel.: +34-868884994 (D.R.)

Abstract

The genus *Sonchus* (Asteraceae) comprises 98 species, including 17 predominantly herbaceous taxa native to the Mediterranean region. These plants have long been utilized as traditional wild food sources due to their high nutritional value, as they are rich in vitamins A, C, and K, essential minerals, and bioactive compounds with antioxidant and anti-inflammatory properties. This review aims to provide a comprehensive synthesis of the taxonomy, geographic distribution, phytochemical composition, traditional uses, historical significance, and pharmacological properties of *Sonchus* species. A systematic literature search was conducted using PubMed, Scopus, Web of Science, and Google Scholar, focusing on studies from 1980 to 2024. Inclusion and exclusion criteria were applied, and methodological quality was assessed using standardized tools. A bibliometric analysis of 440 publications (from 1856 to 2025) reveals evolving research trends, with *S. oleraceus*, *S. arvensis*, and *S. asper* being the most extensively studied species. The review provides detailed taxonomic insights into 17 species and 14 subspecies, emphasizing their ecological adaptations and biogeographical patterns. Additionally, it highlights the cultural and medicinal relevance of *Sonchus* since antiquity while underscoring the threats posed by environmental degradation and changing dietary habits. *Sonchus oleraceus* and *S. tenerrimus* dominate the culinary applications of the genus, likely due to favorable taste, wide accessibility, and longstanding cultural importance. The comprehensive nutritional profile of *Sonchus* species positions these plants as valuable contributors to dietary diversity and food security. Finally, the study identifies current knowledge gaps and proposes future research directions to support the conservation and sustainable utilization of *Sonchus* species.

Keywords: culinary tradition; diet; domestication; ethnobotany; ethnopharmacology; local food; modern cuisine; nutraceuticals; weeds; Asteraceae; sow thistle; functional foods; traditional knowledge; sustainable agriculture; bioactive compounds; wild edibles

1. Introduction

Currently, global food systems exhibit an excessive reliance on a limited number of plant species, a trend that poses significant risks to food security, human health, and environmental sustainability. Broadening the scope of both agricultural production and dietary consumption to incorporate a wider array of plant taxa—particularly those considered underutilized—holds considerable potential for enhancing nutritional quality, supporting livelihoods, and promoting ecological resilience [1]. In this regard, wild plant species have historically played a vital role in complementing staple foods by providing essential micronutrients, including trace elements, vitamins, and minerals, necessary for a balanced diet. Their reintegration into contemporary diets may once again prove indispensable in addressing nutritional challenges [1].

The Mediterranean region, with its rich ethnobotanical heritage, offers a particularly valuable reservoir of such underexploited species. Many plants traditionally consumed as part of the Mediterranean diet may serve as important genetic resources for the innovation and development of novel food products [2,3]. In this context, the systematic review of various species within the genus *Sonchus*, which have been historically utilized in this region, becomes particularly relevant. Such an examination may shed light on their ethnobotanical significance, nutritional potential, and applicability in sustainable food systems.

The genus *Sonchus* (Asteraceae) comprises 98 species, of which 17 are predominantly herbaceous and native to the Mediterranean region [4–10]. These species include 12 monotypic (*Sonchus briquetianus* Gand., *S. crassifolius* Willd., *S. erzincanicus* V. A. Matthews, *S. fragilis* Ball, *S. macrocarpus* Boulos & C. Jeffrey, *S. maculigerus* H.Lindb., *S. masguindalii* Pau & Font Quer, *S. mauritanicus* Boiss. & Reut., *S. microcephalus* Mejías, *S. oleraceus* L., *S. pustulatus* Willk., *S. suberosus* Zohary & P.H.Davis) and 5 polytypic (*Sonchus arvensis* L., *S. asper* (L.) Hill, *S. bulbosus* (L.) N. Kilian & Greuter (≡*Aetheorhiza bulbosa* (L.) Cass. in Cuvier), *S. maritimus* L., *S. palustris* L., *S. pustulatus* Willk., *S. suberosus* Zohary & P. H. Davis, and *S. tenerrimus* L.) with 14 subspecies. These species exhibit remarkable adaptability, thriving in a wide range of habitats, from natural environments such as rocky outcrops, wetlands, and springs to seminatural and anthropized landscapes, including cultivated fields where they often grow as weeds.

Sonchus species have served as valued wild food plants in Mediterranean rural communities for centuries, consumed fresh in salads, cooked in stews, or preserved as brined products [11]. These plants possess significant nutritional value, containing high levels of vitamins A, C, and K, essential minerals, including calcium, magnesium, and iron, and bioactive compounds such as polyphenols and flavonoids that confer antioxidant and anti-inflammatory properties [12].

Historical documentation of *Sonchus* consumption dates to ancient Greek and Roman civilizations. The Greek physician Dioscorides recorded their medicinal applications in De Materia Medica [13], while contemporary traditional dishes such as Greek "horta" and Italian "pistic" demonstrate their enduring cultural relevance [14]. Beyond culinary applications, these species contribute to Mediterranean ecosystem stability by colonizing disturbed habitats and providing soil stabilization in challenging environmental conditions [15].

However, urbanization and evolving dietary patterns threaten the preservation of traditional knowledge surrounding *Sonchus* species, necessitating urgent documentation efforts to maintain Mediterranean cultural heritage and biodiversity [16,17].

Contemporary research has expanded understanding of the genus *Sonchus*, which comprises over 50 recognized species distributed across the Mediterranean and beyond, exhibiting considerable morphological and ecological diversity [5–10,18]. Phytochemical analyses have identified abundant bioactive compounds, including polyphenols, flavonoids, sesquiterpene lactones, and triterpenoids, which underpin their antioxidant, anti-inflammatory, and antimicrobial properties [19].

Recent studies have confirmed traditional medicinal applications for treating digestive disorders, liver conditions, and skin infections [20] while exploring their potential in functional foods and nutraceuticals due to exceptional nutritional profiles and health-promoting compounds [21]. Despite these advances, current reviews remain limited in scope, addressing either individual species or specific aspects rather than providing comprehensive analytical frameworks [21–23].

Experimental cultivation trials have investigated the agronomic requirements, yield potential, and environmental adaptability of *Sonchus* species [24]. These studies reveal promising prospects for integration into sustainable agricultural systems, particularly for diversified crop production. However, significant challenges persist, including the need for optimized cultivation protocols, enhanced understanding of genetic diversity, and development of strategies to control potential invasive behavior in certain environments [25,26].

A significant scientific gap persists in our understanding of Mediterranean species within the genus *Sonchus*. There is a pressing need for a comprehensive review that integrates detailed data on species taxonomy and distribution, ethnobotanical and ethnopharmacological knowledge, phytochemical composition, and nutritional and gastronomic value, as well as information on the natural habitats and ecosystems in which these species grow. Such ecological insights are essential for informing potential cultivation practices. This integrative approach would provide a clear synthesis of the current state of knowledge and highlight critical gaps that warrant further scientific investigation. Further research is essential to fully characterize the ecological, chemical, and pharmacological diversity within this genus and realize its potential for modern applications.

This review paper aims to critically synthesize the current body of knowledge on *Sonchus* species, with a particular focus on their horticultural development. The specific objectives are as follows:

- To provide a comprehensive overview of the taxonomy, distribution, and ecological adaptability of *Sonchus* species, highlighting their diversity and the environmental conditions under which they thrive.
- To review the traditional uses of *Sonchus* species as wild food plants but also as medicinal resources for humans and livestock, emphasizing their cultural significance and the ways in which they have been utilized by local communities.
- To analyze the phytochemical composition and nutritional value of *Sonchus* species, with a focus on their bioactive compounds and potential health benefits.
- To evaluate the results of experimental cultivation trials, assessing the agronomic potential of *Sonchus* species as novel crops and identifying key factors influencing their growth and yield.
- To discuss the prospects and challenges associated with the horticultural development of *Sonchus* species, including opportunities for their integration into sustainable agricultural systems and strategies to address potential limitations.
- To identify gaps in the current research and propose future directions, with the aim of advancing the understanding and utilization of *Sonchus* species in horticulture and beyond.

By addressing these objectives, this review seeks to provide a holistic perspective on the potential of *Sonchus* species as valuable resources for food, nutrition, and sustainable agriculture while also highlighting the need for further research to fully realize their benefits.

2. Materials and Methods

2.1. Research Design and Approach

To conduct a comprehensive review of the diversity, phytochemistry, historical context, and uses of *Sonchus* species, a systematic approach was employed to gather and analyze relevant data. We manually performed a systematic review complemented by bibliometric analysis. The methodology included the following steps:

Literature Search: A thorough search of scientific databases (e.g., PubMed, Scopus, Web of Science, and Google Scholar) was conducted using keywords such as "*Sonchus*", "sow thistle", "phytochemistry", "traditional uses", "ethnobotany", "horticulture", "nutritional value", and "bioactive compounds". The search was limited to peer-reviewed articles, books, and conference proceedings published in English, Spanish, French, Italian, and other languages, covering the period from 1980 to 2024.

A bibliometric analysis was performed by means of a search in Scopus [27] using "*Sonchus*" as a search term in the titles, and the 440 records were exported in a CSV file, which was processed using the tools of Bibliometrix 4.1 [28,29] and VOSviewer 1.6.20 [30]. VOSviewer is a software tool for creating maps based on network data and visualizing and exploring these maps [31,32].

The bibliometric analysis identified the most taxonomically relevant species and primary research domains, facilitating the selection of appropriate keywords for the systematic literature search. The comprehensive search encompassed all *Sonchus* taxa examined in this review, including recognized synonyms.

Figures were elaborated using different methods. In the case of Figure 1, Figure 4, Figure 7, and Figure 8, the images were obtained from nature using conventional digital photo cameras. Figure 2 was elaborated using the tools of VOSviewer for the construction and visualization of a co-occurrence network of terms extracted from the titles and abstracts of publications [30–32]. Figure 3 is elaborated using the Bibliometrix 4.1 R package [28,29] with Keywords Plus as the analytical parameter and employing six chronological demarcations (1992, 2009, 2013, 2018, 2020, 2022, 2025) to establish discrete temporal intervals [29]. In Figure 5, a heat map was created using Python 3.11.13 with the following libraries: Pandas 2.3.1 (used to structure the dataset into a DataFrame for easy handling before plotting), Matplotlib 3.10.3 (matplotlib.pyplot) (used for the overall plotting framework), Seaborn 0.13.2 (used for generating the heat map via seaborn.heatmap) and Color Map of Matplotlib 3.10.3. Figure 6 was manually elaborated using Excel 365.

Inclusion and Exclusion Criteria: Studies were included if they provided detailed information on the taxonomy, distribution, phytochemical composition, traditional uses, history, or pharmacological properties of *Sonchus* species present in the Mediterranean. Articles lacking proper identification of the material studied or referring to "Sonchus viruses" were excluded.

Data Extraction and Synthesis: Data from selected studies were systematically extracted, including species names, geographical distribution, phytochemical profiles, traditional uses, and biological activities. The information was organized into thematic categories (e.g., diversity, phytochemistry, traditional uses, and modern applications) to facilitate analysis.

2.2. Data Quality and Analysis

Quality Assessment: The methodological quality of included studies was evaluated using standardized tools, adapting the PRISMA guidelines for systematic reviews to ensure reliability and validity [33] to the peculiarities of the available literature. However, given the heterogeneity of fields and styles of publications, it was impossible to obtain a tool that was operative for the different subjects covered. We developed a comprehensive scoring rubric to systematically evaluate references in our *Sonchus* phytochemical literature review. Each reference was meticulously assessed using a standardized scoring system ranging from 0 to 5. The evaluation criteria were strategically designed to assess two primary dimensions of scholarly literature: botanical integrity and either ethnobotanical or chemical merit, depending on the references' specific focus. Detailed scoring criteria are elaborated in Supplementary Table S1, ensuring a rigorous and objective approach to literature evaluation. Given the heterogeneity in the analyzed fields, we have not systematically developed specific quality criteria for all fields. However, we have paid particular attention to the reliability of botanical identification, the clear definition of properties and uses, and the geographical area.

Data Analysis: Descriptive statistics and qualitative synthesis were used to summarize findings. Trends and gaps in the literature were identified, and the potential for future research was discussed.

Verification and Validation: To ensure accuracy, data were cross-referenced with authoritative botanical databases (e.g., Plants of the World Online) [4] and validated by consulting experts in ethnobotany and phytochemistry.

3. Sonchus Diversity and Research Fields

3.1. Sonchus Diversity

The genus *Sonchus* exhibits significant taxonomic diversity, encompassing ninety-eight species distributed across temperate and subtropical biomes. The distribution patterns reveal a pronounced Mediterranean–Irano–Turanian center of diversity, with several taxa demonstrating remarkable colonization capacity, as evidenced by *S. oleraceus* and *S. asper* (Table 1) [4–10]. Life-form diversity ranges from annual to perennial habits, including specialized adaptations such as tuberous and rhizomatous geophytes. *S. tenerrimus* presents notable infraspecific diversification in the western Mediterranean region, particularly along the Spanish Mediterranean coast and other Mediterranean regions (Table 1) [4–10]. The biogeographical pattern suggests an evolutionary radiation from ancestral Mediterranean elements, with subsequent adaptation to various ecological niches, reflected in the current distribution of endemic taxa such as *S. crassifolius* in central and eastern Spain, *S. erzincanicus* in eastern Turkey, *S. briquetianus*, *S. fragilis*, *S. maculigerus*, *S. masguindalii* in northern Morocco, *S. suberosus* in Israel and Jordan, and *S. macrocarpus* in Egypt.

Table 1. Mediterranean *Sonchus* taxa, distribution range, types, and biomes [1].

Species and Subspecies	Distribution Range	Types	Biomes
Sonchus arvensis L. [Sp. Pl.: 793 (1753)]	Temperate Eurasia, introduced in N. America. Naturalized in other regions [34]	Perennial	Temperate
Sonchus arvensis subsp. *uliginosus* (M.Bieb.) Nyman [Consp. Fl. Eur.: 433 (1879)]	Temperate Eurasia, introduced in N. America	Perennial	Temperate
Sonchus asper (L.) Hill [Herb. Brit. 1: 47 (1769)] (=*Sonchus oleraceus* var. *asper* L. [Sp. Pl.: 794 (1753)])	Temperate Eurasia, N. Africa, Sahel, Somalia. Naturalized in other regions [34]	Annual	Temperate
Sonchus asper subsp. *glaucescens* (Jord.) Ball [J. Linn. Soc., Bot. 16: 548 (1878)]	Europe to Medit. and Himalaya, Sri Lanka, New Guinea	Biennial	Temperate and Subtropical
Sonchus briquetianus Gand. [Bull. Soc. Bot. France 55: 657 (1909)]	Morocco	Perennial	Subtropical
Sonchus bulbosus (L.) N.Kilian & Greuter [Willdenowia 33: 237 (2003)] (=*Aetheorrhiza bulbosa* (L.) Cass. [G.-F. Cuvier, Dict. Sci. Nat., ed. 2. 48: 426 (1827)])	Mediterranean, Azores	Tuberous geophyte	Temperate

Table 1. *Cont.*

Species and Subspecies	Distribution Range	Types	Biomes
Sonchus bulbosus subsp. *microcephalus* (Rech.f.) N.Kilian & Greuter [Willdenowia 33: 237 (2003)]	E. Mediterranean	Tuberous geophyte	Temperate
Sonchus bulbosus subsp. *willkomii* (Burnat & Barbey) N.Kilian & Greuter [Willdenowia 33: 237 (2003)]	Baleares (Spain)	Tuberous geophyte	Temperate
Sonchus crassifolius Pourr. ex Willd. [Sp. Pl., ed. 4. 3: 1509 (1803)]	Central and E. Spain	Perennial	Temperate
Sonchus erzincanicus V.A.Matthews [Notes Roy. Bot. Gard. Edinburgh 33: 258 (1974)]	E. Turkey	ND	Temperate
Sonchus fragilis Ball [J. Bot. 11: 372 (1873)]	Morocco	Perennial	Subtropical
Sonchus macrocarpus Boulos & C.Jeffrey [Taxon 18: 349 (1969)]	Egypt	Perennial or rhizomatous geophyte	Subtropical
Sonchus maculigerus H.Lindb. [Acta Soc. Sci. Fenn., Ser. B, Opera Biol. 1(2): 169 (1932)]	Morocco	ND	Temperate
Sonchus maritimus L. [Syst. Nat., ed. 10. 2: 1192 (1759)]	Mediterranean to Central Asia, Pakistan, Chad. Naturalized in other regions [34]	Perennial or rhizomatous geophyte	Temperate
Sonchus maritimus subsp. *aquatilis* (Pourr.) Nyman [Consp. Fl. Eur.: 434 (1879)]	W. Mediterranean	Perennial or rhizomatous geophyte	Temperate and subtropical
Sonchus masguindalii Pau & Font Quer [Exsicc. (Iter Marocc.) 1927: n° 732 (1928)]	Morocco		
Sonchus mauritanicus Boiss. & Reut. [Pugill. Pl. Afr. Bor. Hispan.: 70 (1852)]	Morocco, Algeria, Tunisia	Perennial	Temperate and subtropical
Sonchus microcephalus Mejías [Lagascalia 16: 169 (1990)]	E. Mediterranean, Rift Valley, introduced in Spain	ND	Subtropical
Sonchus oleraceus L. [Sp. Pl.: 794 (1753)] (=*Sonchus oleraceus* var. *laevis* L. [Sp. Pl.: 794 (1753)])	Native in Europe, N. Africa Near East and introduced worldwide [8]	Annual or biennial	Temperate and subtropical
Sonchus pustulatus Willk. [M.Willkomm & J.M.C.Lange, Prodr. Fl. Hispan. 2: 242 (1865)]	SE. Spain, NW. Africa	Perennial	Subtropical
Sonchus suberosus Zohary & P.H.Davis [Kew Bull. 2: 87 (1947)]	Israel to Jordan	Perennial	Subtropical
Sonchus tenerrimus L. [Sp. Pl.: 794 (1753)] (=*Sonchus pectinatus* DC.)	Macaronesia, Mauritania, Medit. to Pakistan, Arabian Peninsula, NE. Tropical Africa. Naturalized in other regions [34]	Annual, biennial or perennial	Subtropical
Sonchus tenerrimus subsp. *amicus* (Faure, Maire & Wilczek) Véla [Islands and Pl.: 281 (2013)]	Algeria	Perennial	Subtropical

Table 1. *Cont.*

Species and Subspecies	Distribution Range	Types	Biomes
Sonchus tenerrimus subsp. *dianae* (Lacaita ex Willk.) Ballester, Figuerola, Peris & Stübing [Estud. Multidiscipl. Parque Nat. Montgo: 206 (1991)]	E. Spain	Perennial	Subtropical
Sonchus tenerrimus subsp. *halodianae* Mateo, P.P.Ferrer & R.Roselló [Fl. Montiber. 75: 26 (2019)]	E. Spain	Perennial	Subtropical
Sonchus tenerrimus subsp. *polypodioides* Mateo, P.P.Ferrer & R.Roselló [Fl. Montiber. 75: 26 (2019)]	E. Spain	Perennial	Subtropical

[1] Data according to POWO and Boulos [4–10].

The morphological analysis of Mediterranean *Sonchus* taxa reveals several diagnostic patterns. The genus *Sonchus* exhibits remarkable leaf morphological diversity across its constituent species and subspecies, demonstrating significant variation in shape, texture, margin characteristics, and anatomical features that reflect evolutionary adaptations to diverse ecological niches (Figure 1).

Sonchus species display a range of primary leaf shapes, with lanceolate representing the most prevalent form, occurring in approximately 70% of examined taxa. Oblong configurations are frequent in several species (*S. arvensis*, *S. asper*, *S. crassifolius*, *S. macrocarpus*, *S. maritimus*), while ellipsoid shapes characterize *S. arvensis* and *S. erzincanicus*. Obovate or obovoid forms occur in specialized taxa (*S. oleraceus*, *S. bulbosus*, *S. masguindalii*), representing adaptations to specific ecological niches.

The genus demonstrates a sophisticated gradient of leaf division complexity. Entire margins represent the simplest condition, observed in *S. bulbosus* and certain populations of *S. crassifolius* and *S. maritimus*. Pinnatifid architecture predominates across the genus, with variations including lyrate-pinnatifid (*S. asper*, *S. oleraceus*, *S. masguindalii*) and runcinate-pinnatilobed (*S. suberosus*) patterns.

Advanced division reaches pinnatisect levels in multiple lineages (*S. arvensis* subsp. *uliginosus*, *S. asper*, *S. erzincanicus*, *S. masguindalii*, *S. mauritanicus*, *S. microcephalus*, *S. tenerrimus*), with *S. tenerrimus* achieving the most complex bipinnate architecture, representing the pinnacle of leaf dissection within the genus.

Lobe characteristics exhibit remarkable diversity: linear lobes characterize highly dissected species (*S. briquetianus*, *S. microcephalus*, *S. tenerrimus*), while oblong lobes occur in *S. erzincanicus*. Ovate lobes distinguish *S. fragilis*, and rounded lobes are diagnostic for *S. masguindalii*. Lanceolate lobes appear in several taxa (*S. briquetianus*, *S. mauritanicus*, *S. tenerrimus*), demonstrating convergent morphological solutions.

Spiny margins constitute a major taxonomic character, occurring in the majority of species and representing a presumed ancestral condition. Notable spiny taxa include *S. arvensis*, *S. asper*, *S. crassifolius*, *S. erzincanicus*, *S. macrocarpus*, *S. maculigerus*, *S. maritimus*, and *S. mauritanicus*.

Non-spiny margins represent a derived condition in *S. oleraceus* and the *S. tenerrimus* complex, while reduced spinosity occurs in *S. asper* subsp. *glaucescens* and *S. maritimus* subsp. *aquatilis*. Slightly spiny conditions characterize *S. microcephalus*, suggesting intermediate evolutionary states.

Succulent or fleshy leaf texture represents a significant adaptive syndrome within *Sonchus*, occurring in multiple independent lineages: *S. briquetianus*, *S. crassifolius*, *S. erzincanicus*, *S. masguindalii*, *S. maritimus*, *S. pustulatus*, and *S. tenerrimus* subsp. *halodianae*. This convergent evolution toward succulence reflects adaptation to xeromorphic or halophytic conditions (Figure 1).

Figure 1. *Sonchus* leaf morphology. (**A**) Deeply pinnatisect, *S. microcephalus*; (**B**) entire, *S. maritimus* subsp. *aquatilis*, Arroyo Blanco, Murcia (Spain); (**C**) pinnatisect, spiny, *S. crassifolius*, El Romeral, Toledo (Spain); (**D**) pinnatisect, spiny, *S. crassifolius*, grown at Torretes, Alicante (Spain); (**E**) pinnatifid, *S. pustulatus*, San Telmo, Almería (Spain); (**F**) lyrate pinnatifid, spiny, *S. asper* Columbares, Murcia (Spain); (**G**) somewhat fleshy, adapted to saline environments, *S. tenerrimus* subsp. *halodianae*, San Antonio, Alicante (Spain); (**H**) glaucous, less spiny than *S. asper*, *S. asper* subsp. *glaucescens*, Santo Angel, Murcia (Spain); (**I**) lanceolate to obovoid, entire, *S. bulbosus*, El Saler, Valencia (Spain). Images: (**A**) Jose Luis Cánovas, with permission; (**B**,**F**,**H**) Antonio Robledo, with permission; (**C**,**D**) Segundo Ríos; (**E**) Estefania Mico, with permission; (**G**) Francisco Alcaraz; (**I**) Emilio Laguna, with permission.

Fragile texture characterizes *S. fragilis*, representing a unique mechanical property within the genus. Thick leaves occur in drought-adapted species (*S. crassifolius*, *S. erzincanicus*), while thin leaves distinguish the aquatic adaptation in *S. maritimus* subsp. *aquatilis*.

Glaucous coloration represents a widespread adaptive trait, occurring in approximately 45% of examined taxa (*S. asper* subsp. *glaucescens*, *S. briquetianus*, *S. erzincanicus*, *S. fragilis*, *S. macrocarpus*, *S. maritimus*, *S. masguindalii*, *S. mauritanicus*, *S. microcephalus*, *S. pustulatus*, *S. tenerrimus*). This blue-green pigmentation likely provides photoprotection and reduces water loss.

Green coloration characterizes several species (*S. arvensis*, *S. asper*, *S. crassifolius*, *S. maritimus* subsp. *aquatilis*, *S. oleraceus*, *S. suberosus*), while glossy green surfaces distinguish *S. asper*.

Auriculate leaf bases represent a distinctive character in three major species (*S. arvensis*, *S. asper*, *S. oleraceus*), suggesting possible systematic significance. Pustulate (tuberculate) surface ornamentation uniquely characterizes *S. pustulatus*, representing the most specialized textural modification within the genus.

Size variation ranges from small leaves (*S. briquetianus*, *S. bulbosus*, *S. fragilis*, *S. microcephalus*) to large configurations (*S. macrocarpus*), reflecting diverse resource allocation strategies.

Subspecific taxa demonstrate consistent differentiation through specific morphological parameters: degree of lobation (*S. arvensis* subsp. *uliginosus*, *S. tenerrimus* subsp. *dianae*), lobe width (*S. bulbosus* subsp. *willkomii*, *S. tenerrimus* subsp. *amicus*), division intensity (*S. tenerrimus* subsp. *polypodioides*), spinosity reduction (*S. asper* subsp. *glaucescens*, *S. maritimus* subsp. *aquatilis*), and ecological specialization (*S. tenerrimus* subsp. *halodianae*).

The remarkable morphological diversity in *Sonchus* leaf architecture reflects adaptive radiation responding to varied environmental pressures, including water availability, salinity stress, herbivory pressure, and light conditions. The prevalence of pinnatifid to pinnatisect forms suggests this architecture provides optimal photosynthetic efficiency while maintaining structural integrity across diverse habitats.

The convergent evolution of succulence, glaucescence, and spinosity reduction in multiple lineages indicates strong selective pressures shaping leaf morphology within the genus. The presence of highly specialized forms, such as the bipinnate architecture of *S. tenerrimus* and the pustulate surfaces of *S. pustulatus*, demonstrates the evolutionary potential for morphological innovation within *Sonchus*.

The achene, or cypsela, morphology presents two primary shapes: ellipsoid, predominant across the genus, and claviform, characteristic of *S. asper* and *S. oleraceus*. Achene dimensions exhibit limited variation, typically ranging from 2.5–4 mm, with *S. macrocarpus* displaying the largest fruits (4.5–5 mm) (Table 2). Leaf morphology demonstrates greater plasticity, ranging from soft, non-spiny margins in *S. oleraceus* and *S. tenerrimus* to distinctly spinose margins in *S. asper*. The pappus structure maintains relative uniformity across taxa (5–8 mm), with the presence or absence of an apical collar serving as a taxonomically significant character. Achene indument patterns show limited variation, with most taxa being glabrous or sparsely hairy, except for the distinctive retrorse spinulose hairs in *S. asper* and its subspecies.

Table 2. Mediterranean *Sonchus* leaf and cypsela (achene) characteristics [1].

Species and Subspecies	Leaves	Cypselae Shape	Superficial Processes	Apex Collar and Pappus Length
Sonchus arvensis L.	Lanceolate to oblong or ellipsoid, pinnatifid, green, spiny margins. Sheathing leaf base auriculate	Ellipsoid, 2.5–3.5 × 1–1.5 mm, brown, ribs 4–5(+) on each face	Faces transversely rugulose to tuberculate across and between ribs, glabrous or sparsely hairy	With a short collar. Pappus 8–12(14) mm
Sonchus arvensis subsp. *uliginosus* (M.Bieb.) Nyman	Similar to *S. arvensis* but often more deeply lobed to pinnatisect	Ellipsoid, 2.5–3.5 × 1–1.5 mm, brown, ribbed	Glabrous or sparsely hairy	With a short collar. Pappus 7–10 mm
Sonchus asper (L.) Hill	Oblong to lanceolate, lyrate pinnatifid to pinnatisect, spiny margins, glossy green. Sheathing leaf base auriculate	Claviform to ellipsoid, strongly compressed, 2.5–3 × 1–1.2 mm, whitish brown, each face 3-ribbed	Hairy at ribs and wings only, smooth among them. With minute retrorse spinulose hairs	Without a collar. Pappus (5)7–9 mm

Table 2. *Cont.*

Species and Subspecies	Leaves	Cypselae Shape	Superficial Processes	Apex Collar and Pappus Length
Sonchus asper subsp. *glaucescens* (Jord.) Ball	Glaucous, less spiny than *S. asper*	Claviform, 2.5–3 mm, whitish brown, hairy at ribs and wings	With retrorse spinulose hairs	Without a collar. Pappus 5–7 mm
Sonchus briquetianus Gand.	Leaves lanceolate, pinnatifid, small, with narrow lobes, lanceolate to linear, slightly fleshy, glaucous	Ellipsoid, 3–4 × 1 mm, brown-reddish, smooth	Glabrous	With a short collar. Pappus 6 mm
Sonchus bulbosus (L.) N.Kilian & Greuter	Relatively small leaves, lanceolate to obovoid, entire to slightly lobed	Ellipsoid, 3–4 mm, brown, smooth	Glabrous	With a short collar. Pappus 6–8 mm
Sonchus bulbosus subsp. *microcephalus* (Rech.f.) N.Kilian & Greuter	Similar to subsp. bulbosus but smaller leaves, lobed	Ellipsoid, 3–4 mm, brown, smooth	Glabrous	With a short collar. Pappus 6–8 mm
Sonchus bulbosus subsp. *willkomii* (Burnat & Barbey) N.Kilian & Greuter	Similar to *S. bulbosus* but with narrower lobes	Ellipsoid, 3–4 mm, brown, smooth	Glabrous	With a short collar. Pappus 6–8 mm
Sonchus crassifolius Pourr. ex Willd.	Oblong to lanceolate, entire to lobed or pinnatifid. Green. Thick, fleshy, spiny	Ellipsoid, 2–3.5 × 1–1.4 mm, brown, smooth	Glabrous	With a short collar. Pappus 7–10 mm
Sonchus erzincanicus V.A.Matthews	Widely lanceolate to the elliptic, pinnatisect, lobes oblong, glaucous, margin of leaf serrated, spiny, fleshy and thick	Ellipsoid, 3–4 mm, brown, ribbed	Glabrous or sparsely hairy	With a short collar. Pappus 6–8 mm
Sonchus fragilis Ball	Leaves small, lanceolate, deeply lobed, lobes ovate, fragile, glaucous	Ellipsoid, 3.5–4.25 × 1 mm, brown, smooth	Glabrous	With a short collar. Pappus 6 mm
Sonchus macrocarpus Boulos & C.Jeffrey	Large, oblong to lanceolate, lobate to lyrate pinnatifid, spiny margins, glaucous	Ellipsoid, 4–5.5 × 1.5–1.75 mm, brown to dark brown, smooth	Glabrous	Without a collar, or it is very short. Pappus 7– 8 mm
Sonchus maculigerus H.Lindb.	Lanceolate, pinnatifid, spiny margins	Ellipsoid, 3–4 mm, brown, ribbed	Glabrous or sparsely hairy	With a short collar. Pappus 6–8 mm
Sonchus maritimus L.	Oblong to lanceolate, entire to dentate, somewhat fleshy, spiny margins, glaucous	Ellipsoid, 2.25–3.5 × 1–1.6 mm, brown, smooth	Glabrous	With a short collar. Pappus 6–9 mm
Sonchus maritimus subsp. *aquatilis* (Pourr.) Nyman	Oblong to lanceolate, entire to dentate, rarely spiny, thin, green	Ellipsoid, 3–4 mm, brown, smooth	Glabrous	With a short collar. Pappus 6–8 mm
Sonchus masguindalii Pau & Font Quer	Leaves obovoid in outline, lyrate pinnatisect, with rounded lobes, slightly fleshy, glaucous	Ellipsoid, 2.25–3.25 × 0.6–0.8 mm, brown, smooth	Glabrous	With a short collar. Pappus 6–8 mm, white, bristly
Sonchus mauritanicus Boiss. & Reut.	Lanceolate in outline, pinnatifid to pinnatisect, lobes lanceolate with spiny margins, glaucous	Ellipsoid, 2.5–3 × 0.6–0.8 mm, brown, ribbed	Glabrous or sparsely hairy	With a short collar. Pappus 6–9 mm

Table 2. *Cont.*

Sonchus microcephalus Mejías	Widely lanceolate, small, deeply pinnatisect, lobes almost linear with slightly spiny margins. Somewhat glaucous	Ellipsoid, 2.5–3.5 mm, brown, ribbed	Glabrous or sparsely hairy	With a short collar. Pappus 6–8 mm
Sonchus oleraceus L.	Obovate, deeply lyrate pinnatifid, soft, and non-spiny. Green. Sheathing leaf base auriculate	Claviform, slightly compressed, (2)2.5–3(3.8) × 0.8–1 mm, bright brown, rugose especially at the ribs	Glabrous to minutely tuberculate between ribs	With a short collar. Pappus (5)6–7.5(10) mm
Sonchus pustulatus Willk.	Leaves lanceolate, pinnatifid, lobes fleshy, glaucous, with pustules (small bumps) on the surface	Cylindrical, 3.5–5 × 1.5 mm, brown, ribbed	Glabrous	With a short collar. Pappus 8 mm
Sonchus suberosus Zohary & P.H.Davis	Lanceolate in outline, runcinate-pinnatilobed to pinnatifid; green, lateral lobes irregularly dentate, rarely entire	Oblong, 3.5–4 × 0.8–1 mm, brown, tawny, nearly prismatic, 3–4-ribbed	Glabrous or sparsely hairy, striate between ribs	With a short collar. Pappus 4–5 mm
Sonchus tenerrimus L.	Widely lanceolate. Deeply divided, bipinnate. Lobes linear to lanceolate, glaucous, entire, pinnatifid to pinnatisect, soft and non-spiny	Ellipsoid, 2.5–3.5 × 1.2 mm, brown, ribbed	Glabrous or sparsely hairy	With a short collar. Pappus 6–8 mm
Sonchus tenerrimus subsp. *amicus* (Faure, Maire & Wilczek) Véla	Similar to *S. tenerrimus* but with narrower lobes	Ellipsoid, 2.5–3.5 mm, brown, ribbed	Glabrous or sparsely hairy	With a short collar. Pappus 6–8 mm
Sonchus tenerrimus subsp. *dianae* (Lacaita ex Willk.) Ballester, Figuerola, Peris & Stübing	Similar to *S. tenerrimus* but with more deeply lobed leaves	Ellipsoid, 2.5–3.5 mm, brown, ribbed	Glabrous or sparsely hairy	With a short collar. Pappus 6–8 mm
Sonchus tenerrimus subsp. *halodianae* Mateo, P.P.Ferrer & R.Roselló	Similar to *S. tenerrimus* but somewhat fleshy, adapted to saline environments	Ellipsoid, 2.5–3.5 mm, brown, ribbed	Glabrous or sparsely hairy	With a short collar. Pappus 6–8 mm
Sonchus tenerrimus subsp. *polypodioides* Mateo, P.P.Ferrer & R.Roselló	Similar to *S. tenerrimus* but with more finely divided leaves	Ellipsoid, 2.5–3.5 mm, brown, ribbed	Glabrous or sparsely hairy	With a short collar. Pappus 6–8 mm

[1] Data from [4–10,35–37].

3.2. *Bibliometric Analysis of Sonchus Research Fields*

The bibliometric network visualization map (Figure 2) presents a comprehensive overview of research centered on *Sonchus* species, particularly *Sonchus oleraceus* and *Sonchus asper*, revealing distinct but interconnected research domains. The network structure demonstrates several well-defined thematic clusters, each representing different research orientations and methodological approaches.

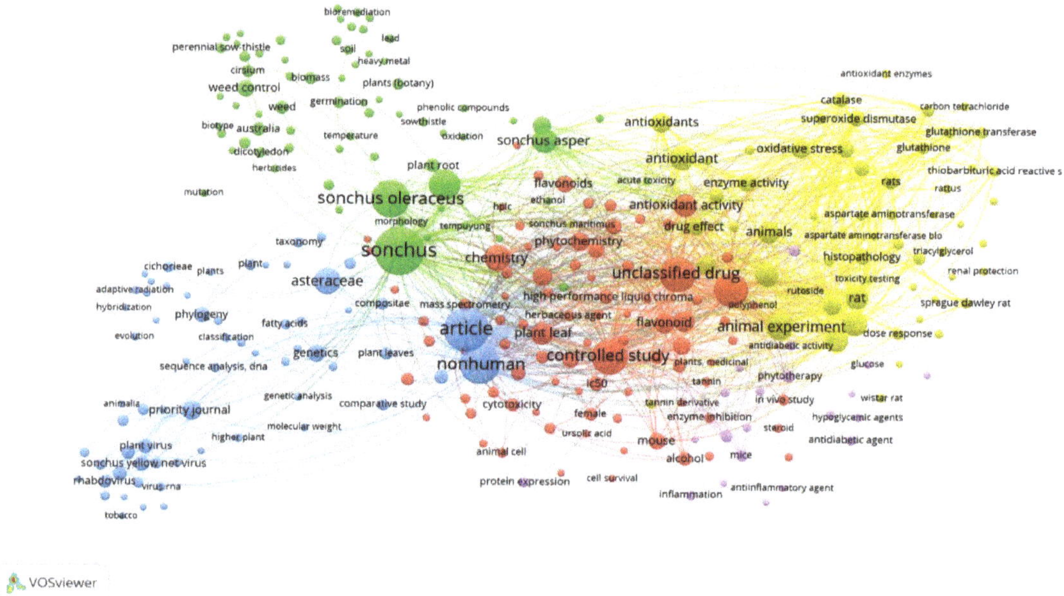

Figure 2. *Sonchus* research, bibliometric network (co-occurrence network of terms extracted from the titles and abstracts of publications) visualization map. Image elaborated by Diego Rivera with data from Scopus [27] processed using VOSviewer [30–32]. In the visualization presented, each circle represents a term. The size of a circle indicates the number of publications that have the corresponding term in their title or abstract. Terms that frequently co-occur tend to be located close to each other in the visualization. VOSviewer has grouped the terms into five clusters, of which four are of significant size. The red cluster, located in the lower central area in the visualization, consists of pharmacology terms. The green cluster, located in the upper left area, covers terms related to taxonomy, plant biology, and environmental factors. The blue cluster, located in the lower left area, consists of terms related to genetics but also to information science and information retrieval. In the right area in the visualization, the yellow cluster covers in vivo experiences in animal models and studies of activities. The small violet cluster, located in the lower central area, covers medicinal properties in terms of pharmacological activities and phytotherapy.

The agricultural and ecological research domain, displayed in green, emphasizes aspects such as weed control, bioremediation, and germination studies, reflecting the significance of these species in agricultural management and environmental applications. This connects naturally to the taxonomic and evolutionary studies shown in the blue cluster, which encompasses phylogenetic research, genetic studies, and viral interactions, providing fundamental understanding of these species within the Asteraceae family.

At the map's center, a prominent red cluster represents phytochemical and analytical research, featuring techniques such as mass spectrometry and HPLC, alongside studies of flavonoids and other chemical constituents. This central position suggests its role as a bridge between basic botanical research and applied studies.

The yellow/olive cluster reveals the significant attention paid to pharmacological and biochemical investigations, particularly focusing on antioxidant properties and oxidative stress studies. This research area demonstrates a substantial connection to experimental work involving animal models, suggesting active investigation of therapeutic potential.

The purple cluster, oriented toward medical applications, shows the translation of basic research into potential therapeutic uses, particularly in areas such as inflammation and diabetes management. This reflects the growing interest in developing plant-based therapeutic agents.

The visualization's structure, with its interconnected nodes of varying sizes, reveals both the breadth of *Sonchus* research and the complex relationships between different research approaches. The larger nodes, representing more frequently occurring terms, highlight the primary research foci, while the connecting lines demonstrate how these various research streams inform and influence each other, creating a rich tapestry of scientific investigation spanning basic science to applied research and potential therapeutic applications.

It can be deduced that the future cultivation and horticultural use of *Sonchus* should not forget its medicinal potential, which makes it a promising nutraceutical.

However, the bibliometric analysis yields ambiguous results in several aspects, which led us to conduct a manual review of the 440 publications selected by Scopus (Table 3). Following the application of exclusion criteria, 36 publications were discarded for referring to species missing in the Mediterranean region (species outside the geographical scope of the study), and 28 publications were excluded due to errors in species identification or, i.e., attribution of data to *Sonchus eruca*, an unknown taxon, or studies focusing specifically on "Sonchus viruses" not *Sonchus* species.

Table 3. Distribution of publications by subject area based on the results of the manual analysis of the *Sonchus* publications recovered by Scopus.

Subject Area	No. of Publications	Comments
Pharmacology	102	Multiple activity models
Phytochemistry	61	Flavonoids, sesquiterpene lactones
Weed Science	58	Weed control
Phytopathology	30	Fungal, viral, and insect-related studies
Ecology	23	Bioremediation
Horticulture	22	Optimal growing conditions
Microbiology	20	Bacteria and viruses
Genetics	17	Phylogenetic studies
Taxonomy	10	Nomenclatural, systematics, and morphological analyses
Nutrition	10	Nutritive properties of *Sonchus* and food processing
Plant Physiology	9	Responses to abiotic and biotic stress
Technology	8	Applications in biotechnology
Toxicology	4	Toxic and nocive substances
Ethnopharmacology	1	Traditional uses
Veterinary Medicine	1	Veterinary applications
Total	376	

Pharmacology, Phytochemistry, and Weed Science are the main research areas (Table 3). Additionally, it should be noted that a substantial portion of the 30 phytopathological studies focused on the development of biological control methods for *Sonchus* species as weeds.

Conversely, the numerous ethnobotanical and ethnopharmacological data documenting traditional uses of *Sonchus* species were excluded from the search results due to the search criteria based on "titles", "keywords", and "abstracts". These data typically appear in comprehensive studies of localities and regions that include dozens or hundreds of species, which authors cannot reflect in their abstracts. This limitation led us to develop specific targeted searches.

Figure 3 depicts a flow diagram representing the thematic evolution of *Sonchus* research from 1856 to 2025, based on 440 studies recorded in Scopus [27], with "*Sonchus*" in their titles and analyzed using Bibliometrix [28,29]. This temporal analysis reveals how research topics have evolved across different periods.

During the initial phase (1856–1992), research concentrated on fundamental aspects such as basic taxonomic studies of the *Sonchus* genus, electron microscopy, *Sonchus* yellow net virus, drug isolation from the Asteraceae family, and early extraction and purification studies.

The period from 1993 to 2009 witnessed an evolution toward specific studies in Cichorieae, research in Australasia, publications in priority journals, and initial work with *Sonchus oleraceus* extracts.

From 2010–2013, research became more specific, focusing on *Sonchus arvensis* (sow thistle), weed control reflecting its consideration as an invasive species, and the first systematic studies in China.

The 2014–2018 timeframe saw an expansion in research areas, including *Sonchus arvensis* extracts, medicinal herb properties, antioxidant studies, ethanol use for extraction, and a greater focus on plant extracts generally.

Between 2019 and 2020, there was notable diversification with studies on herbicides for agricultural control, genetic markers, glyphosate research, pharmacological investigations using rats, analyses of *Sonchus oleraceus* and *Sonchus asper*, and acute toxicity studies.

The 2021–2022 period focused on Asteraceae *Sonchus* taxonomic classification, ascorbic acid, mutation, *S. oleraceus* and *S. arvensis*, and biomass potential as a crop.

Most recent trends (2023–2025) include renewed general *Sonchus* studies, research in China, soil genetics, and pollution, as well as drought, biomass, and oxidative stress studies.

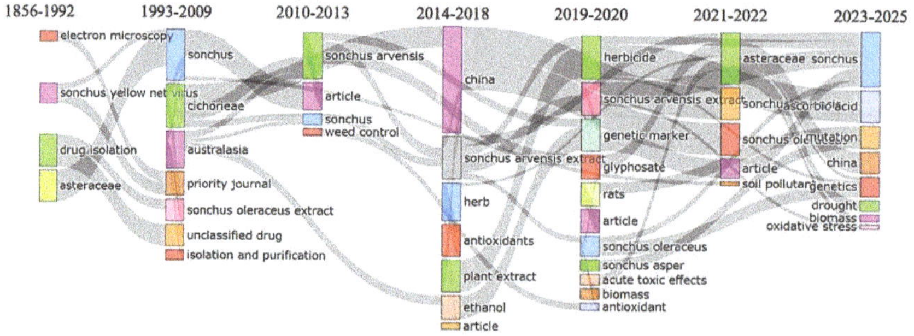

Figure 3. Comprehensive bibliometric analysis of 440 Scopus-indexed publications spanning 1856–2025 reveals the temporal evolution of research themes related to *Sonchus* species [27]. This longitudinal examination, visualized through alluvial flow mapping, demonstrates the shifting focus and emerging trends in *Sonchus* research over nearly two centuries of scientific investigation. The visualization was generated using the Bibliometrix R package [28,29] with Keywords Plus as the analytical parameter and employing six chronological demarcations (1992, 2009, 2013, 2018, 2020, 2022, 2025) to establish discrete temporal intervals [29]. The diagram illustrates significant thematic transitions across seven research periods, revealing the conceptual evolution from fundamental botanical characterization to applied horticultural and pharmacological investigations. Node size corresponds to keyword frequency, while flow width represents the strength of thematic connections between consecutive time periods. This visualization demonstrates the progressive diversification of *Sonchus* research, particularly the transition from taxonomic and ecological perspectives toward phytochemical, stress-response, and cultivation-oriented paradigms in recent years.

In the context of horticultural development, the data shows a clear evolution from purely botanical and taxonomic studies toward practical applications. There has been a gradual shift from considering *Sonchus* as a weed to valuing it as a potential crop, as evidenced by biomass studies in recent years. Relevant phytochemical advances, including the identification of specific compounds, like ascorbic acid, and antioxidant properties, suggest nutraceutical value.

Recent studies on environmental stresses, such as drought and oxidative stress, are particularly valuable for horticultural development under adverse conditions. The emergence of genetic marker research provides tools for improvement, while growing interest in China suggests an emerging potential market.

This thematic evolution supports the horticultural development of *Sonchus*, demonstrating a transition from basic research to more practical applications, with recent emphasis on characteristics relevant to sustainable cultivation and commercial uses.

4. Historical Context

4.1. Earliest Written Records and References to Sonchus Species in Historical Texts from Ancient Greece to Medieval Sources

The archaeological data includes the identification of cypselae from *Sonchus arvensis* L. in *Urartian* sites of the Caucasus, 1st mill BC. This is an example of the occasional presence of *Sonchus* remains in archaeological contexts of Europe, North Africa, and West Asia from Bronze Age levels of Baltic sites of Lithuania to Punic and Byzantine Carthage [38,39].

Campbell-Thompson's (posthumous edition in 1949) [40] interpretation of the Assyrian term ^{šam}HA-HI-IN—rendered as $^{\text{(šam)(is)}}$GIR-HA-AH in logographic form and *puquttu* in syllabic Assyrian—remains contentious within scholarly discourse. This botanical term, referring to a thistle employed medicinally as both a diuretic and lithontriptic agent, was identified by Campbell-Thompson as potentially representing several taxa: *Carduus* sp., *Scolymus* sp., *Sonchus oleraceus* L., or *Silybum marianum* L. While this taxonomic attribution has generated considerable debate, the identification merits serious consideration, as it may constitute the earliest documented reference to a *Sonchus* species in the ancient Near Eastern pharmacological literature [41].

The earliest reliable references to *Sonchus* species can be traced back to ancient Greek and Roman texts. For instance, Dioscorides, in his seminal work De Materia Medica, documented the medicinal properties of *Sonchus*, emphasizing its use for digestive and liver ailments [13]. *Sonchus* species have been traditionally consumed as food and used medicinally across the Mediterranean. Historical accounts describe their incorporation into daily diets and healing practices, often linked to their perceived detoxifying and anti-inflammatory properties [13]. The earliest codices of Dioscorides' work with information on *Sonchus* are the Codex Anicia Juliana of the Wien Library, the Codex Neapolitanus, and the manuscript of the Pierpont Morgan Library.

The preservation of ancient botanical knowledge owes much to the meticulous work of Byzantine scribes who maintained and illustrated crucial medical texts. The most significant repository of early *Sonchus* documentation exists within the manuscript tradition of Dioscorides' De Materia Medica. The Codex Vindobonensis Med. Gr. 1, created circa 512 CE for the imperial princess Anicia Juliana in Constantinople, stands as the oldest and most valuable illustrated herbal manuscript in existence, housed today at the Austrian National Library in Vienna. This illuminated codex contains detailed illustrations of medicinal plants, including *Sonchus* species, representing an invaluable bridge between ancient knowledge and medieval practice.

Contemporary with this work, the Codex Neapolitanus, dating to the late sixth or early seventh century and preserved at the National Library of Naples, provides another crucial early illustrated copy of Dioscorides' botanical descriptions. The manuscript tradition continued through the tenth century with Morgan 652, a Byzantine manuscript held at the Morgan Library & Museum in New York, which contains significant portions of Dioscorides' text with accompanying botanical illustrations.

Dioscorides' systematic approach to botanical documentation established the foundation for understanding *Sonchus* species in medical contexts [13,41]. His De Materia Medica distinguished between two primary types of "*Sonchos*": "*Sonchos trachos*" and "*Sonchos trypheros*". Modern botanical scholarship generally interprets "*Sonchos trachos*", meaning "rough sow thistle", as corresponding to *Sonchus asper*, characterized by its spiny leaves and coarse texture. Conversely, "*Sonchos trypheros*", the "tender sow thistle", is commonly associated with *Sonchus oleraceus*, distinguished by its smoother, more palatable foliage.

Dioscorides' descriptions emphasize both morphological and therapeutic characteristics, noting that both species possessed angular stalks with reddish interiors surrounded by irregularly disposed leaves. Medicinally, he characterized these plants as cooling and moderately astringent, making them suitable for treating gastric inflammation, stomach ailments, and even conditions associated with venomous stings [13,41]. The versatility of these plants extended to gynecological applications, with their juice being employed to alleviate stomach pains, promote lactation, and address inflammatory conditions of the reproductive system.

The textual tradition of Dioscorides reveals interesting variations in interpretation and presentation. While Osbaldeston's modern academic translation emphasizes clarity and systematic organization, Gunther's interpretation preserves the archaic linguistic conventions that reflect the historical context of the original Greek formulations. These variations demonstrate the evolution of scholarly interpretation while maintaining core botanical and therapeutic knowledge.

Pliny the Elder's Natural History provides a complementary perspective on *Sonchus* species that both reinforces and expands upon Dioscoridean knowledge [42]. Writing in Book XXII, Chapter 44, Pliny distinguished between species based on color rather than texture, referring to "white" and "black" kinds of *Sonchus*. This classificatory difference suggests either regional variations in the plants themselves or alternative observational approaches to botanical categorization.

Both classical authors agreed on fundamental morphological characteristics, including angular stems and the presence of milky latex when stems were broken, suggesting consistent identification of *Sonchus oleraceus* and related species across different geographical contexts. Their shared emphasis on the cooling and emollient properties of these plants, particularly for gynecological and gastrointestinal applications, demonstrates remarkable consistency in therapeutic understanding.

Pliny's account distinguished itself through more elaborate pharmacological detail, including specific dosage instructions and preparation methods. His recommendations for combining *Sonchus* with wine, oil, or broth, along with precise measurements, such as "*three cyathi*", reflect a more systematized approach to medicinal preparation. The Roman author's therapeutic applications extended beyond Dioscoridean usage to include respiratory difficulties, urinary calculi, halitosis, and various dermatological conditions, suggesting an expansion of medicinal knowledge through practical experience.

The transition from classical to medieval medicine witnessed the continued recognition of *Sonchus* species across various European medical traditions. The Salernitan school, representing one of the most influential medieval medical centers, incorporated *Sonchus* into its systematic approach to health maintenance. The *Regimen Sanitatis Salernitanum* described "*sonco*" as a bitter herb with blood-cleansing properties that could "purify the vision" while treating fevers through its cooling effects [43].

The *Antidotarium Nicolai*, an early twelfth-century medical text from the School of Salerno, provided comprehensive therapeutic applications for sow thistle, emphasizing its cooling and binding properties [44]. This medieval compilation extended traditional uses to include digestive aid, sleep promotion, and treatment of urinary stones, migraines, and convulsions. The text's systematic organization reflects the growing sophistication of medieval pharmaceutical knowledge while maintaining continuity with classical sources.

Hildegard von Bingen's twelfth-century *Physica* represents a distinct Germanic approach to therapeutics, identifying the plant under *"distel"*, notably *"carduus tam lenis quam hirsutus"* [45].

4.2. Uses of Sonchus in Traditional Medicine, Rituals, and as Food from the Renaissance to the 19th Century

The Renaissance marked a pivotal transition in botanical scholarship, characterized by the systematic reinterpretation of classical herbal knowledge. Herbalists of the period not only preserved ancient texts but also integrated contemporary observations, regional knowledge, and empirical findings into their treatises. The documentation of *Sonchus* species exemplifies this evolution, illustrating how Renaissance botanists expanded upon classical foundations with original insights derived from direct observation and cultural synthesis.

Pietro Andrea Mattioli's 1544 commentary on Dioscorides was foundational to Renaissance herbal literature, setting new standards by integrating classical sources with contemporary medical understanding. His treatment of *Sonchus* emphasized its therapeutic applications within the humoral framework, attributing cooling and moistening properties effective for inflammation [46].

Mattioli focused on medicinal uses relevant to physicians and apothecaries, particularly in women's health and lactation support. This practical orientation reflected a broader Renaissance trend of aligning medical utility with classical theory. His systematic categorization of therapeutic effects represented a shift from the descriptive style of ancient authors toward a more accessible clinical framework.

Andrés Laguna's 1555 Spanish translation and commentary on Dioscorides expanded upon Mattioli's work while introducing original content grounded in empirical observation and cultural context [47]. His documentation of *Sonchus* demonstrated a sophisticated synthesis of classical and contemporary scholarship, addressing both theoretical and practical dimensions.

Laguna identified three distinct *Sonchus* species—the wild and spiny *Sonchus asperus*, the tender *Sonchus laevis*, and a tree-like variety with broad, divided leaves—demonstrating detailed observational classification. His descriptions, consistent with classical sources, included angular, reddish stems and dissected leaves, while reaffirming traditional uses such as cooling, moderate astringency, and treatments for gastrointestinal discomfort, inflammation, lactation, and venomous stings.

Laguna's major contribution lay in his extensive cultural and linguistic contextualization. By documenting regional names across Romance and Germanic languages—such as *Cicerbita*, *Latucella*, *Latteron*, and *Gensz Distel*—he emphasized the widespread recognition of these plants, advancing botanical knowledge across linguistic boundaries.

His empirical observations further distinguished his work. Notably, he recorded that all *Sonchus* species are spineless in early growth, acquiring spines upon maturity—a developmental insight absent in classical texts. This phenological detail exemplifies the Renaissance shift from static to dynamic botanical descriptions.

Laguna also incorporated folk knowledge, referencing Apuleius' term *"lactuca leporina"* (hare's lettuce), tied to the belief that hares used the plant for cooling. His etymological explanation of *Latucella*, linked to its milky latex and lettuce-like appearance, demonstrated interpretive depth and alignment with vernacular associations.

He further contributed practical applications, including preparation methods like applying it with wool (*"puesto con un poco de lana"*), remedies for scorpion bites, and treatments mixing roots with pear juice for wounds, offering concrete therapeutic protocols largely absent from earlier sources.

Leonhart Fuchs' herbal observations [48], translated into Spanish by Jaraba in 1557 [49], also enriched *Sonchus* documentation by combining classical descriptions with folk knowledge. Drawing on Pliny, Fuchs identified two primary types: one more pungent and rustic and another milder and edible. His focus on morphological and sensory characteristics supported both medicinal and culinary uses.

Fuchs confirmed traditional cooling and constricting properties, with applications for inflammation, gastric heat, and lactation, aligning with classical therapeutics but in more accessible language. He preserved folklore around "hare's lettuce", underscoring his commitment to blending scholarly and popular botanical wisdom. His note on *Sonchus* juice for urinary issues suggested continued empirical refinement in therapeutic application.

Mattioli's 1557 edition, featuring *Soncho asper* and *Soncho liscio* with illustrations, epitomized Renaissance efforts to integrate visual and textual documentation [50]. These illustrations, alongside detailed descriptions, reflected the period's emphasis on precise identification and effective therapeutic use through multimodal representation.

Carolus Clusius in 1601 [51] continued the tradition of citing Dioscorides while adding his own observations on the plant's uses for stomach ailments, inflammations, and scorpion bites, demonstrating the persistence of classical medical knowledge through the Renaissance period.

By the seventeenth century, *Sonchus* had become so familiar across the Iberian Peninsula that it transcended medicinal contexts to enter popular language. Sebastián de Covarrubias' *Tesoro de la lengua castellana o española* (1674) captures this cultural integration, noting how the phrase *"agua de cerrajas"* (sow thistle water) became synonymous with weak or inconsequential arguments [52,53]. While acknowledging *Sonchus'* classical therapeutic value, particularly in Dioscorides, Covarrubias highlighted how the plant's perceived mildness led to its association with ineffectiveness in common discourse. His observation that people would dismiss feeble claims by saying *"todo es agua de cerrajas"* illustrates the integration of botanical references into colloquial speech, albeit with diminished regard for their medicinal efficacy.

This shift suggests that by the seventeenth century, *Sonchus* preparations were viewed more as gentle remedies for minor ailments than potent medicinal agents. The divergence between the medical literature, which upheld classical uses, and popular perception, which regarded the plant as ineffectual, reflects broader transformations in medical expectations and practice during the early modern era.

In the eighteenth century, *Sonchus* species saw a notable decline in pharmaceutical use. José Quer's extensive botanical survey noted that by the late 1700s, *cerraja* was seldom employed in Spanish apothecaries, indicating waning professional interest despite continued inclusion in materia medica literature [54].

Nonetheless, eighteenth-century authors largely upheld Dioscoridean assessments. *Sonchus oleraceus*, in particular, was consistently attributed cooling properties and mild astringency, aligning with classical humoral principles. Medicinal broths were still prepared to relieve heat and inflammation, while external applications treated skin conditions. The milky sap retained its use for gastric irritation and lactation support, underscoring the durability of classical concepts through the Enlightenment.

Quer also preserved traditional preparation methods, noting that crushing the plant with its root remained effective for treating scorpion stings in direct continuity with Dioscorides. Culinary uses persisted as well; young, tender leaves harvested before spine development continued to appear in salads, reflecting enduring knowledge about optimal harvesting and usage.

This eighteenth-century perspective reveals a complex relationship between evolving professional practice and persistent folk knowledge. Even as *Sonchus* lost favor in formal pharmacopoeias, traditional applications remained well-documented and widely understood, indicating that scholarly respect for classical therapeutic frameworks endured.

The nineteenth century brought a decisive transformation in the scientific treatment of *Sonchus* species, mirroring broader shifts in botanical science and medical research. Focus shifted from therapeutic descriptions to botanical classification and phytochemical analysis, as taxonomy and laboratory methods gained prominence [55].

This new scientific orientation enabled researchers to investigate the chemical constituents behind traditional effects, while systematic classification clarified interspecies relationships. Such developments marked a departure from earlier therapeutic models toward empirical frameworks rooted in analytical chemistry.

Despite this shift, ethnobotanical studies continued to document traditional uses across Europe. The persistence of local knowledge alongside scientific inquiry underscores the resilience of folk practices and the growing interest in validating them through modern research. This coexistence highlights a transitional phase in which traditional and scientific epistemologies were not mutually exclusive but mutually informative.

The documentation of continued ethnobotanical usage across Europe during this period provides important evidence for the enduring relevance of traditional knowledge systems. This parallel development of scientific investigation and traditional practice created opportunities for cross-validation between empirical folk knowledge and emerging scientific methodologies, laying the groundwork for modern approaches to natural product research and ethnopharmacology.

4.3. Cross-Cultural Adoption

The medicinal and culinary applications of *Sonchus* species eventually transcended their Mediterranean origins. Following European colonization of the Americas, Native American communities incorporated the plant into their pharmacopeia, employing it for diverse purposes, including as a vegetable, medicine, and notably as a treatment for opium addiction [56–58]. The *Biblioteca Digital de la Medicina Tradicional Mexicana* documents its integration into traditional Mexican healing practices, where *Sonchus oleraceus* was adopted alongside indigenous medicinal plants [58].

This cross-cultural adoption represents a significant example of medicinal plant knowledge exchange and adaptation, as *Sonchus* species introduced to the Americas were recognized for properties similar to those documented by ancient Mediterranean writers nearly two millennia earlier.

5. Ethnopharmacological Evidence

5.1. Medicinal Applications of Sonchus Species in the Mediterranean: A Quantitative and Qualitative Review

The traditional use of *Sonchus* species across Mediterranean cultures in late 20th and early 21st centuries (Table 4), through the analysis of references published from 1981 to 2024 [59–108], reflects a multifaceted medicinal profile, with evidence pointing to both high frequency and broad therapeutic scope (Supplementary Tables S2 and S3).

Quantitative analysis of ethnobotanical data reveals that the most common applications relate to the digestive system, accounting for approximately 28.8% of recorded uses. Treatments for injuries, including wounds, burns, and stings, follow at 16.8%, while respiratory disorders are addressed in 7.1% of cases. Uses pertaining to the circulatory system, skin diseases, and infectious or parasitic conditions each constitute around 3.3% of mentions. Less frequently, *Sonchus* is employed in the treatment of nervous system complaints (2.7%), as well as endocrine and nutritional disorders (1.6%). An additional 33.2% of uses pertain to various other conditions, including genitourinary issues, ear problems, and menstrual irregularities.

Table 4. Records for *Sonchus* ethnobotanical, ethnopharmacological, and ethnoveterinary uses by country.

Country	No. of Publications [1]	Records	% of Records	Average Records Per Publication	References
Spain	24(21)	463	65.8	19.3	[59–83]
Italy	15	169	23.0	11.3	[84–98]
France	3	27	3.8	9.0	[97,99,100]
Cyprus	1	13	1.8	13.0	[101]
Morocco	2	10	1.4	5.0	[102,103]
Greece	1	8	1.1	8.0	[104]
Turkey	3	7	1.0	2.3	[105–107]
Egypt	1	3	0.4	3.0	[108]
Libya	1	3	0.4	3.0	[109]
Croatia	2	3	0.4	1.5	[110,111]
Czechia	1	2	0.3	2.0	[112]
Pakistan	1	2	0.3	2.0	[113]
Algeria	1	1	0.1	1.0	[114]

[1] No. of publications: The number of publications in the consulted bibliography, including material from that particular country.

When examining specific medicinal functions, *Sonchus* emerges most prominently as a digestive aid, with 14 separate mentions supporting its role in facilitating gastrointestinal health (Supplementary Table S3). It is also frequently described as a depurative, used for systemic detoxification, and as a diuretic. Applications targeting liver ailments are similarly prevalent. The plant is repeatedly cited in wound healing contexts and is employed in the treatment of cutaneous abscesses, thermal injuries, and inflammatory skin conditions. Respiratory ailments such as upper tract infections, as well as anxiety and pain management, are also among its documented uses.

In terms of categorical emphasis, gastrointestinal support comprises roughly one-quarter of all recorded uses. Wound healing and injury-related applications follow at 17%, while detoxification and hepatoprotective treatments represent about 10%. Anti-inflammatory functions span approximately 8% of uses, and respiratory disorders account for 7%. These figures highlight the plant's predominant roles in digestive, dermatological, and anti-inflammatory therapeutic contexts.

Qualitative insights further underscore the versatility and cultural integration of *Sonchus* in traditional medicine. Its primary functions are centered around digestive regulation, hepatic support, and systemic detoxification. The plant is also widely recognized for its external applications in treating wounds, insect bites, scorpion stings, and burns. These practices are often grounded in their perceived anti-inflammatory and antimicrobial properties, which are reflected across multiple therapeutic categories.

Traditional methods of preparation vary depending on the intended use. Internal disorders are typically treated using decoctions and infusions, while external ailments are addressed through the application of poultices or direct topical treatments. In some cases, leaves are chewed for immediate relief of digestive discomfort, indicating a reliance on the plant's efficacy in both processed and raw forms.

From a phytochemical standpoint, the frequent use of *Sonchus* for hepatic conditions suggests the presence of hepatoprotective compounds. The consistency with which it is used for inflammatory and infectious conditions across body systems also indicates anti-inflammatory, antimicrobial, and potentially analgesic activities. These properties underscore the plant's pharmacological potential and warrant further biochemical investigation.

The medicinal use of *Sonchus* species within ethnobotanical contexts reveals a clear hierarchical pattern of utilization, with *Sonchus oleraceus* L. emerging as the primary species of therapeutic significance (Supplementary Table S4). Accounting for nearly half of all documented medicinal applications, *S. oleraceus* demonstrates exceptional versatility, followed by *S. asper* (L.) Hill subsp. *asper* and *S. tenerrimus* L. subsp. *tenerrimus*, which also contribute notably but on a reduced scale. In contrast, *S. arvensis* L., *S. maritimus* L., and *S. macrocarpus* exhibit limited-to-marginal medicinal relevance, either due to low intrinsic activity or underdocumentation.

Across the genus, medicinal applications are most concentrated in the treatment of digestive system disorders (Supplementary Table S3). This category alone encompasses over a third of all reported uses, with *S. oleraceus* showing particularly high frequency, suggesting a strong ethnopharmacological reputation for gastrointestinal relief. This widespread use across multiple species may reflect either shared phytochemical properties or convergent empirical validation across cultural groups. Injury treatment also emerges as a significant category, where *S. oleraceus*, *S. asper*, and *S. tenerrimus* are repeatedly employed. This suggests that certain *Sonchus* taxa may possess bioactive compounds with wound-healing potential, including anti-inflammatory, antimicrobial, or tissue-regenerative effects.

The application of *Sonchus* species to respiratory ailments appears more restricted and notably centered around *S. oleraceus*, implying a possible concentration of relevant phytoconstituents in this taxon not commonly present in its congeners. Similarly, the genus shows some involvement in treating infections, anxiety, and parasitic conditions. The presence of *Sonchus* species in contexts such as pyogenic infections and candidiasis supports the hypothesis of general antimicrobial potential across the genus. The use of *S. oleraceus* for anxiety introduces the possibility of neuroactive compounds, a hypothesis further supported by its unique presence in this therapeutic domain.

When examined through a species-specific lens, *S. oleraceus* possesses the most extensive medicinal profile, with documented use for eighteen distinct conditions spanning multiple physiological systems. It is especially prominent in the management of digestive and respiratory issues, circulatory concerns, and physical trauma, and is uniquely cited for treating anxiety and other specialized conditions. *S. asper* presents a moderately broad therapeutic scope, most strongly represented in digestive and genitourinary applications and distinguished as the only species employed in the treatment of candidiasis. *S. tenerrimus*, while comparable in breadth to *S. asper*, is especially relevant in wound care and is uniquely associated with the treatment of anorexia, nervous system disorders, and mite infestations such as sarcoptes. Meanwhile, *S. arvensis* shows modest medicinal engagement yet stands out for its association with anemia treatment, pointing to potential hematinic properties. *S. macrocarpus* and *S. maritimus* remain almost absent from the medicinal corpus, which may reflect either genuinely limited pharmacological utility or gaps in ethnomedical documentation.

The phytopharmacological implications of these patterns are significant. The multifunctionality of the more widely used species, particularly *S. oleraceus*, implies the presence of a suite of bioactive secondary metabolites with effects on diverse biological systems. This biochemical diversity may be underpinned by both common phytochemical frameworks shared across the genus and species-specific compounds that drive particular therapeutic uses. The consistent recurrence of certain uses—such as treatment of digestive ailments and injuries—across multiple species reinforces the credibility of these applications and suggests their entrenchment in traditional medical systems. Such patterns are likely the result of long-term empirical observation and cultural transmission rather than random or incidental use.

The widespread use of *Sonchus* in everyday contexts reflects its accessibility and cultural familiarity (Supplementary Tables S2–S4). Many of the conditions it addresses—such as indigestion, minor injuries, and respiratory discomfort—are common and recurrent, suggesting the plant played a regular role in household healthcare. Notably, its use in pediatric and maternal contexts (e.g., as a galactagogue or child sedative) points to its trusted status within the domestic sphere.

5.2. Veterinary Applications of Sonchus *Species in the Mediterranean: A Quantitative and Qualitative Review*

Although less documented than its use in human medicine, the ethnoveterinary application of *Sonchus* species offers valuable insights into traditional animal healthcare within the Mediterranean context (Supplementary Table S5). A notable feature is the clear parallel between human and veterinary uses—particularly in treating wounds, inflammation, and insect or scorpion stings—suggesting a shared recognition of the plant's therapeutic efficacy based on empirical observation [82,96,97].

Sonchus is especially prominent in livestock care, where it is employed to treat mastitis and promote lactation in dairy animals. These uses reflect its integration into traditional husbandry practices and its role in supporting agricultural productivity. Applications also extend to pigs, rabbits, and pigeons, animals of economic significance in rural Mediterranean settings, highlighting the plant's versatility in sustaining animal health.

Certain treatments appear unique to veterinary practice, such as remedies for pig rheumatism and rabbit dermatomycosis. These species-specific uses indicate a depth of traditional knowledge that goes beyond analogies with human medicine, reflecting a nuanced understanding of animal physiology and disease.

The practical focus of these applications is evident in their orientation toward conditions that directly impact productivity and welfare. The plant's use across both productive and companion animals illustrates its broad perceived value within traditional care systems. Viewed through a wider ethnomedical lens, this pattern aligns with a "One Health" perspective, wherein human, animal, and environmental health are interlinked. The shared use of *Sonchus* suggests an integrated knowledge system shaped by observation and transmitted across generations and domains.

Economically, the plant's role in treating animals is vital to household subsistence—particularly dairy livestock, pigs, and rabbits—and reinforces its significance in agro-medicine. Its contribution extends beyond health to the resilience of rural livelihoods.

Finally, the recurrence of certain applications—especially for inflammation, infection, and lactation—supports phytopharmacological hypotheses. The evidence for galactagogue, antimicrobial, and anti-inflammatory effects across species strengthens the case for bioactive compounds within *Sonchus*.

6. Ethnobotanical Evidence for Sonchus Uses as Food

6.1. Documentation of Local Knowledge and Practices Surrounding Sonchus *Gathering and Preparation by Rural Communities*

Over time, the perception of *Sonchus* species has evolved from being a staple wild food to a plant of marginal economic importance. However, recent interest in wild edible plants has revived their cultural and culinary significance [24,60,115].

Rural communities have preserved extensive knowledge about gathering and preparing *Sonchus* species, often passed down through generations. These practices reflect a deep connection to local ecosystems and cultural heritage [115]. Contemporary ethnobotanical research has documented the ongoing use of *Sonchus* species in both rural and urban settings. While traditional practices persist in some areas, there are notable differences in consumption patterns between rural and urban populations [60,115]. The data highlights (Supplementary Table S2) the widespread consumption of *Sonchus* as food, its significant role in traditional medicine, and its value as livestock feed.

An examination of the geographic distribution of *Sonchus* ethnobotanical and ethnopharmacological data (Table 4) reveals a highly uneven landscape in both the quantity and quality of available records. Across twelve countries, a total of 714 use records were extracted from 55 academic references. However, this apparent breadth in geographic scope is deceptive, as a deeper statistical analysis underscores significant disparities in research intensity and documentation. The numerical distribution reveals a profound skew. While the mean number of records per country stands at approximately 55, the median is only 5.5, and the standard deviation is exceptionally high at 130.4. This statistical profile highlights a data landscape marked by concentration rather than balance. Indeed, Spain alone contributes 464 records, accounting for 65% of the entire dataset, followed by Italy with 169 records (23.7%) and France with 71 (9.9%). Together, these three Western European nations account for nearly 99% of all documented uses, while the remaining nine countries are responsible for just over 1%.

The geographic distribution further emphasizes this imbalance. The Mediterranean Basin dominates, with 84.3% of all records originating from this region, while the remainder is split between North Africa (2.4%) and a sparse collection of non-Mediterranean countries, accounting for 13.4% of the dataset. North Africa, a region historically rich in traditional medicinal and culinary practices, is severely underrepresented, with just 17 records across three countries and only four references. The Middle East is similarly marginal, with only a single reference from Pakistan. Other regions, such as Eastern Europe, Sub-Saharan Africa, the broader Asian continent, and the Americas, are either absent or nearly invisible in the corpus, despite the known presence or naturalization of *Sonchus* species in these areas.

The roots of this imbalance are multifaceted, beginning with the geographical limits of this review but also linked to academic infrastructure. Countries such as Spain and Italy benefit from robust ethnobotanical research programs, systematic regional surveys, and an institutional culture that values the documentation of traditional knowledge. Their high record counts are likely a result not only of cultural emphasis on traditional plant use but also of access to European Union research funding streams that prioritize biodiversity and heritage studies. In contrast, countries with less research infrastructure, lower publication output, or more fragmented knowledge transmission systems—particularly those that rely on oral traditions—are systematically underrepresented.

The complete absence of data from the Americas, much of Africa, and most of Asia—despite the presence of naturalized or historically used *Sonchus* species—is due to the geographical limitations defined for this review, but it does not mean true ethnobotanical insignificance in these areas.

Historically and culturally, the high concentration of records in Mediterranean countries aligns with longstanding traditions of *Sonchus* consumption and medicinal use (Table 4). These regions are deeply embedded in folk medicine practices and the broader framework of the Mediterranean diet, where wild greens, including *Sonchus*, play both nutritional and therapeutic roles.

In conclusion, the geographic patterns evident in the current *Sonchus* ethnobotanical literature underscore a critical need for more inclusive and globally representative research

(Table 4). Future studies should prioritize underrepresented regions—particularly the Middle East, Sub-Saharan Africa, Asia, and Latin America—where traditional plant knowledge exists but remains poorly documented. Collaborative research models that engage local scholars, community practitioners, and indigenous knowledge holders will be essential for addressing these gaps and constructing a more accurate global portrait of *Sonchus* uses.

6.2. Comparative Ethnobotanical Analysis Across Mediterranean Regions

Comparative ethnobotanical studies across Mediterranean regions have highlighted the diverse uses of *Sonchus* species in different cultures. These studies provide a comprehensive understanding of the plant's importance in traditional medicine and cuisine throughout the Mediterranean basin [115].

The analysis of total observations per country (Table 4 and Supplementary Table S2) demonstrates that Spain accounts for the highest number of *Sonchus* observations, with 463 records, representing approximately 65% of the total 704 observations. This substantial figure suggests that Spain may serve as a critical biodiversity hotspot for the genus, potentially due to its favorable Mediterranean climate, rich biodiversity, and extensive research endeavors. Italy follows with 169 observations, further reinforcing the notion of a strong *Sonchus* presence within Mediterranean regions. Countries such as France (27), Morocco (10), Greece (8), and Turkey (7) exhibit moderate observation numbers, while nations like Algeria (1), Pakistan (2), and Czechia (2) show very limited records, which could be attributed to either constrained research efforts or unsuitable ecological conditions.

Regarding species-specific distribution, *Sonchus oleraceus* emerges as the most widespread species, with 311 observations, being particularly prevalent in Spain (197) and Italy (74). Similarly, *S. tenerrimus* subsp. *tenerrimus* is also abundant, primarily observed in Spain (167) and Italy (20). *Sonchus asper* subsp. *asper* is frequently recorded in both Spain (72) and Italy (46). In contrast, several species exhibit highly restricted distributions. *Sonchus macrocarpus*, for instance, is exclusively found in Egypt (three observations), while *Sonchus crassifolius* (four observations) and *Sonchus maritimus* subsp. *aquatilis* (six observations) are confined to Spain. *Sonchus asper* subsp. *glaucescens* shows a scattered presence across Croatia (one), Turkey (two), and France (one). These patterns suggest a dichotomy between widespread generalist species and specialized species with narrow ecological niches or geographic ranges.

The favorable mild winters and hot summers characteristic of the Mediterranean climate appear to support a higher diversity and abundance of these species. Conversely, arid or African regions such as Morocco, Egypt, and Libya exhibit few observations (Table 4), likely due to less hospitable drier conditions. Similarly, the minimal presence in northern and central European countries like Czechia and Croatia suggests a preference for warmer climates.

The distinction between generalist and specialist species is evident. *Sonchus oleraceus* and *S. asper* are clear generalists, thriving across multiple countries, while species like *S. macrocarpus* and *S. crassifolius* are specialists, possibly adapted to unique local environmental conditions. Furthermore, the concentration of *Sonchus bulbosus* and *S. maritimus* observations predominantly within Spain and Italy hints at potential endemism or strong habitat preferences, such as coastal environments.

The high number of observations in Spain does not solely reflect actual abundance but could also be a consequence of more intensive botanical surveys in the region. Conversely, countries with limited records, such as Algeria, Libya, and Pakistan, may be considerably understudied. The influence of climate is also a significant factor, with *Sonchus* species generally thriving in temperate and Mediterranean zones but showing limited presence in deserts or colder climates.

In conclusion, Spain emerges as the predominant region for *Sonchus* diversity, closely followed by Italy, largely driven by the conducive Mediterranean climate. *Sonchus oleraceus* is identified as the most widespread species, contrasting with the localized distribution of species such as *S. macrocarpus* and *S. crassifolius*. The distribution patterns strongly suggest that Mediterranean climates favor *Sonchus* growth, while arid and colder regions present limitations. To obtain a more comprehensive understanding of *Sonchus* distribution, further research in currently underrepresented countries would be invaluable.

6.3. Food Applications Versus Other Traditional Uses: A Comparative Analysis of Contemporary Ethnobotanical Relevance in Sonchus Species

The analysis of total uses per category reveals that food is the most prevalent application of *Sonchus* species, accounting for 409 observations of traditional uses. This widespread consumption underscores the importance of *Sonchus* as a dietary component. Medicinal uses constitute the second most frequent category, with 158 observations, indicating the ethnobotanical significance of these plants in traditional remedies. Livestock feed also represents a substantial use, with 114 records, particularly for species such as *S. oleraceus* (54 observations) and *S. asper* (26 observations). In contrast, proverbs, sayings, and colloquial expressions (9 observations) and veterinary uses (12 observations) are less common but suggest a notable cultural integration of *Sonchus*. The use of *Sonchus* as an insecticide is negligible, with only a single record attributed to *S. oleraceus*.

Certain *Sonchus* species exhibit greater versatility in their applications. *Sonchus oleraceus* is identified as the most utilized species, demonstrating dominance across food (170 observations), medicinal (77 observations), and livestock feed (54 observations) categories. *S. tenerrimus* and *S. asper* subsp. *asper* follow, with strong applications in both food and medicinal contexts.

Niche uses are also observed. *Sonchus asper*, *S. bulbosus*, *S. oleraceus*, and *S. tenerrimus* are referenced in proverbs and sayings, suggesting their cultural recognition in folklore. Veterinary applications are predominantly associated with *S. oleraceus* (eight observations), with minor mentions for *S. asper* (one observation) and *S. tenerrimus* (three observations). The singular insecticide use is exclusively linked to *S. oleraceus*. Conversely, species such as *S. crassifolius* (four food uses only), *S. maritimus* subsp. *aquatilis* (six food uses only), and *S. macrocarpus* demonstrate minimal usage across food, feed, and medicinal categories.

The dominance of food uses suggests that *Sonchus* species are commonly consumed as leafy greens, akin to other wild edibles like dandelions or wild lettuce. The high value placed on *S. oleraceus* for food consumption may be due to its palatability and widespread availability. The significant number of medicinal uses indicates their integration into traditional remedies for various ailments, including inflammation, digestive issues, or wound healing (Supplementary Tables S3 and S4). Notably, *S. arvensis* has 12 recorded medicinal uses despite its less common application as food or livestock feed, warranting further investigation into its specific chemical properties. The presence of *Sonchus* in proverbs and sayings points to its symbolic significance within certain cultures, particularly in Mediterranean regions where it might be referenced in folk wisdom. The limited veterinary applications may relate to traditional animal health remedies, while the single insecticide use of *S. oleraceus* suggests experimental or highly localized pest control methods.

In summary, the primary uses of *Sonchus* species are for food, medicine, and livestock feed, with *Sonchus oleraceus* emerging as the most versatile species. While proverbs and veterinary applications are less common, they highlight noteworthy cultural and niche uses. Conversely, *S. crassifolius*, *S. macrocarpus*, and *S. maritimus* appear to be underutilized. Further ethnobotanical studies could explore the reasons behind the preference for certain species, and chemical analyses might elucidate the specific properties contributing to *S. arvensis*'s medicinal value. Additionally, the significant use of *Sonchus* as livestock feed suggests its potential for agricultural applications as a forage crop.

6.4. Traditional Agricultural Applications of Sonchus Species for Animal Feed

Analysis of *Sonchus* species utilization as animal fodder reveals their significant role in traditional small-scale and subsistence farming systems (Supplementary Table S2). A clear taxonomic hierarchy emerges in utilization frequency: *Sonchus oleraceus* dominates with 54 recorded uses, followed by *S. asper* subsp. *asper* (26 uses) and *S. tenerrimus* subsp. *tenerrimus* (21 uses). Other species (*S. maritimus*, *S. arvensis*, *S. bulbosus*, and *S. macrocarpus*) show minimal use, indicating marked variation in fodder relevance across the genus.

Animal-Specific Utilization Patterns: Feed applications demonstrate clear livestock preferences. Rabbits represent the most frequent recipients (31 instances), followed by general backyard animals (26 instances), partridges (16), and chickens (12). Goats, sheep, pigs, and various bird species appear less frequently (5–7 instances each), while waterfowl (geese and ducks) show a rare but notable association exclusively with *S. tenerrimus*.

Species-animal affinities reveal distinct utilization patterns. *S. oleraceus* exhibits the broadest utility across nine animal categories, serving primarily as general fodder (14 instances), rabbit feed (11), and partridge feed (9). *S. asper* shows specialized use in rabbit husbandry (13 instances), while *S. tenerrimus* functions primarily as general fodder (6 instances), occupying an intermediate position without dominating any single category.

Ecological and Agricultural Implications: The predominant use of *Sonchus* species as general fodder reflects their importance in small-scale agricultural systems where flexible, accessible feed resources are essential. The prominence of rabbit feeding suggests both physiological compatibility and farmer recognition of diverse dietary options for lagomorphs. Partridge utilization indicates a role in game bird management spanning both domesticated and wild contexts.

The specific association between *S. tenerrimus* and waterfowl may reflect unique palatability or digestibility characteristics. These patterns underscore the influence of animal preference, plant palatability, and regional ecological knowledge in shaping ethnoveterinary practices.

Taxonomic Hierarchy and Specialization: *S. oleraceus* emerges as the dominant multipurpose species, mirroring its prominence in culinary and medicinal applications and reinforcing its status as a cornerstone plant in Mediterranean ethnobotany. This broad applicability contrasts with the more specialized roles of other taxa: *S. asper* shows concentrated use in rabbit husbandry, suggesting specific biochemical properties favoring lagomorph consumption.

The focus on backyard and small-scale livestock operations highlights Sonchus species' particular value within traditional agricultural systems, supporting household food security through contributions to animal nutrition. The inclusion of wild and semi-domesticated species extends utility beyond conventional livestock to encompass broader wildlife management within traditional ecological systems.

Research Gaps and Future Potential: Limited documentation of several *Sonchus* species in livestock applications suggests significant unexplored potential within this genus. This gap presents opportunities for integrating these resources into sustainable livestock feeding strategies, particularly within contemporary ecological farming and agroforestry systems that combine traditional knowledge with modern sustainable practices.

The livestock applications of *Sonchus* species complement their established culinary and medicinal roles, demonstrating fundamental multifunctionality. This convergent utilization pattern establishes these species as integral components of the ecological and economic systems sustaining traditional Mediterranean communities, where their value transcends individual use categories to encompass comprehensive resource utilization strategies.

7. Gastronomical Value

7.1. Plant Parts

Our findings reveal a selective preference for young plants and tender plant parts for culinary consumption, attributed to their enhanced palatability and reduced bitterness. However, this pattern does not extend to medicinal applications, as active compounds typically accumulate throughout the plant's developmental stages, as discussed subsequently. For gastronomic purposes, young leaves, shoots, and bulbs are consistently favored due to their superior organoleptic properties (Figure 4).

Figure 4. Relevant edible and/or medicinal *Sonchus* plant parts. (**A**) Young leaves along the stem but also in a basal rosette *Sonchus aquatilis*. (**B**) Inmature capitulum, *S. oleraceus*. (**C**) Underground stem and rhizome, *S. crassifolius*. (**D**) "Bulbs"—root tubers, *S. bulbosus*. Images: (**A**,**B**) Antonio Robledo, with permission. (**C**) Javier Tardío, with permission. (**D**) S. Ríos.

The analysis of plant part utilization in *Sonchus* species reveals distinct patterns of exploitation for food and medicinal purposes. Aerial parts demonstrate the highest overall usage (338 records) (Table 5), with leaves representing the most frequently utilized component (233 records), followed by whole plant preparations (72 records). This pattern reflects the practical accessibility and abundance of above-ground tissues, particularly the tender foliage that constitutes the primary edible portion of these species. The preference for aerial parts aligns with traditional harvesting practices that prioritize easily accessible plant materials while allowing for sustainable collection without destroying the entire plant.

The distribution between food and medicinal applications shows notable differences in plant part selection. While aerial parts and leaves dominate both categories, medicinal uses show greater diversity in plant-part exploitation, incorporating specialized tissues such as latex (seven records), flowers (six records), and cypselae (two records) that are absent or minimal in food applications. Roots, including both standard and tuberous forms, contribute modestly to both categories but show relatively higher representation in medicinal uses (18 total records) compared to their food applications. This pattern suggests that medicinal preparations often target specific bioactive compounds concentrated in particular plant tissues, whereas culinary uses focus primarily on palatability and nutritional content found in commonly consumed aerial parts.

Table 5. Relevance of the different *Sonchus* plant parts in terms of frequency among the analyzed records.

Plant Part	Records of *Sonchus* Traditional Uses	Food	Medicinal Plants
Aerial parts	338	190	85
Leaf	233	173	40
Whole plant	72	15	11
Root	18	11	7
Stem	15	14	1
Flower	9	0	6
Latex	7	0	7
Tuberous root	6	3	0
Cypselae	3	0	2

7.2. Traditional Culinary Uses

Sonchus species have been incorporated into various culinary traditions, particularly in Mediterranean cuisine. These plants are often used in salads, stews, and as brined products, contributing to the diverse gastronomic landscape of the region [22,24] (Table 6). Regional recipes, such as Greek "horta" variants, including *Bombotópita* or *Plastós* [98,104] and Italian "pistic" [11,96], highlight the adaptability of *Sonchus* species to local culinary traditions [12,24].

Table 6. Records of traditional food uses and *Sonchus* species.

N° Food Uses	*S. arv*	*S. aspasp*	*S. aspgla*	*S. bul*	*S. cras*	*S. mac*	*S. mar*	*S. maqu*	*S. ole*	*S. ten*
Boiled	4	22	2	1	1	1	0	2	63	43
Salads	1	34	2	4	3	0	0	4	68	81
Bombotópita, is a traditional Greek cornbread pie or cornmeal cake	0	1	0	0	0	0	0	0	1	0
"Braised" refers to a cooking method where food, typically meat or vegetables, is first seared at a high temperature and then slowly cooked in a covered pot with a small amount of liquid, such as broth or wine	0	0	0	0	0	0	0	0	1	0
Cooked in a variety of preparations, including soupy rice dishes	0	0	0	0	0	0	0	0	1	0
Cooked in a variety of preparations, including soupy rice dishes or pan-fried with garlic	0	1	0	0	0	0	0	0	9	2
Lake fish-based soups	0	0	0	0	0	0	1	0	0	0
Pistic refers to a traditional dish from Western Friuli, Italy, made from a collection of over 50 wild herbaceous meadow and wood plants that are boiled and then sautéed together	0	1	0	0	0	0	0	0	1	0
Plastós is a traditional pie from various regions of Greece, particularly Thessaly and parts of central Greece	0	1	0	0	0	0	0	0	1	0
Regional Greek wild green sauté	0	1	0	0	0	0	0	0	1	0
Seasoning or sauce suitable for all varieties of pasta	0	2	0	0	0	0	0	0	1	1
Tender spring leaves blanched and subsequently sautéed, finished with smoked ricotta cheese for added depth and aroma	0	0	0	0	0	0	0	0	1	0
The plants are initially boiled and subsequently sautéed	2	5	0	0	0	0	0	0	3	1

Table 6. *Cont.*

N° Food Uses	*S. arv*	*S. aspasp*	*S. aspgla*	*S. bul*	*S. cras*	*S. mac*	*S. mar*	*S. maqu*	*S. ole*	*S. ten*
The plants are initially boiled and subsequently sautéed with garlic and green peppers to enhance flavor and aroma	0	0	0	0	0	0	0	0	4	0
The plants are initially boiled and subsequently sautéed with garlic to enhance flavor and aroma	0	0	0	0	0	0	0	0	1	2
The plants are initially boiled and subsequently sautéed with ñora peppers and salted sardines to enhance flavor and aroma	0	1	0	0	0	0	0	0	1	0
The plants are initially boiled with Swiss chard and subsequently sautéed with blood sausages to enhance flavor and aroma	0	0	0	0	0	0	0	0	1	0
The plants are initially boiled with Swiss chard and subsequently sautéed with garlic to enhance flavor and aroma	0	0	0	0	0	0	0	0	2	0
The root undergoes roasting and grinding processes to be served as a coffee surrogate	0	1	0	0	0	0	0	0	1	0
The roots are incised to extract latex, which is subsequently chewed	0	0	0	0	0	0	0	0	1	0
These are prepared as Italian vegetable frittata, in which diced seasonal vegetables are incorporated into an egg matrix and cooked until set	0	1	0	0	0	0	0	0	2	0
These are prepared as Spanish vegetable omelets, in which diced seasonal vegetables are incorporated into an egg matrix and cooked until set	0	1	0	0	0	0	0	0	3	8
These leafy greens are commonly employed as substitutes for spinach in culinary preparations such as pizzas and soups	0	0	0	0	0	0	0	0	1	1
Typically roasted and then cooked with rice to enhance flavor and texture	0	0	0	0	0	0	0	0	1	0
Used as a coffee substitute	0	0	0	0	0	0	0	0	1	0
	7	72	4	5	4	1	1	6	170	139

Abbreviations: S. arv, Sonchus arvensis L.; S. aspasp, S. asper (L.) Hill subsp. asper; S. aspgla, S. asper subsp. glaucescens (Jord.) Ball; S. bul, S. bulbosus (L.) N.Kilian & Greuter; S. cras, S. crassifolius Pourr. ex Willd.; S. mac, S. macrocarpus Boulos & C.Jeffrey; S. mar, S. maritimus L.; S. maqu, S. maritimus subsp. aquatilis (Pourr.) Nyman; S. ole, S. oleraceus L.; S. ten, S. tenerrimus L. subsp. tenerrimus. Data from Supplementary Table S2.

An analysis of culinary applications of *Sonchus* species in Mediterranean ethnobotany reveals a strong preference for certain taxa and preparation methods, underscoring their cultural and gastronomic importance (Table 6). Among the documented species, *Sonchus oleraceus* and *Sonchus tenerrimus* subsp. *tenerrimus* emerge as the most frequently utilized, accounting for 170 and 139 recorded uses, respectively. These species dominate the culinary landscape, far surpassing others in frequency of use. In contrast, *Sonchus asper* subsp. *asper* appears significantly less often, with 72 recorded uses, while all remaining taxa exhibit markedly limited culinary relevance. Species such as *S. maritimus* subsp. *aquatilis*, *S. bulbosus*, *S. arvensis*, *S. crassifolius*, *S. asper* subsp. *glaucescens*, *S. maritimus*, and *S. macrocarpus* collectively account for only a handful of instances, reflecting their marginal presence in traditional foodways.

In terms of preparation methods, raw consumption in salads constitutes the predominant mode of culinary use, with 197 instances recorded (Table 6). Boiling, either on its own or mixed with other wild greens, is the second most common preparation technique, cited 139 times. These two methods together represent the vast majority of culinary applications, while other techniques—such as sautéing or frying—appear only sporadically, typically with fewer than ten mentions. This pattern suggests a culinary preference for simplicity, likely intended to preserve or subtly enhance the natural characteristics of the plant.

A closer examination of species-specific culinary combinations reinforces the centrality of *S. oleraceus*, *S. tenerrimus*, and *S. asper*. The most frequently recorded preparations include *S. tenerrimus* consumed in salads (81 instances) and *S. oleraceus* similarly used (68 instances), followed by their respective boiled preparations. *S. asper* is also present in both categories, though with a significantly lower frequency. The high rate of repetition across these core preparations highlights their integration into daily food practices and suggests they are the most palatable or culturally valued members of the genus.

Qualitative data provides further insight into the culinary versatility of *Sonchus* species. Their frequent raw consumption, particularly in salads, indicates that certain species are palatable without the need for cooking, a trait not universally common among wild leafy greens. Where cooking is involved, boiling dominates, potentially serving to reduce bitterness or soften fibrous textures. In some regional traditions, additional techniques such as sautéing follow boiling, suggesting layered preparation methods aimed at refining flavor or texture.

Distinct regional cuisines further reflect the depth of culinary integration. In the western Mediterranean, *Sonchus* is incorporated into Spanish omelets [62,67,73,75,81,82] and traditional Catalan or Valencian dishes. Greek gastronomy includes preparations such as *plastós* and *bombotópita* [104], while in Italy, *Sonchus* is used in frittatas and pasta-based dishes [96]. These examples suggest that the plant is not only accepted but embedded within long-standing culinary repertoires.

Flavor enhancement is a notable theme across these preparations. Garlic is the most consistently cited additive [52,67,82,83], followed by regionally specific ingredients such as *ñora* peppers [67], salted sardines [67], blood sausages [67], and smoked ricotta [96]. These accompaniments imply that *Sonchus* is frequently paired with strong or savory elements, perhaps to balance bitterness or augment its earthy flavors.

Beyond their role in primary dishes, *Sonchus* species also serve alternative culinary and quasi-culinary functions. In some traditions, the roots are roasted and used as a coffee substitute [88], while its latex has been noted for both medicinal [83,93,97] and recreational purposes [107], suggesting a broader spectrum of plant utility within local knowledge systems.

Certain species exhibit unique or highly specialized culinary applications. *S. maritimus* subsp. *maritimus*, for instance, is exclusively associated with fish-based lake soups, potentially indicating a specific flavor compatibility [86]. While *S. asper* is employed across several methods, it appears less frequently in complex or culturally specific recipes compared to *S. oleraceus* and *S. tenerrimus*, reinforcing the latter's greater culinary status.

Overall, the data affirms the significant cultural embeddedness of *Sonchus* species within traditional Mediterranean cuisine. The specificity of dishes and preparation techniques points to a deep-rooted culinary heritage rather than opportunistic or incidental use. Moreover, the predominance of just a few species suggests that palatability, availability, or cultural preference has historically shaped selection practices (Table 6).

8. Phytochemical and Nutritional Evidence

8.1. Phytochemical Composition

8.1.1. Overview of Phytochemical Evidence

Sonchus species contain a diverse array of bioactive compounds, including polyphenols, flavonoids, and sesquiterpene lactones. These phytochemicals contribute to the plant's medicinal properties and potential health benefits [22,24]. In the review (Table 7) of the 25 publications selected, *S. oleraceus* exhibited the highest value with 302 chemical records distributed across 11 publications, followed by *S. maritimus* (293 records primarily from 2 publications) and *S. arvensis* (210 records across 8 publications). Less studied species include *S. crassifolius*, *S. macrocarpus*, and *S. erzincanicus*, with only 5–7 records each from single publications. These data highlight significant research concentration on select *Sonchus* species while revealing substantial knowledge gaps for the less-studied taxa.

Table 7. Distribution of chemical compounds in *Sonchus* species across the published literature. This table presents the frequency of chemical compound records across 25 representative publications (columns 1–25) for 11 *Sonchus* taxa. Numbers within cells indicate the quantity of distinct chemical compounds reported in each publication for the corresponding taxon.

Taxa/Publication [1]	a	b	c	d	e	f	g	h	i	j	k	l	m	n	o	p	q	r	s	t	u	v	w	x	y	N° Records
S. oleraceus L.	39	25	0	23	0	9	4	64	32	69 *	0	0	0	13	0	0	0	16	0	8	0	0	0	0	0	302
S. maritimus L.	0	0	288 *	0	0	0	5	0	0	0	0	0	0	0	0	0	0	0	0	0	0	0	0	0	0	293
S. arvensis L.	0	0	0	0	0	0	0	0	0	0	0	0	108	0	12	9	3	0	10	0	0	0	25	17	26	210
S. tenerrimus L.	0	25	0	0	0	0	4	0	0	0	0	0	0	0	0	0	0	0	0	5	0	50	0	0	0	84
S. asper (L.) Hill	0	0	0	0	5	9	0	0	0	0	0	35	0	0	0	0	0	0	0	9	10	0	0	0	0	68
S. asper subsp. *glaucescens* (Jord.) Ball	0	0	0	0	0	0	5	0	0	0	0	0	0	0	0	0	0	0	0	9	0	0	0	0	0	14
S. maritimus subsp. *aquatilis* (Pourr.) Nyman	0	0	0	0	0	0	0	0	0	0	0	0	0	0	0	0	0	0	0	10	0	0	0	0	0	10
S. erzincanicus V.A.Matthews	0	0	0	0	0	0	0	0	0	0	7	0	0	0	0	0	0	0	0	0	0	0	0	0	0	7
S. arvensis subsp. *uliginosus* (M.Bieb.) Nyman	0	0	0	0	0	0	6	0	0	0	0	0	0	0	0	0	0	0	0	0	0	0	0	0	0	6
S. macrocarpus Boulos & C.Jeffrey	0	0	0	0	0	0	5	0	0	0	0	0	0	0	0	0	0	0	0	0	0	0	0	0	0	5
S. crassifolius Pourr. ex Willd.	0	0	0	0	0	0	5	0	0	0	0	0	0	0	0	0	0	0	0	0	0	0	0	0	0	5
Total Chemicals	39	50	288	23	5	18	34	64	32	69	7	35	108	13	12	9	3	16	10	41	10	50	25	17	26	1004
Botanical score	5	2.7	4	5	1.8	1.6	1.5	2	2.4	3.7	3.3	3.5	0	2.5	3.2	3.4	1.4	0	2.7	1.3	2.7	2.7	3	1.2	0.1	-
Chemical score	4.9	4.2	2.5	4.1	4	1	1.2	4.2	3.7	1.9	4.5	4.2	4.5	3	4.4	4.4	3	0	4.2	3.6	3.1	2.2	2.5	2.6	0.8	-

[1] References and countries: a [116] Algeria; b [24] Spain; c [117] Algeria; d [118] Egypt; e [119] Egypt; f [120] Pakistan; g [121] Egypt and other countries; h [122] Egypt; i [123] South Africa; j [124] Tunisia; k [125] Turkey; l [126] Italy; m [127] Indonesia; n [128] India; o [129] China; p [130] China; q [131] India; r [132] Kazakhstan; s [133] China; t [134] Spain; u [135] Pakistan; v [136] Libya; w [137] Indonesia; x [138] Indonesia; y [139] Ukraine. (*) Caution: Some of the reported compounds potentially represent analytical artifacts or possess marginal evidential support for their structural identification.

Table 7 reveals a striking shift in the geographical focus of *Sonchus* chemical research, expanding far beyond the traditional Mediterranean centers that historically dominated ethnobotanical documentation. The data encompasses 1004 distinct chemical compound records across 11 *Sonchus* taxa from 25 publications spanning 15 countries, representing a significant diversification from the previously Spain-, Italy-, and France-centric research landscape.

The distribution of research effort shows pronounced taxonomic bias, with *Sonchus oleraceus* dominating the literature, with 302 compound records (30.1% of all records), followed by *S. maritimus* with 293 records (29.2%) and *S. arvensis* with 210 records (20.9%). These three species account for approximately 80% of all documented chemical compounds, while the remaining eight taxa collectively contribute only 199 records. This concentration suggests either greater chemical diversity in the dominant species or research preference for more widely distributed taxa.

Geographically, the research landscape has undergone a remarkable transformation. North African countries now contribute substantially, with Algeria [116,117] providing the highest single-study chemical diversity (288 compounds) [117] and Egypt contributing across multiple studies [118,119,121,122]. Asian representation has expanded significantly, including contributions from China [129,130,133], Indonesia [127,137,138], India [128,131], Pakistan [120], and Kazakhstan [132]. African research extends beyond the Mediterranean to include South Africa, while the traditional European strongholds are represented by only two Spanish studies [24,134] and one Italian study [126] among the 25 references.

The temporal analysis reveals interesting patterns in research intensity. Recent publications (2020–2025) account for 12 of the 25 references, indicating accelerated research activity. The chemical scores, ranging from 0.8 to 4.9, suggest varying analytical rigor or scope across studies, with several African studies achieving high scores despite being geographically distant from traditional research centers.

The geographical redistribution of *Sonchus* research reflects broader trends in global scientific capacity building and the recognition of traditional knowledge systems outside Europe. The emergence of substantial research contributions from Algeria, Egypt, Indonesia, and other developing nations suggests successful technology transfer and local expertise development in phytochemical analysis. This shift is particularly significant given that these regions often harbor greater plant biodiversity and traditional medicinal knowledge than the historically dominant research centers.

The taxonomic focus reveals potential biases in species selection that may not reflect actual chemical diversity. *S. oleraceus* and *S. maritimus* show remarkably high compound counts in individual studies, with some studies reporting over 60 compounds from a single species. This could indicate either exceptional chemical richness or methodological differences in compound identification and reporting standards. The cautionary note in Table 7 regarding potential analytical artifacts in some highly prolific studies suggests that quantity may not always correlate with quality in phytochemical research.

The emergence of specialized regional research programs is evident from the data. Indonesian researchers appear to be developing systematic approaches to *Sonchus* chemistry, with multiple recent publications focusing on *S. arvensis*. Similarly, North African research centers have demonstrated sophisticated analytical capabilities, particularly in Algeria, where a single study [117] identified 288 compounds from *S. maritimus*.

The botanical and chemical scoring systems reveal interesting disparities between traditional botanical knowledge and modern chemical analysis capabilities. Some studies with lower botanical scores achieve high chemical scores, suggesting that chemical analysis expertise may now exist independently of classical taxonomic expertise in some regions. However, this casts doubt on the reliability of botanical identification of such analyses.

This geographical diversification represents a democratization of phytochemical research that contrasts sharply with the historical concentration in Mediterranean Europe. The data suggests that the center of *Sonchus* chemical research has effectively shifted toward regions where these plants may have greater ethnobotanical significance and where modern analytical techniques are being applied to traditional knowledge systems. However, the wide variation in compound numbers per study and the cautionary notes about analytical artifacts highlight the need for standardized methodologies and quality control measures as this research field continues to expand globally.

The evaluation of the 25 papers on *Sonchus* phytochemistry (Table 7), based on botanical and chemical scoring criteria (Supplementary Table S1), reveals notable disparities in methodological rigor. The botanical scores, which assess critical aspects of plant identification, taxonomy, and documentation, exhibit significant variability, ranging from 0 to 5, with a mean of approximately 2.3. This indicates that a substantial proportion of studies lack thorough botanical validation, as nearly 28% of papers scored below 1.5, while only a minority (28%) achieved scores above 3. In contrast, chemical scores—reflecting the robustness of extraction, analytical procedures, and compound identification—demonstrate comparatively stronger performance, with a mean of 3.2 and 32% of studies scoring above 4. However, this relative strength in chemical methodology is undermined by inconsistencies, as 20% of papers received chemical scores below 2, suggesting deficiencies in experimental reproducibility or analytical precision.

A striking observation is the frequent dissociation between botanical and chemical rigor. Several studies (e.g., with botanical scores of 0 paired with chemical scores of 4.5) [127] exhibit advanced laboratory techniques but neglect fundamental botanical documentation, potentially compromising the reproducibility and ecological relevance of their findings. Conversely, a subset of papers with high botanical scores (e.g., 5) but moderate chemical scores (e.g., 2.5) suggests rigorous plant science that is undermined by less stringent analytical protocols. Only a single study [116] achieved high marks in both domains (botanical = 5, chemical = 4.9), serving as a benchmark for comprehensive phytochemical research. These disparities highlight an urgent need for more balanced methodologies, integrating meticulous botanical practices—such as voucher specimen deposition and precise taxonomic identification—with standardized chemical analyses to ensure both accuracy and ecological validity.

The findings underscore a broader trend in phytochemical research, where technical advancements in chemical analysis often outpace foundational botanical rigor. To address this, future studies should prioritize harmonizing these dimensions, adopting established guidelines for plant collection and identification alongside validated analytical techniques. Such an approach would enhance the reliability, reproducibility, and applicability of phytochemical data, particularly in understudied genera like *Sonchus*. Further statistical exploration, including correlation analyses between botanical and chemical scores, could yield additional insights into the interdependence of these research domains.

8.1.2. Comparative Analysis of Significant Compounds Across Different Sonchus Species

Comparative studies reveal significant variations in the phytochemical profiles of different *Sonchus* species, influenced by environmental and genetic factors [121,122].

The phytochemical profile, organized by main chemical classes, presents significant chemical distribution patterns within the genus while also acknowledging the influence of research intensity disparities across species (Figure 5).

Volatile organic compounds (VOCs) emerge as the most abundantly documented chemical class in *Sonchus*, with 224 total records. Their distribution is notably uneven, with *S. maritimus* containing the highest concentration (165 records), followed by *S. tenerrimus* (24), *S. arvensis* (21), and *S. oleraceus* (14). The overwhelming concentration of VOCs in *S. maritimus* (Figure 5) represents a striking chemotaxonomic feature, suggesting specialized ecological adaptations possibly related to its coastal habitat but also differential intensity in the analyses among species.

Flavonoids constitute the second most prevalent class with 172 records (Figure 5), showing a markedly different distribution pattern. *S. arvensis* demonstrates exceptional flavonoid diversity (61 records), followed by *S. oleraceus* (52), with substantially lower but still significant numbers in *S. asper* (17), *S. tenerrimus* (6), *S. maritimus* subsp. *aquatilis* (6), and *S. asper* subsp. *glaucescens* (5). This distribution points to *S. arvensis* as being particularly rich in flavonoids but also establishes *S. oleraceus* as another major flavonoid-containing species.

Phenolic compounds (142 records) display an even more pronounced concentration in *S. oleraceus* (87 records) (Figure 5), distantly followed by *S. arvensis* (29) and *S. asper* (11). This dramatic predominance in *S. oleraceus,* in parallel with the minimal phenolic presence in other species, underscores *S. oleraceus'* distinctive phenolic profile.

Terpenoids, comprising 100 records, show a relatively balanced distribution across *S. oleraceus* (28), *S. arvensis* (25), *S. maritimus* (24), and *S. tenerrimus* (13), with *S. asper* (10) containing a moderate number. This substantial terpenoid presence is relevant, particularly given the ecological and pharmacological importance of terpenoids in the Asteraceae family.

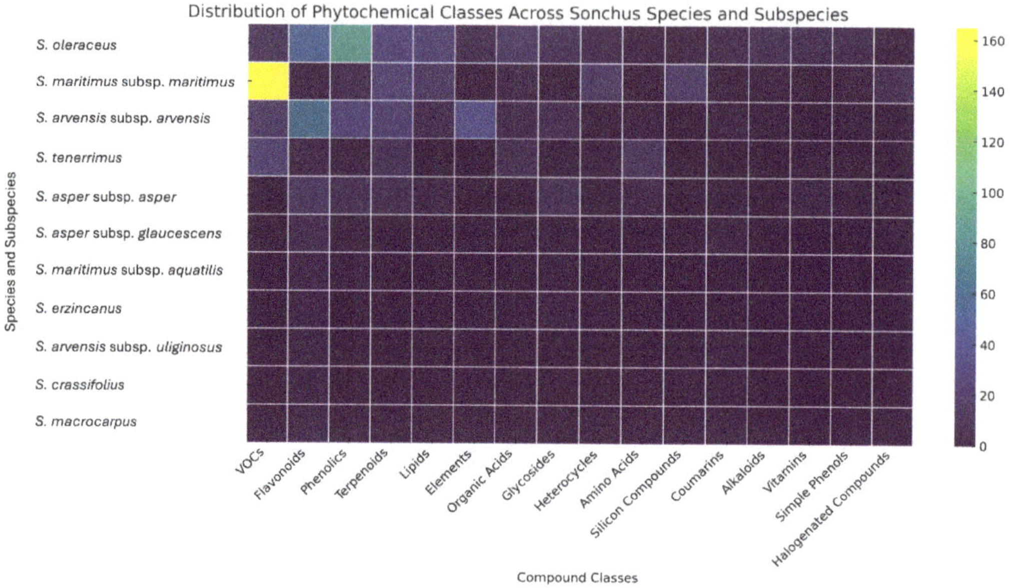

Figure 5. Heatmap displaying the distribution of 16 major phytochemical compound classes across selected *Sonchus* species and subspecies. The data represent the presence and relative abundance (as count of reported compounds) for each class, including volatile organic compounds (VOCs), flavonoids, phenolics, terpenoids, lipids, glycosides, alkaloids, and others. Notably, species such as *S. oleraceus* and *S. maritimus* exhibit high phytochemical diversity, while other taxa show limited chemical representation based on the published literature. Data from Supplementary Table S6.

The lipid profile (56 records) reveals concentrations in *S. oleraceus* (24) and *S. maritimus* (18), with modest representation in *S. arvensis* (9) and *S. asper* (5). Elements (44 records) demonstrate an overwhelming concentration in *S. arvensis* (36), with the remainder found in *S. oleraceus* (8), highlighting the lack of a systematic analysis of elements along the entire genus.

Organic acids (40 records) and glycosides (40 records) show distinctive distribution patterns. Organic acids appear predominantly in *S. oleraceus* (18) and *S. tenerrimus* (10), with modest presence in *S. arvensis* (7) and *S. maritimus* (5). Glycosides concentrate in *S. oleraceus* (13), *S. asper* (12), and *S. arvensis* (10), with minimal representation elsewhere.

Several additional chemical classes reveal species-specific accumulation patterns. Heterocycles (29 records) appear concentrated in *S. maritimus* (18). Amino acids (27 records) show their highest concentration in *S. tenerrimus* (16). Silicon compounds (22 records) demonstrate a striking concentration in *S. maritimus* (20) and are most likely artifacts of the analysis. Coumarins (16 records) appear most abundant in *S. oleraceus* (6). Alkaloids (14 records) and vitamins (13 records) both concentrate in *S. oleraceus* (10 and 8, respectively). Simple phenols (11 records), likewise, demonstrate the highest representation in *S. oleraceus* (7).

The comprehensive dataset reveals distinct chemotaxonomic patterns that substantially revise previous interpretations.

S. maritimus emerges as remarkably VOC-rich and the primary reservoir within the genus, suggesting specialized adaptations possibly related to its wetland/coastal habitat, but this requires further confirmation.

S. arvensis maintains its status as flavonoid-rich but also demonstrates exceptional elemental accumulation capacity, suggesting distinctive nutritional strategies.

S. oleraceus reinforces its phenomenal phenolic diversity while additionally revealing substantial flavonoid, terpenoid, lipid, organic acid, alkaloid, and vitamin content, establishing it as perhaps the most phytochemically diverse but also the most widely analyzed (Figures 5 and 6) *Sonchus* species.

S. tenerrimus demonstrates a distinctive amino acid profile alongside significant terpenoid and organic acid content, suggesting unique biosynthetic pathways.

S. asper exhibits moderate diversity across flavonoids, phenolics, terpenoids, and glycosides but lacks the pronounced chemical specialization observed in other species.

The subspecies, including *S. arvensis* subsp. *uliginosus*, *S. maritimus* subsp. *aquatilis*, and *S. asper* subsp. *glaucescens*, show markedly reduced chemical diversity compared to their parent species, though this may partially reflect research bias rather than biological reality.

Figure 6 addresses compounds reported in at least five samples within the genus, representing only a fraction of the approximately 580 different compounds identified across all samples. The chemical constituents can be categorized into seven primary classes: flavonoids, phenolic acids, coumarins, simple phenolics, other organic acids, carbohydrates, and minerals or elements.

The notable absence of volatile organic compound (VOC) representatives in Figure 6 appears inconsistent with their apparent significance in *Sonchus*, as indicated in Figure 5. Analysis of Supplementary Table S6 reveals that while VOCs are reported with high frequency, they occur in fewer than five individual cases. This discrepancy necessitates consideration of two critical factors in future investigations: first, whether the observed high diversity of VOCs reflects genuine genetic and environmental variability within the genus, or second, whether this diversity may be partially attributable to methodological artifacts inherent in the analytical approaches employed.

Flavonoids that constitute a predominant class within *Sonchus* species (Figure 5) are represented by compounds such as luteolin derivatives, apigenin derivatives, quercetin, and rutin, differentially distributed across taxa. Notably, *Sonchus arvensis* demonstrates exceptional flavonoid diversity, particularly with respect to luteolin (seven refs), apigenin (five refs), and is the sole species with significant quercetin and rutin presence in this high-frequency dataset (Figure 6). This distinctive flavonoid profile suggests potential species-specific biosynthetic pathways or ecological adaptations. *Sonchus oleraceus* similarly exhibits substantial flavonoid diversity, though with a different compositional pattern than *S. arvensis*.

Phenolic acids represent another significant chemical class in *Sonchus* (Figure 5), with caffeic acid, chlorogenic acid, gallic acid, chicoric acid, and ferulic acid being the most frequently detected compounds (Figure 6). *S. oleraceus* demonstrates remarkable accumulation of these compounds, containing the highest observed frequencies of caffeic acid (five), chlorogenic acid (six), and gallic acid (six). This extraordinary phenolic acid frequency in *S. oleraceus* suggests enhanced antioxidant defense mechanisms, potentially reflecting evolutionary adaptations to specific environmental stressors or herbivory pressures. The absence or minimal presence of these compounds in species such as *S. erzincanicus*, *S. crassifolius*, and *S. macrocarpus* could further highlight the taxonomic significance of phenolic acid distribution patterns or the lack of exhaustive analysis for these species.

Compound and Chemical Class / Taxa	*S. arvensis* subsp. *arvensis*	*S. arvensis* subsp. *uliginosus*	*S. asper* subsp. *asper*	*S. asper* subsp. *glaucescens*	*S. crassifolius*	*S. erzincanicus*	*S. macrocarpus*	*S. maritimus* subsp. *maritimus*	*S. oleraceus*	*S. tenerrimus*	*S. maritimus* subsp. *aquatilis*
Flavonoids											
Luteolin 7-O-glucoside	2	1	2	1	1	0	1	1	2	2	1
Apigenin 7-O-glucuronide	3	1	0	1	1	1	1	1	2	1	0
Luteolin	7	0	0	0	0	0	0	0	1	1	1
Luteolin 7-O-glucuronide	2	1	0	1	1	1	1	1	2	1	0
Apigenin	5	0	1	0	0	0	0	0	2	0	1
Luteolin 7-O-glucosylglucuronide	0	1	0	1	1	0	1	1	1	1	0
Apigenin 7-O-glucoside	1	0	2	1	0	0	0	0	1	0	1
Quercetin	4	0	0	0	0	0	0	0	0	0	1
Rutin	4	0	0	0	0	0	0	0	1	0	0
Phenolic Acids											
Caffeic acid	3	0	1	1	0	0	0	0	5	1	1
Chlorogenic acid	1	0	2	1	0	0	0	0	6	1	1
Gallic acid	0	0	0	0	0	0	0	0	6	0	0
Chicoric acid	2	0	1	0	0	0	0	0	2	0	0
Ferulic acid	1	0	0	0	0	0	0	0	4	0	0
Coumarins											
Esculetin	2	1	1	2	1	0	1	1	0	0	0
Simple Phenolics											
Catechol	0	0	0	0	0	0	0	0	5	0	0
Pyrogallol	0	0	0	0	0	0	0	0	5	0	0
Other Organic Acids											
Quinic acid	1	0	0	0	0	0	0	0	5	1	0
Azelaic acid	4	0	0	0	0	0	0	0	1	0	0
L-Ascorbic acid	0	0	1	0	0	0	0	0	3	1	0
Carbohydrates											
1-Butyl β-D-glucopyranoside	5	0	0	0	0	0	0	0	1	0	0
Minerals											
Calcium	4	0	0	0	0	0	0	0	1	0	0
Magnesium	4	0	0	0	0	0	0	0	1	0	0
Manganese	4	0	0	0	0	0	0	0	1	0	0
Phosphorus	4	0	0	0	0	0	0	0	1	0	0
Potassium	4	0	0	0	0	0	0	0	1	0	0
Sodium	4	0	0	0	0	0	0	0	1	0	0
Publications	8	1	5	2	1	1	1	2	11	4	1

Figure 6. Heatmap showing the frequency of the more commonly reported chemical compounds in different *Sonchus* taxa species. Numbers represent the frequency of reports in the scientific literature. Publications row (green) shows the total number of papers analyzed for each taxon. Low presence (1), medium presence (3–4), high presence (6+). It is important to note that among the 580 chemicals recorded in the revised literature from *Sonchus*, only those recorded 5 times or more are included. Generated using Excel.

The coumarin esculetin displays an interesting distribution pattern across the genus, being present in multiple species but conspicuously absent in *S. erzincanicus*, *S. oleraceus*, and *S. tenerrimus*. This differential distribution may serve as a valuable taxonomic marker and warrants further investigation regarding its ecological significance. Simple phenolics, specifically catechol and pyrogallol, appear exclusively concentrated in *S. oleraceus* (frequency of five for both compounds), further distinguishing this species' chemical profile.

Regarding other organic acids, quinic acid shows notable frequency in *S. oleraceus* (five), while azelaic acid appears predominantly in *S. arvensis* (four). L-ascorbic acid (vitamin C) demonstrates a more restricted distribution, with the highest frequencies in *S. oleraceus* (three). The carbohydrate 1-butyl β-D-glucopyranoside appears concentrated in *S. arvensis* (five), with limited presence elsewhere in the genus.

Mineral accumulation patterns reveal another dimension of chemotaxonomic differentiation, with calcium, magnesium, manganese, phosphorus, potassium, and sodium all showing significantly higher frequencies in *S. arvensis* (four for each) and modest presence in *S. oleraceus* (one for each). This mineral distribution pattern may reflect differential soil adaptation strategies or nutrient uptake mechanisms between species but also deficiencies in the systematic study of minerals along the different *Sonchus* species.

The sampling bias inherent in examining only compounds detected in five or more samples necessitates acknowledging significant limitations in this analysis. The complete phytochemical diversity of *Sonchus*, comprising approximately 580 compounds, likely includes numerous species-specific or rare compounds that could serve as important biomarkers or possess unique biological activities. Notably absent from this high-frequency dataset are compound classes typically abundant in Asteraceae, including terpenoids, sesquiterpene lactones, alkaloids, saponins, and essential oils, which might be represented in the broader dataset.

The chemical profiles observed provide valuable insights into taxonomic relationships within *Sonchus*. Subspecies, such as *S. arvensis* subsp. *uliginosus*, exhibit distinct chemical signatures from their parent species, lending phytochemical support to their taxonomic differentiation. Furthermore, the distribution of defensive compounds, particularly flavonoids and phenolics, likely reflects adaptation to varying ecological pressures across species' native ranges.

From an applied perspective, the distinctive phytochemical profiles suggest potential applications in natural product research. *S. oleraceus* demonstrates considerable promise for antioxidant applications due to its exceptional phenolic frequencies, while *S. arvensis* could be explored for flavonoid-related bioactivities. These unique chemical signatures could guide the strategic selection of Sonchus species for further phytopharmaceutical investigation.

This analysis, while necessarily constrained by the limited dataset of high-frequency compounds, provides valuable chemotaxonomic insights into *Sonchus*. A comprehensive analysis incorporating the full complement of 580 identified compounds would likely reveal additional species-specific patterns and potentially discover bioactive compounds of significant pharmacological interest. The current findings nonetheless establish a foundation for understanding the chemical ecology and potential applications of this diverse plant genus. The substantial discrepancies in compound numbers across species likely reflect, in great part, differences in research intensity rather than solely biological differences (Table 7, Figures 5 and 6). *S. oleraceus, S. arvensis, S. maritimus,* and *S. asper* appear to be most extensively investigated, while subspecies and less common species have received comparatively minimal phytochemical attention. This research bias must temper interpretations of apparent chemical absence in less-studied taxa.

8.1.3. Ecological and Pharmacological Implications

The revised chemical patterns suggest distinct ecological adaptations. *S. maritimus*' VOC and silicon-rich profile likely reflects adaptations to coastal/wetland environments, potentially functioning in salt tolerance or herbivore deterrence. *S. oleraceus*' exceptional phenolic and alkaloid diversity suggests strong selection for chemical defense against herbivores and pathogens, while *S. arvensis*' mineral accumulation capacity may represent an adaptation to nutrient-poor soils, but this could be expected in other *Sonchus* species.

From a pharmacological perspective, this comprehensive analysis highlights *S. oleraceus* as an outstanding candidate for broad-spectrum bioactivity investigation, with exceptional diversity across multiple bioactive compound classes. *S. maritimus* emerges as a potentially valuable source of novel VOC compounds, with possible applications in agricultural chemistry and materials science. *S. arvensis*, with its unique flavonoid and elemental profile, warrants investigation for nutritional applications.

8.2. Nutritional Value and Health Benefits

8.2.1. Macronutrient and Micronutrient Composition

Sonchus species are a good source of fiber, vitamins (A, C, and K), and minerals (calcium, magnesium, and iron), making them a nutrient-dense food [23].

The genus *Sonchus* comprises several species of unconventional food plants that have garnered attention for their potential nutritional value. Multiple studies have investigated the chemical composition of these species, particularly focusing on their application as alternative food sources for rural populations and their contribution to food security.

Nutritional analyses of *Sonchus* species employed standardized methodologies to ensure reliable and comparable results. Proximate composition analyses followed the protocols established by the Association of Official Analytical Chemists (AOAC), while mineral content determination utilized atomic emission spectrometry with inductively coupled plasma. High-Performance Liquid Chromatography (HPLC) has been the preferred method for quantifying carotenoids and vitamins, providing accurate measurements of these bioactive compounds [23,24,140].

The four primary *Sonchus* species studied (*S. oleraceus*, *S. asper*, *S. tenerrimus*, and *S. arvensis*) exhibit distinct macronutrient profiles. *S. asper* demonstrates superior lipid content (1.32 g/100 g) and available carbohydrate concentrations (0.34 g/100 g) compared to its congeners. Notably, all species maintain characteristically low proportions of available carbohydrates, a feature that may contribute to their potential dietary applications. Fiber content consistently exceeds 30 g/kg across all species, indicating substantial dietary fiber availability [23,140].

Vitamin C emerges as the most extensively studied and abundant vitamin component across *Sonchus* species. Concentrations range from 457 mg/kg in *S. tenerrimus* to 779 mg/kg in *S. oleraceus*, representing a 2-3-fold increase compared to conventional leafy vegetables such as lettuce [140]. These concentrations are nutritionally significant, as consumption of 142 g of *S. oleraceus* or 172 g of *S. tenerrimus* would satisfy the average daily recommended vitamin C requirements for women (75 mg). *S. arvensis* exhibits the highest vitamin E content (72.98 μg/100 g) and exceptional vitamin K concentrations (604.85 mg/100 g), distinguishing it from other species in the genus [24,140].

Carotenoid concentrations vary substantially among species, with *S. oleraceus* demonstrating the highest levels at 158 mg/kg fresh weight, while *S. asper* contains 5.58 mg/100 g total carotenoids. These compounds contribute to the antioxidant capacity of the plants and represent important provitamin A sources [140].

The mineral profiles of *Sonchus* species align with those of conventional green leafy vegetables, though specific concentrations vary among species. *S. asper* exhibits superior calcium content (96.25 mg/100 g), while *S. oleraceus* contains the highest iron concentrations (23.74 mg/100 g). Essential minerals, including sodium, potassium, magnesium, phosphorus, copper, zinc, and manganese, are present in nutritionally relevant quantities across all species [23,140].

The fatty acid composition of *Sonchus* species reveals notable concentrations of essential omega-3 fatty acids, with *S. oleraceus* containing the highest proportion (44.97%) of ω-3 series fatty acids. This characteristic enhances the nutritional value of these species, particularly considering the limited availability of plant-based omega-3 sources [140].

Studies conducted in different geographic regions, including southeastern Spain and the Brazilian Atlantic Forest, demonstrate the adaptability of *Sonchus* species to various environmental conditions while maintaining their nutritional integrity. This geographic distribution supports their potential as accessible food resources for diverse rural populations.

The comprehensive nutritional profile of *Sonchus* species, characterized by high concentrations of essential vitamins, minerals, fiber, and beneficial fatty acids, positions these plants as valuable contributors to dietary diversity and food security. Their widespread availability in rural environments, combined with their superior nutritional density compared to conventional vegetables, makes them particularly relevant for populations with limited access to commercial food sources.

The collective evidence supports the classification of *Sonchus* species as nutritionally valuable unconventional food plants. Their exceptional vitamin C content, substantial fiber levels, balanced mineral profiles, and presence of essential fatty acids demonstrate their potential for integration into human diets. The species-specific variations in nutritional composition suggest that different *Sonchus* species may serve complementary roles in addressing various nutritional requirements, thereby contributing to enhanced food security and dietary quality in rural communities.

8.2.2. Health Benefits Associated with Consumption of Sonchus

The *Sonchus* genus, comprising species such as *S. arvensis*, *S. oleraceus*, and *S. asper*, has been extensively studied for its diverse pharmacological properties. These plants are rich in bioactive compounds, including polyphenols, flavonoids, and terpenoids, which contribute to their antioxidant, anti-inflammatory, antidiabetic, hepatoprotective, and hypolipidemic effects [141,142]. These findings support the traditional use of these plants in folk medicine and highlight the potential of *Sonchus* species in preventing chronic diseases due to their bioactive compounds [59].

The antioxidant capacity of *Sonchus* species is primarily attributed to their high phenolic and flavonoid content, including kaempferol, quercetin, and rutin. Methanolic extracts of *S. arvensis* demonstrated potent free radical scavenging activity (420 ± 6.9 mg GAE/g total phenolics), with significant correlations between flavonoid content and antioxidant efficacy [141]. Similarly, *S. oleraceus* ethanolic extracts exhibited 43.46% DPPH radical inhibition, while *S. asper* extracts from discarded leaves showed stronger antioxidant activity than edible parts, suggesting potential applications in nutraceutical formulations [121,137]. These properties mitigate oxidative stress, reducing cellular damage linked to chronic diseases such as diabetes, cancer, and atherosclerosis [143].

Anti-Inflammatory Effects: *Sonchus oleraceus* aqueous extract has been shown to suppress pro-inflammatory cytokines (TNF-α, IL-6) by inhibiting TLR-4, NF-κB, and COX-2 pathways in LPS-stimulated macrophages [61,144]. In vivo studies confirmed its efficacy, with doses of 125–250 mg/kg significantly reducing ear edema in mice without adverse effects. HPLC-MS analysis identified villosol, ferulic acid, and ursolic acid as key anti-inflammatory constituents [61,144]. These findings suggest therapeutic potential for inflammatory conditions such as arthritis and inflammatory bowel disease.

Antidiabetic and Antiglycation Activity: *Sonchus* extracts exhibit notable α-glucosidase and α-amylase inhibitory effects, modulating postprandial glucose absorption [143]. In diabetic rats, *S. oleraceus* extract (200–300 mg/kg) reduced blood glucose, LDL cholesterol, and improved pancreatic histology, comparable to glibenclamide [142]. Additionally, antiglycation properties were observed, with extracts preventing advanced glycation end-products (AGEs) and preserving protein thiols, indicating protective effects against diabetic complications [143].

Hepatoprotective and Hypolipidemic Effects: *S. asper* has traditionally been used for liver detoxification, with studies demonstrating its ability to reduce hepatotoxicity markers (AST, ALT) and lipid peroxidation in CCl4-induced liver injury models [145]. Furthermore, *S. oleraceus* extract significantly lowered total cholesterol and LDL while increasing HDL in hyperlipidemic rats, supporting its role in managing metabolic disorders [142].

Anti-Inflammatory and Gastroprotective Effects: Preclinical studies confirm the anti-inflammatory and gastroprotective properties of *Sonchus oleraceus* hydroalcoholic extracts (SoHE). In rat models, oral SoHE (100–300 mg/kg) reduced carrageenan-induced paw edema, inhibited leukocyte migration, and attenuated lipopolysaccharide-induced fever [146]. SoHE also suppressed cotton pellet-induced granuloma formation, confirming chronic anti-inflammatory activity.

Gastroprotective mechanisms involve multiple pathways. SoHE (30–300 mg/kg) reduced ethanol-induced gastric ulceration by 71.5–76.2% while increasing mucin production up to 477.5% [147]. These effects resulted from enhanced antioxidant defenses (elevated glutathione, superoxide dismutase, catalase, and glutathione-S-transferase), reduced myeloperoxidase activity and TNF levels, and inhibited gastric acid secretion and pepsin activity. Phytochemical analysis identified luteolin-7-O-β-D-glucoside as a key bioactive constituent with notable free radical scavenging capacity (IC_{50} = 15.41 µg/mL against DPPH). These findings support *S. oleraceus* as a therapeutic candidate for inflammatory and gastrointestinal disorders.

The pharmacological potential of *Sonchus* species is well-supported by in vitro and in vivo evidence, highlighting their antioxidant, anti-inflammatory, antidiabetic, and hepatoprotective properties. Their rich phytochemical profile and nutritional value position them as promising candidates for functional foods and therapeutic applications.

9. Ecological and Habitat Analysis

9.1. Habitat Preferences of Sonchus Species

9.1.1. Natural Habitats: Rocky Outcrops, Wetlands, Springs, etc.

Sonchus species exhibit diverse habitat preferences, including rocky outcrops, wetlands, and springs (Table 8). Understanding these natural habitats is crucial for conservation efforts and potential cultivation strategies [25].

Table 8. Habitat types and ecological distribution of *Sonchus* species across Mediterranean environments.

Taxa	Habitat	Ref.
	Rocky habitat	
S. briquetianus	Coastal rocks and cliffs	[8]
S. bulbosus subsp. *microcephalus*	Rocky habitats	[8]
S. fragilis	Rocky, sometimes nitrified walls	[8,148,149]
S. masguindalii	Coastal calcareous rocks	[8,148,149]
S. pustulatus	Cliffs with a steep slope	[8,148,149]
S. suberosus	Limestone rocky slopes	[150]
S. tenerrimus subsp. *amicus*	Rheolitic volcanic rocks	[151]
S. tenerrimus subsp. *dianae*	Cliffs near the sea	[151]
S. tenerrimus subsp. *halodianae*	Coastal cliffs and coastal sands	[151]
S. tenerrimus subsp. *polypodioides*	Steep limestone rocks	[151]
	Scree slopes	
S. bulbosus subsp. *willkomii*	Carbonate screes and scrub on stony soils	[152]
	Wet soils and wetlands	
S. macrocarpus	Moist ground, Nile, and canal banks	[8]
S. maculigerus	Meadows on wet soils	
S. maritimus subsp. *aquatilis*	Wet grasslands on the banks of freshwater streams	[8]
S. mauritanicus	Natural and artificial wetlands	[8]
	Sandy soils	
S. bulbosus subsp. *bulbosus*	Sandy shores, occasionally on roadsides and embankments	[152]
	Salinized habitats	

Table 8. *Cont.*

Taxa	Habitat	Ref.
S. crassifolius	Salt marshes and grasslands on saline soils, as well as the margins of brackish streams and lagoons	[8]
S. erzincanicus	Halophytic grasslands and reedbeds	[37,153]
S. maritimus subsp. *maritimus*	Salt steppes and reed beds	[8]
	Anthropized habitats	
S. arvensis subsp. *arvensis*	Crops, gardens, parks, roadsides, and other anthropized areas	[8]
S. arvensis subsp. *uliginosus*	Crops, gardens, parks, roadsides, and other anthropized areas	[8]
S. asper	Crops, gardens, parks, roadsides, and other anthropized areas	[8]
S. microcephalus	Crops, gardens, parks, roadsides, and other anthropized areas	[154]
S. oleraceus	Crops, gardens, parks, roadsides, and other anthropized areas	[8]
S. tenerrimus subsp. *tenerrimus*	Crops, roadsides, walls, palm trunks, and other anthropized areas	[8]

Abbreviations: Ref., references.

The *Sonchus* taxa exhibits a clear dichotomy in habitat preference, with distinct ecological niches separating natural and anthropic environments (Table 8, Figure 7). A quantitative assessment reveals that out of the 21 entries, 15 taxa (approximately 71%) are associated with natural habitats, while 6 taxa (29%) predominantly occupy anthropized areas. This distribution underscores the genus's ecological versatility, with a majority of species adapted to undisturbed ecosystems, while a smaller subset thrives in human-modified landscapes.

Figure 7. *Sonchus* habitats. (**A**) Sand dunes, *S. bulbosus*, Mar Menor, Murcia (Spain); (**B**) anthropized, *S. tenerrimus* subsp. *tenerrimus*, Saccargia, Sardinia (Italy); (**C**) coastal cliff, *S. tenerrimus* subsp. *halodianae*, San Antonio, Alicante (Spain); (**D**) grassland in the park La Benamá, *S. tenerrimus*, Albalat de la Ribera, Valencia (Spain); (**E**) anthropized, crack at the edge of an asphalt parking lot, next to a wall, *S. asper*, Saccargia, Sardinia (Italy); (**F**) coastal cliff, *S. pustulatus*, San Telmo, Almería (Spain); (**G**) anthropized wet habitats of Albufera, *S. oleraceus*, Catarroja, Valencia (Spain); (**H**) anthropized, *S. tenerrimus* subsp. *tenerrimus*, in a street of Cagliari, Sardinia (Spain); (**I**) grassland on saline soil, *S. oleraceus*, Nora archeological park, Sardinia (Italy). Images: (**A**) Antonio Robledo, with permission; (**B**,**E**,**H**,**I**) Diego Rivera; (**C**) Francisco Alcaraz; (**F**) Estefania Mico, with permission; (**D**,**G**) Emilio Laguna, with permission.

Natural habitats are further divisible into three primary categories: rocky or coastal environments (Figure 7), wetlands, and saline or halophytic ecosystems. Rocky and coastal habitats are the most prevalent, hosting nine taxa, including *Sonchus briquetianus*, *S. bulbosus* subsp. *microcephalus*, and *S. tenerrimus* subsp. *dianae*. These species demonstrate adaptations to xeric conditions, steep slopes, and mineral-rich substrates, suggesting a specialization in low-competition, high-stress environments. Wetland-affiliated taxa, such as *Sonchus macrocarpus*, *S. maritimus* subsp. *aquatilis*, and *S. mauritanicus*, occupy hydric soils along riverbanks, canals, and marshes, indicating a tolerance for waterlogged conditions. Halophytic species, including *Sonchus crassifolius* and *S. maritimus* subsp. *maritimus*, are restricted to saline grasslands and brackish wetlands, reflecting niche specialization in high-salinity environments.

9.1.2. Anthropized Environments: Cultivated Fields, Roadside Verges, Urban Landscapes

Anthropic habitats are dominated by ruderal species such as *Sonchus asper*, *S. oleraceus*, and *S. arvensis* subsp. *arvensis*, which colonize disturbed areas, including croplands, roadsides, and urban spaces. These taxa exhibit broad ecological amplitudes, enabling them to exploit nutrient-rich, frequently disturbed soils [25,26]. The presence of multiple subspecies in both natural and anthropized settings (e.g., *Sonchus tenerrimus* subsp. *tenerrimus* in anthropogenic areas versus subsp. *amicus* in volcanic rocks) suggests that certain lineages within the genus have diverged to occupy contrasting ecological niches [151].

9.2. Role in Ecosystem Services

Sonchus species play important roles in ecosystem services, contributing to biodiversity, soil health, and pollinator support. Their presence in various habitats underscores their ecological significance (Table 8 and Figure 7).

The *Sonchus* genus contributes variably to ecosystem services depending on habitat specialization. Species occupying natural habitats play critical roles in erosion control, particularly those colonizing rocky slopes and coastal cliffs, where root systems stabilize loose substrates. Wetland-associated taxa contribute to water filtration and flood mitigation by reinforcing riparian vegetation structure. Halophytic species, such as *Sonchus crassifolius*, enhance soil formation in saline environments while providing forage for specialized fauna.

Anthropic-adapted *Sonchus* species, though often regarded as weeds, contribute to provisional and regulatory services. Their rapid colonization of disturbed soils aids in early succession, preventing further degradation. Some, like *Sonchus oleraceus*, serve as food sources for pollinators, while others may contribute to nutrient cycling in agroecosystems. However, their presence in cultivated areas may also entail disservices, such as competition with crops. Overall, the genus demonstrates a dual role across ecosystems, with natural-habitat species supporting stability and resilience, while anthropic taxa influence urban and agricultural ecological dynamics. The potential contributions of *Sonchus* species to food security and sustainable agriculture warrant further investigation. Their adaptability to various environments and nutritional value makes them promising candidates for addressing global food challenges. Furthermore, *Sonchus* species that have adapted to anthropogenic habitats may be incorporated into seed mixtures of wild greens sown in organically cultivated Mediterranean systems, such as olive and almond orchards, vineyards, and other agroforestry settings. When young, these plants provide a valuable yield of tender, edible leaves, serve as fodder for livestock, and later contribute to soil fertility as green manure. These were named "crypto crops" [155].

10. Cultivation and Commercialization

10.1. Agronomic Performance and Cultivation Requirements

Sonchus plants have shown remarkable adaptability to anthropized environments, often thriving in cultivated fields, along roadside verges, and in urban landscapes. This adaptability contributes to their widespread distribution and potential for sustainable harvesting [156,157].

Scientific research on cultivating *Sonchus* species has generated substantial insights into their agronomic potential, nutritional characteristics, and commercial viability as domesticated crops. These investigations have established a foundation for understanding both the opportunities and challenges associated with transitioning these wild edible plants into cultivated food systems [14,24,158].

Recent studies have demonstrated that *Sonchus* species, particularly *S. oleraceus* and *S. arvensis*, exhibit considerable adaptability to various cultivation conditions. Research conducted in Greece revealed that *S. oleraceus* displays moderate tolerance to salinity stress, maintaining growth performance under irrigation water containing 40–60 mM NaCl concentrations [159]. While high salinity levels adversely affected plant growth, the species showed promise for cultivation in regions where irrigation water quality is compromised, though the health implications of consuming salt-stressed plants require further investigation.

Fertilization studies have indicated that organic fertilization approaches, particularly those utilizing vinasse-based fertilizers, can achieve comparable results to conventional fertilization methods in promoting growth characteristics and nutrient accumulation [160]. Organic fertilization treatments produced the highest plant height in common sow thistle, while combination treatments of conventional fertilizers with side-dressing applications resulted in leaf number increases of up to 41.6% compared to conventional fertilization alone. Notably, organic fertilization enhanced dry matter content by up to 20.9% relative to other treatments, suggesting potential benefits for crop quality.

Hydroponic cultivation systems have shown particular promise for *Sonchus* production. Research by [161] demonstrated that sow thistle can be successfully cultivated in soilless systems, with optimal nitrogen supply concentrations of 100–200 mg L^{-1} promoting both plant growth and nutritional quality. These intermediate nitrogen concentrations enhanced plant defense mechanisms against oxidative stress while achieving high agronomic efficiency and irrigation water productivity.

10.2. Challenges and Solutions in Domesticating Wild Sonchus Species

Cultivation significantly influences the nutritional profile of *Sonchus* species compared to their wild counterparts. Botella et al. [24] found that cultivated *S. oleraceus* exhibited reduced phenolic compound concentrations but increased vitamin C content relative to wild-harvested plants. Similarly, cultivation of *S. tenerrimus* enhanced concentrations of sugars, organic acids, and β-carotene, which are compounds with important nutritional and health-promoting properties. These compositional changes suggest that cultivation practices can be optimized to enhance specific nutritional attributes depending on intended applications.

Domestication efforts face challenges, such as variability in seed germination and susceptibility to pests [24,25].

The transition from wild harvesting to controlled cultivation presents several technical challenges that require systematic solutions. Seed germination represents a primary obstacle, with field germination rates significantly lower than those in laboratory conditions. Botella et al. [24] reported field germination rates of 49.9% for *S. oleraceus* and only 17.8% for *S. tenerrimus*, compared to laboratory rates of 75.8% and 82.1%, respectively. To address this inconsistency, researchers have developed pre-germination protocols using Petri dishes, followed by seedbed transplantation during autumn–winter cycles to achieve greater uniformity in cultivation trials.

Comparative germination studies have revealed species-specific adaptations that influence cultivation success. Lee et al. [25]. demonstrated that *S. asper* exhibits superior tolerance to environmental stresses compared to *S. oleraceus*, including better performance under low temperatures, drought conditions, and soil burial. These findings have important implications for species selection and cultivation system design.

The rapid maturation characteristics of *Sonchus* species present both opportunities and challenges for sustained production. While quick growth cycles enable multiple harvests, they also limit the collection period for optimal young leaves and aerial parts [162]. Additionally, the weedy nature of these species and their potential for developing herbicide resistance, as documented in glyphosate-resistant populations of *S. oleraceus* in New South Wales [163], requires careful management strategies to prevent unwanted spread.

10.3. Commercial Cultivation

Although large-scale commercial cultivation of *Sonchus* species remains limited, growing interest in functional foods and sustainable agriculture has created market opportunities for these nutrient-dense plants [24,164].

Small-scale cultivation trials across various regions, including Cyprus and other Mediterranean regions and Brazil, have demonstrated the feasibility of incorporating *Sonchus* species into organic farming systems [160]. These initiatives suggest that with appropriate cultivation protocols, *Sonchus* species can be successfully integrated into commercial agricultural operations.

The market appeal of *Sonchus* species stems from their exceptional nutritional profile, including high concentrations of carotenoids, phenolic compounds, and antioxidants, which align with consumer demand for functional foods [24]. Their culinary versatility and sensory acceptability further enhance their commercial potential, particularly in haute cuisine and health-focused market segments.

Sustainability considerations provide additional market advantages, as organic cultivation methods have proven effective for *Sonchus* production, aligning with eco-conscious consumer preferences [161]. However, achieving commercial scalability requires addressing challenges related to germination consistency and yield stability across diverse growing conditions.

Beyond food applications, *Sonchus masguindalii* has been identified as a candidate for ornamental use in Morocco, reflecting broader biodiversity valorization efforts [165]. However, economic feasibility hinges on addressing gaps in supply chain development and consumer awareness.

11. Future Prospects and Research Directions

11.1. Opportunities for Horticultural Development

Scientific research has established that *Sonchus* species (e.g., *S. oleraceus, S. asper, S. arvensis*) possess significant potential as cultivated crops due to their adaptability to diverse agricultural systems and high nutritional value. Studies demonstrate their resilience in varied climates, from Mediterranean regions to subtropical zones, with *S. asper* showing superior drought tolerance and *S. oleraceus* thriving in disturbed soils. Their nutritional profiles—rich in vitamins (A, C, E, K), minerals (iron, calcium), and antioxidants—position them as functional foods, with *S. arvensis* containing notably high vitamin K (604.85 mg/100 g) and *S. oleraceus* providing iron (23.74 mg/100 g) [23–25].

While challenges such as rapid maturation, seed germination variability (e.g., light-dependent germination in *S. oleraceus*), and compositional fluctuations under cultivation persist, targeted agronomic strategies can mitigate these issues. For instance, cultivation trials of *S. oleraceus* and *S. tenerrimus* in Mediterranean climates revealed that controlled environments enhance vitamin C content, though phenolic compounds may decrease. Future research should prioritize standardized cultivation protocols, nutritional optimization, and sustainability assessments.

11.2. Areas for Further Research

Standardized cultivation protocols should focus on optimizing planting density, irrigation, and soil management to balance yield and nutrient retention [24,162]. Nutritional optimization should investigate fertilization regimes to enhance bioactive compounds (e.g., carotenoids, polyphenols) [23,126]. Sustainability assessments should evaluate long-term impacts on soil health and ecosystem biodiversity in agroecological systems [156].

Breeding programs could further enhance commercial viability by selecting traits like delayed bolting (*S. oleraceus* matures in 60–90 days), improved pest resistance, and uniformity in leaf morphology [25]. Integration into agroecological systems—such as intercropping with grains—could reduce herbicide reliance, as *Sonchus* competes effectively with weeds but is suppressed by competitive crops like barley [156].

Despite growing interest, critical knowledge gaps remain. Phytochemical variability, including compositional differences between wild and cultivated populations, is poorly documented. For example, cultivated *S. oleraceus* shows reduced phenolics but higher sugars compared to wild counterparts [21,120]. Germination ecology studies indicate that *S. asper* outperforms *S. oleraceus* under stress (e.g., salinity up to 40 mM NaCl, osmotic potential of −0.4 MPa), suggesting species-specific cultivation niches [25].

Socioeconomic barriers, such as farmer adoption and consumer acceptance, require study, particularly in regions where *Sonchus* is traditionally foraged (e.g., Mediterranean, Brazil). Pilot projects could test marketability, as seen with *S. asper*'s use in gourmet dishes in Italy [23,126,166,167].

Digital ethnobotanical platforms such as CONECT-e represent a transformative advancement in the preservation and democratization of traditional ecological knowledge (TEK) [168]. These platforms function as vital repositories, safeguarding ancestral wisdom from irreversible erosion while fostering cross-cultural knowledge dissemination and scientific inquiry. The establishment of analogous collaborative databases in other regions would substantially bolster global initiatives aimed at documenting endangered traditional practices, particularly as elder knowledge custodians pass away and rural communities experience accelerated modernization.

Although several international efforts exist—including the traditional knowledge section of the Encyclopedia of Life and various regional databases across Europe and Latin America—few adopt CONECT-e's integrative wiki-based framework, which enables direct community contributions. EOL does not have a dedicated "Traditional Knowledge section" as a centralized feature, but it systematically incorporates TK where data exists, particularly through trait annotations and partnerships [169]. The extensive documentation of *Sonchus* species within CONECT-e, notably the 402 records for *S. oleraceus* [170] and 159 for *S. tenerrimus* [171], underscores the platform's efficacy in elucidating the depth of traditional use associated with even ostensibly common flora. Such data provide researchers with unparalleled access to geographically and culturally varied applications, offering valuable insights for ethnobotanical research and potential bioprospecting endeavors.

This robust corpus of crowdsourced information not only reaffirms the cultural importance of *Sonchus* species across Mediterranean societies but also establishes a foundation for systematic comparative analyses of traditional uses. Such analyses may reveal previously unrecorded applications while supporting evidence-based strategies for integrating TEK into contemporary sustainable agriculture and healthcare systems.

11.3. Broader Applications

Sonchus species have ethnomedicinal applications, including anti-inflammatory and antimicrobial properties attributed to sesquiterpene lactones and flavonoids [126,172]. Nanoformulations of *S. asper* extracts (e.g., liposome encapsulation) are being explored for enhanced bioavailability in nutraceuticals [126]. Industrial uses could include latex-derived biopolymers, though toxicity to livestock (e.g., lambs) necessitates caution [173].

Their adaptability to marginal lands and climate resilience make *Sonchus* species candidates for addressing food insecurity. In Brazil's Atlantic Forest, rural communities rely on wild *Sonchus* for nutrition, highlighting its role in food sovereignty [23].

High antioxidant levels (4× red wine) and essential fatty acids support *Sonchus'* use in health-promoting products. Clinical studies on hypoglycemic effects (e.g., GLP-1 modulation by *S. asper*) are nascent but promising.

11.4. Culinary Innovation

Chefs and food innovators are increasingly incorporating *Sonchus* into haute cuisine, experimenting with its flavors and textures in creative dishes. Chefs in avant-garde restaurants incorporate *Sonchus* into salads, soups, and functional foods (e.g., *S. oleraceus* as a bitter green in Mediterranean cuisine). Traditional dishes like *zuppa delle streghe* (Italy) and *Māori puha* (New Zealand) exemplify cultural integration [24].

The unique nutritional profile and flavor characteristics of *Sonchus* species present opportunities for their use in functional and gourmet foods. Ongoing research and culinary experimentation are exploring the potential of these plants as novel ingredients in the food industry. The nutritional profile of *Sonchus* species makes them a promising ingredient for functional foods aimed at promoting health and wellness [24,126].

The incorporation of various species from the genus *Sonchus* in modern cuisine in southern and southeastern Spain originates from traditional cooking, featuring dishes such as dry or brothy rice (typically using short-grain rice varieties like "Bomba" or "Bahía", *Oryza sativa* var. *japonica*, known for their high broth absorption capacity), omelets, *cocas*, and *minchos* (Spanish-style flatbreads with wild greens and herbs, similar to Italian pizza but without cheese).

Today, numerous restaurants utilize different species of sow thistles (e.g., *Sonchus oleraceus, S. tenerrimus, S. asper*) in these traditional recipes with varying degrees of innovation. Many of these dishes are documented by Ruiz in 2022 [174]. Additionally, new culinary creations can be found in the work of the amateur cook Orengo [175], such as the "*Quitxe de Llicsons amb alls tendres i anacardos*" (*Sonchus* tart with tender garlic and cashews) described by the author.

However, one of the most prolific chefs and educators in the use of wild plants is David Ariza, who currently manages Restaurante Rices & Bones in Alicante [176] (Figure 8). His culinary creations frequently incorporate *Sonchus* species in three main preparations.

Vegetable Broth: Ingredients include dried tomato (*Solanum lycopersicum* L.), wild rocket leaves (*Diplotaxis erucoides* (L.) DC.), dandelion (*Taraxacum* sect. *Taraxacum* F.H.Wigg.), sow thistles (*Sonchus oleraceus, S. tenerrimus,* and/or *S. asper*), hedge mustard (*Sisymbrium officinale* (L.) Scop. and related species), saffron (*Crocus sativus* L.), onions (*Allium cepa* L.), garlic (*Allium sativum* L.), carrot (*Daucus carota* L.), ginger (*Zingiber officinale* Roscoe), honey, and salt.

Figure 8. Creative modern recipes made with wild *Sonchus* and other wild greens as main ingredients: (**A**) Spanish rice with marinated weeds ("escabeche"); (**B**) marinated eggplant and weeds ("escabeche"); (**C**) marinated mullet fish with weeds ("escabeche"); (**D**) Spanish rice with fresh weeds. Images by David Ariza, Rest. Rices & Bones Alicante, with permission.

Fermented Weeds: A blend of wild rocket (*Diplotaxis erucoides*), hedge mustard (*Sisymbrium officinale* and related species), dandelion (*Taraxacum* sect. *Taraxacum* F.H.Wigg.), and sow thistles (*Sonchus oleraceus*, *S. tenerrimus*, and/or *S. asper*).

Stir-Fry: Comprising onion (*Allium cepa*), garlic (*Allium sativum*), saffron (*Crocus sativus*), dandelion (*Taraxacum* gr. *officinale*), tomato (*Solanum lycopersicum*), and sow thistles (*Sonchus oleraceus*, *S. tenerrimus*, and/or *S. asper*).

12. Conclusions

The comprehensive examination of *Sonchus* species across botanical, phytochemical, and agricultural dimensions reveals both significant potential and critical research needs. The key findings from each section demonstrate that while these species possess notable nutritional and bioactive properties, substantial gaps remain in our understanding of their phytochemical composition, optimal cultivation methods, and commercial viability. By bridging the gap between botanical and chemical rigor, future studies can not only elevate the quality of *Sonchus* phytochemical research but also serve as a model for interdisciplinary standardization in ethnopharmacology and natural product chemistry.

The potential of *Sonchus* species as novel crops extends beyond traditional food applications, encompassing medicinal and industrial uses that could unlock new economic opportunities. Future research directions should prioritize breeding programs aimed at optimizing yield, enhancing pest resistance, and improving nutritional value to develop cultivars better suited for commercial production. The integration of these species into agroecological systems presents particularly promising opportunities for sustainable agriculture and biodiversity conservation, potentially contributing to more resilient and diverse farming practices that align with contemporary environmental priorities.

However, realizing this potential requires addressing several critical knowledge gaps and inherent challenges. Despite growing scientific interest, our understanding of optimal cultivation methods remains limited, hindering the transition from wild harvesting to systematic agricultural production. The weedy nature of many *Sonchus* species presents a particular challenge for domestication efforts, as their aggressive growth habits and dispersal mechanisms could pose risks in agricultural settings if not properly managed through careful breeding and cultivation protocols. Moreover, the variability in phytochemical profiles across different *Sonchus* species and growing conditions necessitates standardized analytical approaches to ensure consistent quality and efficacy of derived products. Socioeconomic research examining farmer adoption patterns and consumer acceptance is equally essential for assessing commercial viability and informing market development strategies.

The path forward demands multidisciplinary collaboration that combines expertise from agronomy, ethnobotany, food science, and policy development. Such interdisciplinary approaches should focus on optimizing cultivation practices while exploring diverse applications across food, pharmaceutical, and industrial sectors. Policy support mechanisms that incentivize *Sonchus* cultivation within broader agroecological programs, similar to existing EU wild plant initiatives, could provide the institutional framework necessary for scaling research efforts and commercial implementation. Only through this coordinated, multifaceted approach can the scientific community fully unlock the potential of *Sonchus* species as valuable contributors to sustainable agriculture and human nutrition.

Supplementary Materials: The following supporting information can be downloaded at: https://www.mdpi.com/article/10.3390/horticulturae11080893/s1, Table S1: Scoring rubric for *Sonchus* phytochemical literature review. Table S2: Ethnopharmacological and ethnobotanical *Sonchus* data in the Mediterranean. Table S3: Synthesis of traditional medicinal uses of *Sonchus* in the Mediterranean. Table S4: Synthesis of traditional medicinal uses of the different *Sonchus* species. Table S5: Traditional veterinary uses of *Sonchus*. Table S6: Phytochemical *Sonchus* data.

Author Contributions: Conceptualization, D.R. and A.R.-R.; methodology, D.R. and C.O.; software, D.R.; validation, A.R.-R.; formal analysis, D.R. and A.R.-R.; investigation, D.R., A.R.-R., S.R., F.A. and C.O.; resources, D.R. and S.R.; data curation, D.R.; writing—original draft preparation, D.R.; writing—review and editing, C.O.; visualization, D.R., S.R., F.A. and A.R.-R.; supervision, D.R.; project administration, D.R.; funding acquisition, D.R. All authors have read and agreed to the published version of the manuscript.

Funding: This research in an early phase was funded by Ministerio de Educación y Universidades (Spain), Estudio de indicadores de procesos de coevolución entre las poblaciones humanas y las especies vegetales recolectadas como alimento (gfps) en el Mediterráneo, CGL2008-04635/BOS.

Data Availability Statement: The original contributions presented in the study are included in the article and Supplementary Materials; further inquiries can be directed to the corresponding authors.

Acknowledgments: We gratefully acknowledge the contributions of Rodrigo Roldán, for his assistance in Castilla La Mancha. We also extend our gratitude to Antonio Robledo, José Luis Cánovas, Javier Tardío, Estefania Mico, and Emilio Laguna for allowing us to publish their photographs.

Conflicts of Interest: The authors declare no conflicts of interest.

References

1. Romojaro, A.; Botella, M.Á.; Obón, C.; Pretel, M.T. Nutritional and antioxidant properties of wild edible plants and their use as potential ingredients in the modern diet. *Int. J. Food Sci. Nutr.* **2013**, *64*, 944–952. [CrossRef]
2. Tardío, J.; Pardo de Santayana, M.; Morales, R. Ethnobotanical review of wild edible plants in Spain. *Bot. J. Linn. Soc.* **2006**, *152*, 27–71. [CrossRef]
3. Rivera, D.; Obón, C.; Inocencio, C.; Heinrich, M.; Verde, A.; Fajardo, J.; Llorach, R. The Ethnobotanical Study of Local Mediterranean Food Plants as Medicinal Resources in Southern Spain. *J. Physiol. Pharmacol.* **2006**, *57*, 43–59.
4. POWO. Plants of the World Online. Royal Botanic Gardens, Kew. Available online: http://www.plantsoftheworldonline.org (accessed on 20 February 2025).
5. Boulos, L. Révision Systématique du Genre *Sonchus*. I. Introduction et Classification. *Bot. Not.* **1972**, *125*, 287–304.
6. Boulos, L. Révision Systématique du Genre *Sonchus*. II. Etude Caryologique. *Bot. Not.* **1972**, *125*, 306–309.
7. Boulos, L. Révision Systématique du Genre *Sonchus*. III. Etude Palynologique. *Bot. Not.* **1972**, *125*, 310–319.
8. Boulos, L. Révision Systématique du Genre *Sonchus*. IV. Sous Genre 1. *Sonchus*. *Bot. Not.* **1973**, *126*, 155–196.
9. Boulos, L. Révision Systématique du Genre *Sonchus*. V. Sous Genre 2. *Dendrosonchus*. *Bot. Not.* **1974**, *127*, 7–37.
10. Boulos, L. Révision Systématique du Genre *Sonchus*. VI. Sous Genre 3. *Origosonchus*. *Bot. Not.* **1974**, *127*, 402–451.
11. Guarrera, P.M.; Savo, V. Wild food plants used in traditional vegetable mixtures in Italy. *J. Ethnopharmacol.* **2016**, *185*, 202–234. [CrossRef]
12. Morales, P.; Ferreira, I.C.F.R.; Carvalho, A.M.; Sánchez-Mata, C.; Cámara, M.; Fernández-Ruiz, V.; Pardo-de-Santayana, M.; Tardío, J. Mediterranean non-cultivated vegetables as dietary sources of compounds with antioxidant and biological activity. *LWT Food Sci. Technol.* **2014**, *55*, 389–396. [CrossRef]
13. Dioscorides, P. *De Materia Medica*; Osbaldeston, T.A., Translator; Ibidis Press: Johannesburg, South Africa, 2000.
14. Hadjichambis, A.C.; Paraskeva-Hadjichambi, D.; Della, A.; Giusti, M.E.; De Pasquale, C.; Lenzarini, C.; Censorii, E.; Gonzales-Tejero, M.R.; Sanchez-Rojas, C.P.; Ramiro-Gutierrez, J.M.; et al. Wild and semi-domesticated food plant consumption in seven circum-Mediterranean areas. *Int. J. Food Sci. Nutr.* **2008**, *59*, 383–414. [CrossRef]
15. Pasta, S.; La Mantia, T.; Rühl, J. The impact of anthropogenic disturbance on the population structure and dynamics of *Sonchus* species in the Mediterranean basin. *Plant Biosyst.* **2012**, *146*, 392–401. [CrossRef]
16. Pieroni, A.; Quave, C.L. Functional foods or food medicines? On the consumption of wild plants among Albanians and Southern Italians in Lucania. *Eat Tradit. Foods* **2005**, *8*, 1–25.
17. González-Tejero, M.R.; Casares-Porcel, M.; Sánchez-Rojas, C.P.; Ramiro-Gutiérrez, J.M.; Molero-Mesa, J.; Pieroni, A.; Giusti, M.E.; Censorii, E.; de Pasquale, C.; Della, A.; et al. Medicinal plants in the Mediterranean area: Synthesis of the results of the project Rubia. *J. Ethnopharmacol.* **2008**, *116*, 341–357. [CrossRef] [PubMed]
18. Kilian, N.; Gemeinholzer, B.; Lack, H.W. Sonchus. In *Flora of the Mediterranean*; Springer: Berlin/Heidelberg, Germany, 2009; pp. 325–340.
19. Li, X.; Yang, P. Research progress of *Sonchus* species. *Int. J. Food Prop.* **2018**, *21*, 147–157. [CrossRef]
20. Alothman, E.; Awaad, A.; Safhi, A.; Almoqren, S.; El-Meligy, R.; Zain, Y.; Alasmary, F.; Alqasoumi, S.I. Evaluation of anti-ulcer and ulcerative colitis of *Sonchus oleraceus* L. *Saudi Pharm. J.* **2018**, *26*, 956–959. [CrossRef] [PubMed]
21. Zubair, M.; Gul, Z.; Awais, M.; Saeed, S.; Akbar, A. Nutritional and medicinal importance of *Sonchus asper* (L.) hill plant-a review. *Pak-Euro J. Med. Life Sci.* **2023**, *6*, 219–224.
22. Heba, F.; Eman, S.; Safwat, A. Phytochemistry and pharmacological effects of plants in genus *Sonchus* (Asteraceae). *Rec. Pharm. Biomed. Sci.* **2020**, *4*, 40–50. [CrossRef]
23. de Paula, G.; Barreira, T.; Pinheiro-Sant'Ana, H. Chemical composition and nutritional value of three *Sonchus* species. *Int. J. Food Sci.* **2022**, *2022*, 4181656. [CrossRef]
24. Botella, M.Á.; Hellín, P.; Hernández, V.; Dabauza, M.; Robledo, A.; Sánchez, A.; Fenoll, J.; Flores, P. Chemical Composition of Wild Collected and Cultivated Edible Plants (*Sonchus oleraceus* L. and *Sonchus tenerrimus* L.). *Plants* **2024**, *13*, 269. [CrossRef] [PubMed]
25. Lee, Y.; Mahajan, G.; Beregszaszi, R.; Chauhan, B.S. Seed Germination Ecology of *Sonchus asper* and *Sonchus oleraceus* in Queensland Australia. *Plants* **2024**, *13*, 3451. [CrossRef]
26. Peerzada, A.M.; O'Donnell, C.; Adkins, S. Biology, impact, and management of common sowthistle (*Sonchus oleraceus* L.). *Acta Physiol. Plant* **2019**, *41*, 136. [CrossRef]
27. SCOPUS. Scopus Document Search. Available online: https://www.scopus.com/search/form.uri?display=basic#basic (accessed on 17 February 2025).
28. Aria, M.; Cuccurullo, C. Bibliometrix: An R-tool for comprehensive science mapping analysis. *J. Informetr.* **2017**, *11*, 959–975. [CrossRef]
29. Aria, M.; Cuccurullo, C. Bibliometrix and Biblioshiny 4.1. Available online: https://www.bibliometrix.org/home/index.php (accessed on 17 February 2025).

30. VOS. VOSviewer: Visualizing Scientific Landscapes. Available online: https://www.vosviewer.com/ (accessed on 17 February 2025).
31. van Eck, N.J. VOSviewer Manual. Waltman. L. Available online: https://www.vosviewer.com/documentation/Manual_VOSviewer_1.6.20.pdf (accessed on 17 February 2025).
32. van Eck, N.J.; Waltman, L. Visualizing bibliometric networks. In *Measuring Scholarly Impact*; Ding, Y., Rousseau, R., Wolfram, D., Eds.; Springer: Cham, Switzerland, 2014; pp. 285–320. [CrossRef]
33. Moher, D.; Liberati, A.; Tetzlaff, J.; Altman, D.G. Preferred Reporting Items for Systematic Reviews and Meta-Analyses: The PRISMA Statement. *PLoS Med.* **2009**, *6*, e1000097. [CrossRef]
34. GBIF. Sonchus in Global Biodiversity Information Facility Free and Open Access to Biodiversity Data. Available online: https://www.gbif.org/species/3105646/ (accessed on 17 February 2025).
35. Shamso, E.; Hosni, H.; Ahmed, D.; Shaltout, K. Achene Characteristics of Some Taxa of Asteraceae from the Northwestern Mediterranean Coast of Egypt. *Egypt. J. Bot.* **2021**, *61*, 1–31. [CrossRef]
36. Wahyuni, D.K.; Rahayu, S.; Purnama, P.R.; Saputro, T.B.; Suharyanto; Wijayanti, N.; Purnobasuki, H. Morpho-anatomical structure and DNA barcode of *Sonchus arvensis* L. *Biodiversitas* **2019**, *20*, 2417–2426. [CrossRef]
37. Kandemir, A.; Makbul, S.; Türkmen, Z.; Yilmaz, M. Morphological, Anatomical and Palynological Investigation on Sonchus erzincanicus Matthews (Asteraceae). *Turk. J. Bot.* **2006**, *30*, 405–411. Available online: https://journals.tubitak.gov.tr/botany/vol30/iss5/8 (accessed on 3 April 2025).
38. Lisitsina, G. Main types of ancient farming on the Caucasus—On the basis of palaeoethnobotanical research. *Ber. Deutsch. Bot. Ges.* **1978**, *91*, 47–57. [CrossRef]
39. Rivera, D.; Matilla, G.; Obón, C.; Alcaraz, F. *Plants and Humans in The Near East and the Caucasus, Volume 2. The Plants: Angiosperms, Ancient and Traditional Uses of Plants as Food and Medicine, a Diachronic Ethnobotanical Review (Armenia, Azerbaijan, Georgia, Iran, Iraq, Lebanon, Syria, and Turkey)*; EDITUM–Plants & Humans: Murcia, Spain, 2012.
40. Campbell-Thompson, R. *A Dictionary of Assyrian Botany*; British Academy: London, UK, 1949.
41. Gunther, R.T. *The Greek Herbal of Dioscorides (Translated by John Goodyer in 1655)*; Hafner Publishing Co.: New York, NY, USA, 1968.
42. Bostock, J.; Riley, J. *The Natural History. Pliny the Elder*; Taylor and Francis: London, UK, 1855; Volume 4, Available online: https://www.perseus.tufts.edu/hopper/text?doc=plin.+nat.+toc (accessed on 10 April 2025).
43. de Frutos González, V. *Edición Crítica del Regimen Sanitatis Salernitanum Transmitido por los Manuscritos Add. 12190 y Sloane 351 de la British Library de Londres*; Universidad de Valladolid: Valladolid, Spain, 2010.
44. González Blanco, M.I. An Edition of the Middle English Translation of the Antidotarium Nicolai. Master's Thesis, University of Glasgow, Glasgow, UK, 2018. Available online: https://theses.gla.ac.uk/8965/7/2018GonzalezBlancoMPhilR.pdf (accessed on 10 April 2025).
45. von Bingen, H. *Physica: Liber Subtilitatum Diversarum Naturarum Creaturarum Edition der Florentiner Handschrift (Cod. Laur. Ashb. 1323, ca. 1300) in Vergleich mit der Textkonstitution der Patrologia Latina (MIGNE)*; Müller, I., Schulze, C., Eds.; Olms-Weidmann: Hildesheim, Germany, 2008.
46. Mattioli, P.A. *Di Pedacio Dioscoride Anazarbeo Libri Cinque Della Historia, & Materia Medicinale*; Niccolò de Bascarini: Venice, Italy, 1544.
47. Laguna, A. *Pedacio Dioscorides Anazarbeo, Acerca de la Materia Medicinal y de los Venenos Mortiferos*; Iuan Latio: Antwerp, Belgium, 1555; Available online: https://bdh-rd.bne.es/viewer.vm?id=0000037225&page=1 (accessed on 10 April 2025).
48. Fuchs, L. *De Historia Stirpium Commentarii Insignes (Notable Commentaries on the History of Plants)*; Michael Isingrin: Basel, Switzerland, 1542.
49. Fuchs, L. *Historia de Yervas y Plantas*; Jaraba, J., Translator; Iuan Lacio: Antwerp, Belgium, 1557.
50. Matthioli, A. *I Discorsi di M. Pietro Andrea Matthioli Médico Sanese ne i Sei Libri della Materia Medicinale di Pedacio Dioscoride Anazarbeo*; Vincenzo Valgrisi: Venice, Italy, 1557; Book 2, Chapter CXX; p. 180.
51. Clusius, C. *Rariorum Plantarum Historia*; Plantin (Ex Officina Plantiniana apud Ioannem Moretum): Antwerp, Belgium, 1601.
52. Covarruvias, S. *Parte Primera del Tesoro de la Lengua Castellana*; Melchor Sánchez: Madrid, Spain, 1674.
53. Morales, R.; Tardío, J.; Aceituno, L.; Molina, M.; Pardo de Santayana, M. Biodiversidad y Etnobotánica en España. *Mem. R. Soc. Esp. Hist. Nat.* **2011**, *9*, 157–207.
54. Quer, J. *Continuación de la Flora Española o Historia de las Plantas de España*; Joachin Ibarra: Madrid, Spain, 1784; Volume 6, Available online: https://www.google.es/books/edition/Continuacion_de_la_flora_espanola_o_hist/w7rH-jP5PBcC?hl=es&gbpv=1&dq=Quer+flora+espa%C3%B1ola+tomo+6&pg=PP5&printsec=frontcover (accessed on 10 April 2025).
55. Font Quer, P. *Plantas Medicinales: El Dioscorides Renovado*; Labor: Barcelona, Spain, 1981.
56. Moerman, D. Native American Ethnobotany: Sonchus. Available online: http://naeb.brit.org/uses/search/?string=Sonchus (accessed on 16 February 2025).
57. Cocarico, S.; Rivera, D.; Beck, S.; Obón, C. Agrobiodiversity as a Reservoir of Medicinal Resources: Ethnobotanical Insights from Aymara Communities in the Bolivian Andean Altiplano. *Horticulturae* **2025**, *11*, 50. [CrossRef]

58. BDMTM. Biblioteca Digital de la Medicina Tradicional Mexicana. Lechuguilla. Available online: http://www.medicinatradicionalmexicana.unam.mx/apmtm/termino.php?l=3&t=sonchus-oleraceus (accessed on 16 February 2025).
59. Sánchez-Aguirre, O.A.; Sánchez-Medina, A.; Juárez-Aguilar, E.; Barreda-Castillo, J.M.; Cano-Asseleih, L.M. *Sonchus oleraceus* L.: Ethnomedical, phytochemical and pharmacological aspects. *Naunyn Schmiedebergs Arch. Pharmacol.* **2024**, *397*, 4555–4578. [CrossRef] [PubMed]
60. Rivera, D.; Rivera-Obón, D.J.; Palazón, J.A.; Obón, C. From Weeds to Feeds: Exploring the Potential of Wild Plants in Horticulture from a Centuries-Long Journey to an AI-Driven Future. *Horticulturae* **2024**, *10*, 1021. [CrossRef]
61. Li, Q.; Dong, D.D.; Huang, Q.P.; Li, J.; Du, Y.Y.; Li, B.; Li, H.Q.; Huyan, T. The anti-inflammatory effect of *Sonchus oleraceus* aqueous extract on lipopolysaccharide stimulated RAW 264.7 cells and mice. *Pharm. Biol.* **2017**, *55*, 799–809. [CrossRef] [PubMed]
62. Ríos, S.; Alcaraz, F. *Flora de las Riberas y Zonas Húmedas de la Cuenca del río Segura*; Servicio de publicaciones; Universidad de Murcia: Murcia, Spain, 1996; 331p.
63. Nicolás, C. *Estudio de las Plantas Comestibles Silvestres del Municipio de Murcia (Murcia)*; Universidad Miguel Hernández: Elche, Spain, 2010; 146p.
64. Selma, C. *Flora y Vegetación Cormofítica del Sector Noroccidental de la Región de Murcia*; Universidad de Murcia: Murcia, Spain, 1990; 702p.
65. Müller, A. *Ethnobotanik im Unteren Segura-Becken und Einige Phytochemische Untersuchungen*; Facultad de Matemáticas y Ciencias Naturales, Universidad Renana Federico-Guillermo de Bonn: Bonn, Germany, 1996; 131p.
66. CLM. *Sonchus* in Castilla La Mancha DB 2024. Ethnobotanical Database. Available online: https://github.com/drivera2001/Castilla-La-Mancha-Ethnobotany-DB (accessed on 25 July 2025).
67. Obón, C.; Nicolás, C.; Rivera, D. Estudio de las plantas comestibles silvestres del municipio de Murcia. In *Actas del III Congreso de la Naturaleza de la Región de Murcia*; García, P., Ed.; ANSE-CEMACAM: Murcia, Spain, 2007; pp. 97–105.
68. Rivera, D.; Obón, C. Etnopharmacology of Murcia. In *Medicaments et Aliments Approche Etnopharmacologique*; Schröder, E., Balansard, G., Cabalion, P., Fleurentin, J., Mazars, G., Eds.; ORSTOM: Paris, France, 1996; pp. 121–128.
69. Murcia. *Sonchus* in Murcia DB 2024. Ethnobotanical Database (including Laboratorio Etnobotánica 1992, Laboratorio Etnobotánica 1998, Laboratorio Etnobotánica 2006). Available online: https://github.com/drivera2001/Murcia-Region-Ethnobotany-DB (accessed on 25 July 2025).
70. Costa, J.B. Una incorporación norteafricana y andalusí a la farmacopea medieval: Los tubérculos comestibles de *Sonchus bulbosus* (L.) N. Kilian & Greuter. *Panace Rev. De Med. Leng. Traducción* **2019**, *20*, 24–33.
71. Alcaraz, F.; Lozano, M.T.; Llimona, X. Flora y vegetaciónde lo arrozales próximos a Calasparra (Murcia, SE. de España). *An. Univ. De Murcia (Cienc.)* **1981**, *37*, 39–61.
72. Castillo, J. *I Certamen de Investigación Etnográfica a Través de la Tradición Oral y Otras Fuentes. Aljaba, Monográfico. Peña Huertana* <<*La Crilla*>>; Consejo Municipal de Cultura y Festejos (Centro Cultural de Puente Tocinos): Murcia, Spain, 1995; p. 62.
73. Rabal, G. Algunas consideraciones sobre el conocimiento etnobotánico en el campo de Cartagena. *Rev. Murc. De Antropol.* **2010**, *10*, 227–239.
74. Blanco, E.; Muñoz, G. 1996. El mundo vegetal en la tradición del Noroeste de Murcia (Comarca del Altiplano). *Yakka Rev. De Estud. Yeclanos* **1996**, *7*, 173–176.
75. Murcia-Riquelme, R.M. *Estudio Etnobotánico y Propiedades Nutritivas y Antioxidantes de Plantas Comestibles de Sierra Espuña*; Universidad Miguel Hernández: Elche, Spain, 2012; pp. 1–171.
76. Belda, A.; Jordán-Nuñez, J.; Micó-Vicent, B.; López-Rodríguez, D. Long-Term Monitoring of the Traditional Knowledge of Plant Species Used for Culinary Purposes in the Valencia Region, South-Eastern Spain. *Plants* **2024**, *13*, 775. [CrossRef]
77. Jiménez, A. Trabajo realizado sobre las plantas de Jumilla. *Juncellus* **1997**, *12*, 39–50.
78. Rabal, G. Cuando la chicoria echa flor... (Etnobotánica en Torre Pacheco). *Rev. Murc. De Antropol.* **2000**, *6*, 1–205.
79. Revelles, L. *El uso de las plantas en Puerto Lumbreras*; Colegio Público Sagrado Corazón: Puerto Lumbreras, Spain, 1998; p. 49.
80. Ruiz-Marín, D. *Vocabulario de las Hablas Murcianas*; Consejería de Presidencia, Gobierno de Murcia: Murcia, Spain, 2000.
81. Rivera, D.; Obón, C. Las plantas y el hombre en el Valle de Ricote. In *Valle de Ricote II Congreso Turístico Cultural*; Mancomunidad de Municipios Valle de Ricote: Murcia, Spain, 2003; pp. 283–316.
82. Pardo-de-Santayana, M.; Morales, R.; Tardío, J.; Aceituno, L.; Molina, M. Conocimientos Tradicionales Relativos a la Biodiversidad. Segunda Fase. Tomo 3. Ministerio para la Transición Ecológica: Madrid, Spain, 2018.
83. Rivera, D.; Alcaraz, F.; Verde, A.; Fajardo, J.; Obón, C. *Las Plantas en la Cultura Popular*; SOMEHN (Sociedad Mediterránea de Historia Natural): Jumilla, Spain, 2008.
84. Sansanelli, S.; Tassoni, A. Wild food plants traditionally consumed in the area of Bologna (Emilia Romagna region, Italy). *J. Ethnobiol. Ethnomedicine* **2014**, *10*, 69. [CrossRef] [PubMed]
85. Sansanelli, S.; Ferri, M.; Salinitro, M.; Tassoni, A. Ethnobotanical survey of wild food plants traditionally collected and consumed in the Middle Agri Valley (Basilicata region, southern Italy). *J. Ethnobiol. Ethnomedicine* **2017**, *13*, 50. [CrossRef] [PubMed]

86. Biscotti, N.; Bonsanto, D.; Del Viscio, G. The traditional food use of wild vegetables in Apulia (Italy) in the light of Italian ethnobotanical literature. *Ital. Bot.* **2018**, *5*, 1–24. [CrossRef]
87. Licata, M.; Tuttolomondo, T.; Leto, C.; Virga, G.; Bonsangue, G.; Cammalleri, I.; La Bella, S. A Survey of Wild Plant Species for Food Use in Sicily (Italy)—Results of a 3-Year Study in Four Regional Parks. *J. Ethnobiol. Ethnomed.* **2016**, *12*, 12. [CrossRef] [PubMed]
88. Ranfa, A.; Bodesmo, M. An Ethnobotanical investigation of traditional knowledge and uses of edible wild plants in the Umbria Region, Central Italy. *J. Appl. Bot. Food Qual.* **2017**, *90*, 246–258. [CrossRef]
89. Pieroni, A.; Cattero, V. Wild vegetables do not lie: Comparative gastronomic ethnobotany and ethnolinguistics on the Greek traces of the Mediterranean Diet of southeastern Italy. *Acta Bot. Bras.* **2019**, *33*, 198–211. [CrossRef]
90. Tuttolomondo, T.; Licata, M.; Leto, C.; Savo, V.; Bonsangue, G.; Gargano, M.L.; La Bella, S. Ethnobotanical investigation on wild medicinal plants in the Monti Sicani Regional Park (Sicily, Italy). *J. Ethnopharmacol.* **2014**, *153*, 568–586. [CrossRef]
91. Motti, R.; Motti, P. An ethnobotanical survey of useful plants in the agro Nocerino Sarnese (Campania, southern Italy). *Human. Ecol.* **2017**, *45*, 865–878. [CrossRef]
92. Guarrera, P.M.; Lucia, L.M. Ethnobotanical remarks on central and southern Italy. *J. Ethnobiol. Ethnomedicine* **2007**, *3*, 23. [CrossRef]
93. Guarrera, P.M.; Forti, G.; Marignoli, S. Ethnobotanical and ethnomedicinal uses of plants in the district of Acquapendente (Latium, Central Italy). *J. Ethnopharmacol.* **2005**, *96*, 429–444. [CrossRef] [PubMed]
94. Signorini, M.A.; Piredda, M.; Bruschi, P. Plants and traditional knowledge: An ethnobotanical investigation on Monte Ortobene (Nuoro, Sardinia). *J. Ethnobiol. Ethnomedicine* **2009**, *5*, 6. [CrossRef]
95. Monreale, F. *Piante Spontanee Alimentari in Sicilia. Guida di Fitoalimurgia*; Natura Sicula: Siracusa, Syracuse, 2017.
96. Guarrera, P.M. *Usi e Tradizioni Della Flora Italiana, Medicina Popolare ed Etnobotanica*; Aracne: Roma, Italy, 2006.
97. Atzei, A.D. *Le Piante Nella Tradizione Popolare Della Sardegna*; Carlo Delfino Editore: Sassari, Italy, 2003.
98. Nebel, S. *Ta Chòrta: Piante Commestibili Tradizionali di Gallicianò*; School of Pharmacy, University of London: London, UK, 2005.
99. Les Ecologistes de l'Euzière. *Les salades sauvages, L'Ensalada champanèla*; Les Ecologistes de l'Euzière: Prades-le-Lez, France, 2003.
100. Paume, M.C. *Sauvages et Comestibles, Herbes, Fleurs & Petites Salades*; Sarl Édisud: Aix-en-Provence, France, 2005.
101. Savvides, L. *Edible Wild Plants of the Cyprus Flora*; Loucas Savvides: Nicosia, Cyprus, 2000.
102. Chaachouay, N.; Benkhnigue, O.; Fadli, M.; El Ibaoui, H.; Zidane, L. Ethnobotanical and ethnopharmacological studies of medicinal and aromatic plants used in the treatment of metabolic diseases in the Moroccan Rif. *Heliyon* **2019**, *5*, e02191. [CrossRef]
103. Bellakhdar, J. *La Pharmacopée Marocaine Traditionnelle*; Editions Le Fennec-Ibis Press: Paris, France, 1997.
104. Lampraki, M. *Ta Jorta*; Ellinika Grammata: Athens, Greece, 2000.
105. Kayabaşı, N.P.; Tümen, G.; Polat, R. Ethnobotanical Studies on Useful Plants in Manyas (Balıkesir/Turkey) Region. *Biol. Divers. Conserv.* **2016**, *9/3*, 58–63.
106. Şenkardeş, İ.; Bulut, G.; Doğan, A.; Tuzlacı, E. An ethnobotanical analysis on wild edible plants of the turkish asteraceae taxa. *Agric. Conspec. Sci.* **2019**, *84*, 17–28. Available online: https://hrcak.srce.hr/218531 (accessed on 23 May 2025).
107. Simsek, I.; Aytekin, F.; Yesilada, E.; Yildirimli, Ş. An Ethnobotanical Survey of the Beypazari, Ayas, and Gudul District Towns of Ankara Province (Turkey). *Econ. Bot.* **2004**, *58*, 705–720. [CrossRef]
108. Shaltout, K.H.; Ahmed, D.A.; Al-Sodany, Y.M.; Haroun, S.A.; El-Khalafy, M.M. Cultural importance indices of the endemic plants in Egypt. *Egypt. J. Bot.* **2023**, *63*, 649–663. [CrossRef]
109. Ahmed, D.A.; El-Khalafy, M.M.; Almushghub, F.; Sharaf El-Din, A.; Shaltout, K. Ethnobotanical importance of wild plants in Wadi Kaam, Northwestern Libya. *Egypt. J. Bot.* **2023**, *63*, 797–812. [CrossRef]
110. Łuczaj, Ł.; Jug-Dujaković, M.; Dolina, K.; Jeričević, M.; Vitasović-Kosić, I. The ethnobotany and biogeography of wild vegetables in the Adriatic islands. *J. Ethnobiol. Ethnomed.* **2019**, *15*, 18. [CrossRef] [PubMed]
111. Łuczaj, Ł.; Zovko Končić, M.; Miličević, T.; Dolina, K.; Pandža, M. Wild Vegetable Mixes Sold in the Markets of Dalmatia (Southern Croatia). *J. Ethnobiol. Ethnomed.* **2013**, *9*, 2. [CrossRef] [PubMed]
112. Simkova, K.; Polesny, Z. Ethnobotanical review of wild edible plants used in the Czech Republic. *J. Appl. Bot. Food Qual.* **2015**, *88*, 9. [CrossRef]
113. Sher, Z.; Khan, Z.; Hussain, F. Ethnobotanical studies of some plants of Chagharzai valley, district Buner, Pakistan. *Pak. J. Bot.* **2011**, *43*, 1445–1452. Available online: http://www.pakbs.org/pjbot/PDFs/43(3)/PJB43(3)1445.pdf (accessed on 23 May 2025).
114. Senouci, F.; Ababou, A.; Chouieb, M. Ethnobotanical survey of the medicinal plants used in the Southern Mediterranean. Case study: The region of Bissa (Northeastern Dahra Mountains, Algeria). *Pharmacogn. J.* **2019**, *11*, 103. [CrossRef]
115. Leonti, M.; Nebel, S.; Rivera, D.; Heinrich, M. Wild gathered food plants in the European mediterranean: A comparative analysis. *Econ. Bot.* **2006**, *60*, 130–142. [CrossRef]
116. Aissani, F.; Grara, N.; Bensouici, C.; Bousbia, A.; Ayed, H.; Idris, M.H.M.; Teh, L.K. Algerian *Sonchus oleraceus* L.: A comparison of diferent extraction solvent on phytochemical composition, antioxidant properties and anti-cholinesterase activity. *Adv. Tradit. Med.* **2022**, *22*, 383–394. [CrossRef]

117. Chetehounaa, S.; Derouichec, S.; Réggamie, Y.; Boulaaresc, I.; Frahtiac, A. Gas Chromatography Analysis, Mineral Contents and Anti-inflammatory Activity of *Sonchus maritimus*. *Trop. J. Nat. Prod. Res.* **2024**, *8*, 6787–6798.
118. El Gendy, A.E.-N.G.; Mohamed, N.A.; Sarker, T.C.; Hassan, E.M.; Garaa, A.H.; Elshamy, A.I.; Abd-ElGawad, A.M. Chemical Composition, Antioxidant, and Cytotoxic Activity of Essential Oils in the Above-Ground Parts of *Sonchus oleraceus* L. *Plants* **2024**, *13*, 1712. [CrossRef]
119. Helal, A.M.; Nakamura, N.; El-Askary, H.; Hattori, M. Sesquiterpene lactone glucosides from *Sonchus asper*. *Phytochemistry* **2000**, *53*, 473–477. [CrossRef] [PubMed]
120. Hussain, J.; Muhammad, Z.; Ullah, R.; Khan, F.U.; Khan, I.U.; Khan, N.; Ali, J.; Jan, S. Evaluation of the chemical composition of *Sonchus eruca* and *Sonchus asper*. *J. Am. Sci.* **2010**, *6*, 231–235.
121. Mansour, R.M.; Saleh, N.A.; Boulos, L. A chemosystematic study of the phenolics of *Sonchus*. *Phytochemistry* **1983**, *22*, 489–492. [CrossRef]
122. Mohasib, R.M.; Nagib, A.; Samad, A.A.; Salama, Z.A.; Gaafar, A.A.; Taie, H.A.; Hussein, S.R. Identification of bioactive ingredients in *Sonchus oleraceus* by HPLC and GC/MS. *Plant Arch.* **2020**, *20*, 9714–9720.
123. Nobelaa, O.; Ndhlalab, A.; Tugizimanac, F.; Njobeha, P.; Raphashab, D.; Ncubed, B.; Madalae, N. Tapping into the realm of underutilised green leafy vegetables: Using LCIT-Tof-MS based methods to explore phytochemical richness of *Sonchus oleraceus* (L.) L. *S. Afr. J. Bot.* **2022**, *145*, 207–212. [CrossRef]
124. Nouidha, S.; Selmi, S.; Guigonis, J.M.; Pourcher, T.; Chekir-Ghedira, L.; Kilani-Jaziri, S. Metabolomics profiling of Tunisian *Sonchus oleraceus* L. extracts and their antioxidant activities. *Chem. Biodivers.* **2023**, *20*, e202300290. [CrossRef]
125. Ozgen, U.; Sevindik, H.; Kazaz, C.; Yigit, D.; Kandemir, A.; Secen, H.; Calis, I. A new sulfated α-ionone glycoside from *Sonchus erzincanicus* Matthews. *Molecules* **2010**, *15*, 2593–2599. [CrossRef]
126. Parisi, V.; Santoro, V.; Faraone, I.; Benedetto, N.; Vassallo, A.; De Tommasi, N.; Milella, L.; Nesticò, A.; Maselli, G.; Fadda, A.M.; et al. *Sonchus asper* (L.) Hill extracts: Phytochemical characterization and exploitation of its biological activities by loading into nanoformulation. *Front. Plant Sci.* **2024**, *15*, 1416539. [CrossRef]
127. Rafi, M.; Karomah, A.H.; Annisa, D.F.; Maharani; Mulyati, A.H.; Septaningsih, D.A.; Nurcholis, W.; Syaftri, U.; Lim, L.; Achmadi, S.; et al. Phytochemical profiling using FTIR-and LC-HRMS-based metabolomics, mineral content, and free radical scavenging activity of *Sonchus arvensis* extracts from different growth locations. *Vegetos* **2025**, *38*, 12. [CrossRef]
128. Shahar, B.; Dolma, N.; Chongtham, N. Phytochemical analysis, antioxidant activity and identification of bioactive constituents from three wild medicinally important underutilized plants of Ladakh, India using GCMS and FTIR based metabolomics approach. *Food Humanit.* **2023**, *1*, 430–439. [CrossRef]
129. Xia, Z.; Yao, J.; Liang, J. Two new sesquiterpene lactones from *Sonchus arvensis*. *Chem. Nat. Compd.* **2012**, *48*, 47–50. [CrossRef]
130. Xu, Y.J.; Sun, S.B.; Sun, L.M.; Qiu, D.F.; Liu, X.J.; Jiang, Z.B.; Yuan, C.S. Quinic acid esters and sesquiterpenes from *Sonchus arvensis*. *Food Chem.* **2008**, *111*, 92–97. [CrossRef]
131. Baruah, P.; Baruah, N.C.; Sharma, R.P.; Baruah, J.N.; Kulanthaivel, P.; Herz, W. A monoacyl galactosylglycerol from *Sonchus arvensis*. *Phytochemistry* **1983**, *22*, 1741–1744. [CrossRef]
132. Amangeldi, N.; Amangeldikyzy, Z.; Abdukerim, R.Z.; Yerezhepova, A. Sow thistle (*Sonchus oleraceus* L.) plants growth and chemical composition from the aboveground part in Kazakhstan. *J. Zhangir Khan Agric. Tech. Univ. Sci.* **2024**, *3*, 28–38. [CrossRef]
133. Xia, Z.; Qu, W.; Lu, H.; Fu, J.; Ren, Y.; Liang, J. Sesquiterpene lactones from *Sonchus arvensis* L. and their antibacterial activity against *Streptococcus mutans* ATCC 25175. *Fitoterapia* **2010**, *81*, 424–428. [CrossRef]
134. Giner, R.M.; Ubeda, A.; Just, M.J.; Serrano, A.; Máñez, S.; Ríos, J.L. A chemotaxonomic survey of *Sonchus* subgenus *Sonchus*. *Biochem. Syst. Ecol.* **1993**, *21*, 617–620. [CrossRef]
135. Shimizu, S.; Miyase, T.; Ueno, A.; Usmanghani, K. Sesquiterpene lactone glycosides and ionone derivative glycosides from *Sonchus asper*. *Phytochem.* **1989**, *28*, 3399–3402. [CrossRef]
136. Ahmida, M.; Fouad, I.; Elbrghathi, A.; Madi, M.; El Raey, M.; Elmabsout, A.; Khadra, I.M. Phytochemical and Biological Investigation of *Sonchus tenerrimus* Growing in Libya. *EC Nutr.* **2018**, *13*, 656–661. Available online: https://ecronicon.net/assets/ecnu/pdf/ECNU-13-00507.pdf (accessed on 23 May 2025).
137. Rafi, M.; Hasanah, T.; Karomah, A.; Mulyati, A.; Trivadila, A.; Rahminiwati, M.; Achmadi, S.; Iswantini, D. FTIR- and UHPLC-Q-Orbitrap HRMS-based Metabolomics of *Sonchus arvensis* Extracts and Evaluation of Their Free Radical Scavenging Activity. *Sains Malays.* **2022**, *51*, 3261–3269. [CrossRef]
138. Maslahat, M.; Mardinata, D.; Surur, S.M.; Lioe, H.N.; Syafitri, U.D.; Rafi, M.; Rohaeti, E. Untargeted Metabolomics Analysis Using FTIR and LC-HRMS for Differentiating Sonchus arvensis Plant Parts and Evaluating Their Biological Activity. *Chem. Biodivers.* **2025**, *22*, e202401537. [CrossRef] [PubMed]
139. Delyan, E. Analysis of component composition of volatile compounds of field sow thistle (*Sonchus arvensis* L.) leaves using the method of gas chromatography with mass-detection. *Pharma Innov. J.* **2016**, *5*, 118–121.
140. Guil-Guerrero, J.L.; Giménez-Giménez, A.; Rodríguez-García, I.; Torija-Isasa, M.E. Nutritional composition of *Sonchus* species (*S. asper* L., *S. oleraceus* L. and *S. tenerrimus* L.). *J. Sci. Food Agric.* **1998**, *76*, 628–632. [CrossRef]

141. Khan, R.A. Evaluation of flavonoids and diverse antioxidant activities of *Sonchus arvensis*. *Chem. Cent. J.* **2012**, *6*, 126. [CrossRef]
142. Salim, N.S.; Abdel-Alim, M.; Said, H.E.M.; Foda, M.F. Phenolic profiles, antihyperglycemic, anti-diabetic, and antioxidant properties of Egyptian *Sonchus oleraceus* leaves extract: An in vivo study. *Molecules* **2023**, *28*, 6389. [CrossRef] [PubMed]
143. El Hamri, A.L.A.; Kabach, I.; El Asri, S.; Faouzi, M.E.A. Investigation of antioxidant, antidiabetic, and antiglycation properties of *Sonchus oleraceus* and *Lobularia maritima* (L.) Desv. extracts from Taza, Morocco. *Food Chem. Adv.* **2025**, *4*, 100912. [CrossRef]
144. Yatoo, M.I.; Gopalakrishnan, A.; Saxena, A.; Parray, O.R.; Tufani, N.A.; Chakraborty, S.; Tiwari, R.; Dhama, K.; Iqbal, H.M. Anti-inflammatory drugs and herbs with special emphasis on herbal medicines for countering inflammatory diseases and disorders—A review. *Recent Pat. Inflamm. Allergy Drug Discov.* **2018**, *12*, 39–58. [CrossRef]
145. Khan, R.A.; Khan, M.R.; Sahreen, S.; Shah, N.A. Hepatoprotective activity of *Sonchus asper* against carbon tetrachloride-induced injuries in male rats: A randomized controlled trial. *BMC Complement. Altern. Med.* **2012**, *12*, 90. [CrossRef]
146. Vilela, F.C.; Bitencourt, A.D.; Cabral, L.D.; Franqui, L.S.; Soncini, R.; Giusti-Paiva, A. Anti-inflammatory and antipyretic effects of *Sonchus oleraceus* in rats. *J. Ethnopharmacol.* **2010**, *127*, 737–741. [CrossRef]
147. Vecchia, C.A.D.; Locateli, G.; Serpa, P.Z.; Gomes, D.B.; Ernetti, J.; Miorando, D.; Zanatta, M.E.D.C.; Nunes, R.K.S.; Wildner, S.M.; Gutie'rrez, M.V.; et al. *Sonchus oleraceus* L. promotes gastroprotection in rodents via antioxidant, anti-inflammatory, and antisecretory activities. *Evid.-Based Complement. Altern. Med.* **2022**, *2022*, 7413231. [CrossRef]
148. Silva Hernández de Santaolalla, J.L. Biología y Conservación de las Especies de Sonchus Sect. Pustulati (Asteráceas): Endemismos Rupícolas del Complejo Bético-Rifeño (Mediterráneo Occidental). Ph.D. Thesis, Universidad de Sevilla, Sevilla, Spain, 2014; p. 279.
149. Silva, J.L.; Mejías, J.A.; García, M.B. Demographic vulnerability in cliff-dwelling *Sonchus* species endemic to the western Mediterranean. *Basic Appl. Ecol.* **2015**, *16*, 316–324. [CrossRef]
150. GBIF. Sonchus suberosus Zohary & P.H.Davis. Available online: https://www.gbif.org/es/species/3105748 (accessed on 26 May 2025).
151. Mateo, G.; Ferrer-Gallego, F.F.; Roselló, R. Sobre *Sonchus tenerrimus* L. (Compositae) y su variabilidad en la flora valenciana. *Fl. Montiber.* **2019**, *75*, 24–4015. Available online: http://www.floramontiberica.org/FM/075/Flora_Montib_075_024-040_2019.pdf (accessed on 26 May 2025).
152. Rechinger, K.H. *Aetheorrhiza bulbosa* (L.) Cass. und ihre geographischen Rassen. *Phyton* **1974**, *16*, 211–220.
153. Doğan, N.Y.; Kurt, P.; Osma, E. Determination of genetic diversity of *Sonchus erzincanicus* Matthews (Asteraceae), a Critically Endangered plant endemic to Turkey using RAPD markers. *Biharean Biol.* **2018**, *12*, 61.
154. Mejías, A.J. Notas sobre la flora de Andalucía y del Rif: 19. *Sonchus microcephalus* Mejías sp. nov. *Lagascalia* **1990**, *16*, 168–169.
155. Rivera, D.; Verde, A.; Fajardo, J.; Obón, C.; Inocencio, C.; Valdés, A. Modelos etnobiológicos como alternativa al control de malas hierbas con agricultura biológica, los criptocultivos. In *La malherbología en los Nuevos Sistemas de Producción Agraria*; Mansilla, J.A., Artigao Monreal, J.A., Eds.; Sociedad Española de Malherbología: Albacete, Spain, 2007; pp. 149–154.
156. GRDC. UNE45—Ecology and Management of the Weed Common Sowthistle (Sonchus oleraceus). Available online: https://grdc.com.au/research/reports/report?id=5685 (accessed on 16 February 2025).
157. Vanhala, P.; Lötjönen, T.; Hurme, T. Managing *Sonchus arvensis* using mechanical and cultural methods. *Agric. Food Sci.* **2006**, *15*, 444–458. [CrossRef]
158. Luczaj, L.; Pieroni, A.; Tardío, J.; Pardo-de-Santayana, M.; Sõukand, R.; Svanberg, I.; Kalle, R. Wild food plant use in 21st century Europe: The disappearance of old traditions and the search for new cuisines involving wild edibles. *Acta Soc. Bot. Pol.* **2012**, *81*, 359–370. [CrossRef]
159. Gkotzamani, A.; Ipsilantis, I.; Menexes, G.; Katsiotis, A.; Mattas, K.; Koukounaras, A. The Impact of Salinity in the Irrigation of a Wild Underutilized Leafy Vegetable, *Sonchus oleraceus* L. *Plants* **2024**, *13*, 1552. [CrossRef]
160. Hajisolomou, E.; Neofytou, G.; Petropoulos, S.A.; Tzortzakis, N. The Application of Conventional and Organic Fertilizers During Wild Edible Species Cultivation: A Case Study of Purslane and Common Sowthistle. *Horticulturae* **2024**, *10*, 1222. [CrossRef]
161. Chrysargyris, A.; Tzortzakis, N. Optimising Fertigation of Hydroponically Grown Sowthistle (*Sonchus oleraceus* L.): The Impact of Nitrogen Source and Supply Concentration. *Agric. Water Manag.* **2023**, *289*, 10852. [CrossRef]
162. Aguilar Rodríguez, A.; Parra Galant, J.; Gamayo Díaz, J.d.D. *Domestication of Wild Edible Plants: Lisón (Sonchus Tenerrimus, var. Tenerrimus)*; Estación Experimental Agraria de Elche, I.V.I.A.: Elche, Spain, 2012. Available online: https://portalagrari.gva.es/documents/366567370/373114974/DOMESTICACIÓN+DE+PLANTAS+SILVESTRES+COMESTIBLES.+EL+LISÓN+(Sonchus+tenerrimus,+var.+tenerrimus).+2011-2012.pdf/9123632a-3015-4324-b669-fe951b14d71b (accessed on 5 June 2025).
163. Cook, T.; Davidson, B.; Miller, R. A New Glyphosate Resistant Weed Species Confirmed for Northern New South Wales and the World: Common Sowthistle (Sonchus oleraceus)', in Nineteenth Australasian Weeds Conference. 2014. Available online: https://caws.org.nz/old-site/awc/2014/awc201412061.pdf (accessed on 25 May 2025).

164. IWM. Common Sowthistle (Sonchus oleraceus L.): Ecology and Management. Available online: https://www.weedsmart.org.au/app/uploads/2018/09/Common-Sowthistle-Sonchus-oleraceus-L_-ecology-and-management-2.pdf (accessed on 16 February 2025).
165. Krigas, N.; Tsoktouridis, G.; Anestis, I.; Khabbach, A.; Libiad, M.; Megdiche-Ksouri, W.; Ghrabi-Gammar, Z.; Lamchouri, F.; Tsiripidis, I.; Tsiafouli, M.A.; et al. Exploring the Potential of Neglected Local Endemic Plants of Three Mediterranean Regions in the Ornamental Sector. *Sustainability* **2021**, *13*, 2539. [CrossRef]
166. Guarrera, P.M.; Savo, V. Perceived health properties of wild and cultivated food plants in local and popular traditions of Italy: A review. *J. Ethnopharmacol.* **2013**, *146*, 659–680. [CrossRef]
167. Molina, M.; Pardo-de-Santayana, M.; Tardío, J. Natural production and cultivation of Mediterranean wild edibles. In *Mediterranean wild edible plants: Ethnobotany and food composition tables, From explanations to feature selection: Assessing SHAP values as feature selection mechanism. In Mediterranean Wild Edible Plants*; Sánchez-Mata, M.C., Tardío, J., Eds.; Springer: Berlin/Heidelberg, Germany, 2016; pp. 81–107. Available online: https://link.springer.com/chapter/10.1007/978-1-4939-3329-7_5 (accessed on 24 September 2024).
168. CONECT-e. Collaborative Platform for Traditional Ecological Knowledge. Available online: https://www.conecte.es/index.php/es/ (accessed on 29 May 2025).
169. EOL. Encyclopedia of Life. Available online: https://www.eol.org/ (accessed on 29 May 2025).
170. CONECT-e. Sonchus oleraceus. Available online: https://www.conecte.es/index.php/es/plantas/2151-sonchus-oleraceus/usos-tradicionales (accessed on 29 May 2025).
171. CONECT-e. Sonchus tenerrimus. Available online: https://www.conecte.es/index.php/es/plantas/2153-sonchus-tenerrimus (accessed on 29 May 2025).
172. Zulkefle, N.N.; Zainal, N.; Wahyuni, D.K.; Mahmood, S.; Ramarao, K.D.R.; Chin, K.L. Antimicrobial Properties of *Sonchus* Species: A Review. *Asian Pac. J. Trop. Biomed.* **2025**, *15*, 177–188. [CrossRef]
173. Khodaie, L.; Sharma, A.; Bhatia, B.; Al-Harrasi, A. *Introduction to Plant-Based Latex, Phytochemistry, Ethnopharmacology, Biological, and Pharmacological Activities*; Jenny Stanford Publishing: New York, NY, USA, 2025.
174. Ruiz, A. *Vademecum de Cocina Alicantina*; Gastronomía Series; Publicacions Universitat d'Alacant: Alicante, Spain, 2022.
175. Orengo, A. *Herbari Mengívol: Guia per a l'us Gastronòmic de la Flora Silvestre*; Ed. Tívoli: Gandía, Spain, 2016.
176. Rice & Bones. Available online: https://ricebonesbar.com/ (accessed on 28 May 2025).

horticulturae

MDPI

Article

Enhancing Soilless Production of *Portulaca oleracea*, *Mesembryanthemum crystallinum* and *Valerianella locusta* Through Nitrogen Form Ratio Optimization and Biostimulant Application

Theodora Ntanasi [1], Ioannis Karavidas [1], Evangelos Giannothanasis [1], George P. Spyrou [1], Theoni Karaviti [1], Sofia Marka [2], Simona Napoli [3], Damianos Neocleous [4] and Georgia Ntatsi [1,*]

[1] Laboratory of Vegetable Production, Department of Crop Science, Agricultural University of Athens, Iera Odos 75, 11855 Athens, Greece; ntanasi@aua.gr (T.N.); karavidas@aua.gr (I.K.); giannothanasis@aua.gr (E.G.); gspyrou@aua.gr (G.P.S.); theonikaraviti@gmail.com (T.K.)

[2] Laboratory of Cell Technology, Department of Biotechnology, Agricultural University of Athens, 11855 Athens, Greece; smarka@aua.gr

[3] Department of Agricultural, Food and Forest Sciences, University of Palermo, 90128 Palermo, Italy; simona.napoli@unipa.it

[4] Agricultural Research Institute, Department of Natural Resources and Environment, P.O. Box 22016, 1516 Nicosia, Cyprus; dneocleous@ari.moa.gov.cy

* Correspondence: ntatsi@aua.gr

Abstract

Underutilized leafy greens are considered as functional plant species primarily due to their resilience to abiotic stress factors, low nutrient requirements, and high nutritional value. Over the past 30 years, many experiments have been conducted to identify nutrient-efficient species, cultivars, landraces, and ecotypes, but few have successfully entered mainstream agriculture. The integration of these species into advanced horticultural systems, such as hydroponics, has the potential to further strengthen their impact on sustainable agriculture by minimizing use of resources, enabling year-round cultivation, and improving the nutritional profile of the harvested produce. As leafy vegetables, a primary food safety concern is the accumulation of nitrates in the leaves. In hydroponics, this issue is usually addressed by balancing the NH_4-N/total-N ratio (N_r) in the nutrient solution. Provided that the plant responses to high ammonia supply are species-dependent, three wild leafy greens, iceplant, corn salad, and common purslane, were grown in a soilless culture, with perlite as the substrate, under low (0.04) and high (0.12) N_r on a molar basis. Additionally, the potential of protein hydrolysates (PH) and seaweed extracts (SW) to alleviate plant tolerance to excess ammonia supply was also investigated. In terms of yield, high N_r led to significant yield restrictions in iceplant that reached 28%, while on corn salad, it had a positive impact, with yield increasing by 18%. Both biostimulant applications enhanced iceplant productivity only under optimal N_r conditions (0.04). Apart from yield responses, biofertilizers had no substantial impact on the plant nutrient profile. In contrast, high N_r suppressed nitrate accumulation in fresh leaves, while enhancing micronutrient uptake in all three plant species. In conclusion, this study highlights the pivotal role of biostimulants as plant stress protectors and growth regulators and identifies the optimal N_r ratio for maximizing the yield and quality performance of corn salad, iceplant, and common purslane in soilless cultivation systems.

Keywords: plant nutrition; hydroponics; purslane; iceplant; corn salad; seaweed extracts; protein hydrolysates; nitrates

1. Introduction

According to the latest EU Agricultural Outlook, European farmers are encouraged to enhance plant diversity in response to consumer preferences, environmental sustainability, and ecosystem balance [1]. Integrating alternative crops, such as underutilized wild leafy vegetables, into soilless cultivation systems could increase biodiversity and improve the sustainability of the agriculture sector [2]. These species are often rich in essential minerals, vitamins, and antioxidants, indicating their high nutritional value and the potential to address "hidden hunger", which affects more than two billion people globally [3,4]. Furthermore, wild leafy vegetables are typically resilient to abiotic stress and require low inputs [5–10], offering a sustainable option for cultivation and inclusion in the daily human diet. However, targeted research is needed to efficiently address their specific cultivation requirements and ensure consistent optimal product quality.

Purslane (*Portulaca oleracea*) is a nutrient-rich wild leafy vegetable containing significant amounts of essential minerals for the human diet, such as potassium, calcium, magnesium, and iron [11]. Its high concentration of bioactive compounds also makes it a valuable functional food with potential pharmacological benefits [12]. Purslane exhibits higher levels of linolenic acid, beta-carotene, and ascorbic acid concentrations than many conventionally cultivated leafy vegetables, further supporting its strong antioxidant properties [13,14]. Moreover, the presence of cholesterol-free omega-3 fatty acids constitutes it as a viable plant-based alternative to fish oils [15,16]. Physiologically, purslane employs C4 photosynthesis, but under stress conditions, it can switch to CAM, enhancing its adaptability to limited water availability and high temperatures [17]. Moreover, purslane demonstrates remarkable stress resilience, maintaining stable growth, net photosynthetic rate, and macronutrient concentrations (Na, K, Mg, and Ca) in leaf tissues under salinity stress up to 50 mM Na^+ in the root zone [18]. Notably, low nitrogen supply enhances purslane's phenolic and flavonoid content, thereby increasing its antioxidant capacity [19].

Iceplant (*Mesembryanthemum crystallinum*) is a halophytic plant species with notable physiological and biochemical adaptations that enhance its survival under saline conditions [20]. The epidermal bladder cells of this plant develop early and cover the leaves, stems, and peduncle. These cells expand rapidly in response to salt stress by accumulating water and sodium, reflecting substantial metabolic alterations in salt-treated tissues [21]. Under salinity stress conditions, up to 200 mM Na^+ in the root zone, iceplant exhibits enhanced growth and increased accumulation of antioxidants, flavonoids, polyphenols, and D-pinitol. It is also rich in phenolic compounds, such as tannins and catechins, further contributing to its stress tolerance and nutritional value [22,23]. Given its resilience to salinity and high nutritional value, iceplant represents a promising alternative food source for arid regions facing challenges related to salinity and water scarcity [24]. Additionally, its mineral profile, primarily Na, K, and Mg content, gives it a natural salty taste, making it a promising candidate as a salt alternative in culinary applications [25].

Corn salad (*Valerianella locusta*) is widely used as a salad green [26] and is cultivated in greenhouses, both in soil and soilless systems [27]. The bioactive compound profile of corn salad enhances its nutritional value as a fresh vegetable in the human diet. The plant's leaves contain high concentrations of carotenoids, phenolics, folic acid, sterols, fatty acids, and glucosinolates [28,29], indicating strong antioxidant capacity and potential antitumor effects [30]. In terms of fertigation practices, corn salad is a high-N-demanding plant. Under reduced N supply, plant growth is negatively affected, an effect that can be mitigated through the application of biostimulants [31]. Furthermore, eustress application in hydroponically grown *V. locusta* using nutrient solutions (NSs) with increased EC levels and different cationic compositions indicates that the species is sensitive to salinity, as yield is notably restricted when Na^+ is used to increase EC instead of Ca^{2+} [10]. These findings highlight the importance of tailored fertigation strategies for optimizing the yield and nutritional quality of *V. locusta*.

Nitrogen (N) is a crucial macronutrient for plant physiology, serving multiple roles in plant physiology and metabolism. As a fundamental component of proteins, nucleic acids and chlorophyll, it directly influences plant growth, development, and metabolism [32]. It supports enzyme functions, cellular structure, and overall metabolic regulation as the key component of proteins. Plants primarily absorb N in two mineral forms, nitrate (NO_3^-) and ammonium (NH_4^+). In soilless crops, the addition of NH_4^+ aims to effectively manage the pH in the root solution [33]. The reduction in root-zone pH was primarily determined by the NH_4^+/total-N ratio, with the absolute NH_4^+ concentration in the supplied nutrient solution having a lesser effect [7]. This NH_4^+-induced pH decline is attributed to the preferential uptake of NH_4^+ by plant roots. NH_4^+ is preferentially absorbed by the plants when both NO_3^- and NH_4^+ are available in the root zone, despite NO_3^- being the predominant form absorbed by most crop species [34]. This preference persists even though NO_3^- assimilation requires higher energetic expenditure. Following uptake, these inorganic N forms are assimilated into amino acids, mainly through the glutamine synthetase/glutamate synthase cycle pathway [35]. The assimilation of mineral N forms in organic compounds is strongly influenced by photosynthetically active radiation, affecting NO_3^- accumulation in leaves [32]. As a critical food safety parameter for human health, NO_3^- concentration in the edible parts of leafy vegetables is strictly regulated, and safety thresholds of 3000 to 5000 mg per kg of fresh weight have been established by the European Union (Regulation No. 2023/915).

Seaweed extracts (SW) and protein hydrolysates (PHs) can be used as effective biostimulants to enhance stress tolerance in vegetable crops. When applied to stressed plants, seaweed extracts boost the activity of antioxidant enzymes such as SOD, CAT, and APX, contributing to oxidative stress mitigation [36,37]. Additionally, these extracts can promote growth and nutrient uptake in crops like tomato and spinach [38–41]. Similarly, PHs function as potent biostimulants by enhancing nutrient use efficiency and activating both enzymatic and non-enzymatic antioxidant defense systems, supporting plant growth and physiological functions under abiotic stress conditions [42–44]. Moreover, PHs modulate phytohormonal balance, helping to maintain photosynthetic efficiency and reduce stress-induced damage [45].

The objective of this research is to develop cultivation protocols that can optimize open soilless (hydroponic) production of three underutilized leafy greens: iceplant (*Mesembryanthemum crystallinum*), corn salad (*Valerianella locusta*), and common purslane (*Portulaca oleracea*). To achieve this, the study examined a novel approach that combines different NH_4^+/total-N ratios (N_r) in the nutrient solution with biostimulant applications (PH, and SW), evaluating their effects on plant growth, nutritional quality, and NO_3^- accumulation. Specifically, the study seeks to (a) determine the optimal NH_4^+/total-N ratio that maximizes yield and minimizes NO_3^- accumulation in leaves while enhancing nutrient uptake, (b) evaluate the potential of biostimulants to alleviate stress caused by high ammonia supply and improve plant productivity, and (c) assess the impact of nitrogen form and biostimulants on the nutritional profile of these species to ensure food safety and quality.

2. Materials and Methods

2.1. Plant Material and Growing Conditions

Three independent experiments were conducted in a free-drainage soilless culture system, at the greenhouse facilities of the Laboratory of Vegetable Production, the Agricultural University of Athens (37°58′57.8″ N, 23°42′14.3″ E). The first experiment focused on purslane (*Portulaca oleracea*) with seeds (Fitotech Industrial and Commercial S.A., Athens, Greece) sown in rockwool trays (AO Plug, Grodan, Roermond, the Netherlands) on 30 March. At the stage of the first 2–3 true leaves, on 10 April, 5 seedlings were transplanted into 33 L perlite bags (Geoflor Hydro, Greece), with harvest occurring at commercial maturity on 27 April. The second experiment followed an identical protocol for corn salad (*Valerianella locusta*, 'Elixir' (HM. CLAUSE Sas, Portes-Les-Valence, France)), with sowing on 30 March and transplanting into the soilless system at the 2–3 true leaf stage on 10 April, while the experiment terminated on 6 May. The third experiment with iceplant (*Mesembryanthemum crystallinum* L.) also followed the same configuration, with sowing of the seeds (Fitotech Industrial and Commercial S.A., Athens, Greece) on 20 April, transplantation at the same developmental stage on 30 April, and final harvest on 18 May. In all three experiments, the perlite substrates were moistened with starter nutrient solution (NS) (Table 1) prior to transplanting. The greenhouse maintained consistent environmental conditions throughout all three experiments (17–25 °C, 50–80% relative humidity) and the plants were grown under natural light (13 h day/11 h night), using the same soilless culture system configuration for comparability between species.

Table 1. Composition of the irrigation water, starter nutrient solution, and the two supplied nutrient solutions (high and low N_r).

Parameter	Unit	Water	Starter	High N_r	Low N_r
EC	dS m^{-1}	0.30	3.00	2.40	2.30
pH		7.55	5.60	5.60	5.60
K^+	mM	0.00	8.72	7.43	7.43
Ca^{2+}	mM	1.00	5.25	4.10	4.10
Mg^{2+}	mM	0.20	3.42	2.29	2.29
NH_4^+	mM	0.00	0.89	1.79	0.64
SO_4^{2-}	mM	0.20	5.02	3.88	2.73
NO_3^-	mM	0.00	15.05	12.63	13.78
P	mM	0.00	1.65	1.40	1.40
Fe	µM	0.00	40.00	25.00	25.00
Mn	µM	0.00	5.00	8.00	8.00
Zn	µM	0.20	7.00	5.00	5.00
Cu	µM	0.00	0.80	0.75	0.75
B	µM	0.00	40.00	30.00	30.00
Cl^-	mM	0.30	0.40	0.40	0.40
Na^+	mM	0.60	0.60	0.60	0.60
HCO_3^-	mM	2.20	0.40	0.40	0.40
N_r	mol:mol		0.06	0.12	0.04

2.2. Experimental Design

All three experiments employed identical two-factorial experimental designs to maintain comparability between the purslane, corn salad, and iceplant trials. The first factor investigated was the NH_4^+/total-N ratio (N_r) in the NS with two treatment levels: a low N_r level of 0.04 and a high N_r level of 0.12 on a molar basis (Table 1). The second factor examined biostimulant application, comparing two products: a seaweed (SW: 'Algastar' liquid extract of *Ascophyllum nodosum* extract) (Table S1) and a protein hydrolysate (PH: Tyson) (Table S2), both supplied by Mugavero Fertilizers, Italy. Both biostimulants were applied via foliar spraying, according to the manufacturer's recommendations. The spraying solution of SW contained Algastar at a concentration of 2 mL L^{-1}, while that of PH contained Tyson at a concentration of 3 mL L^{-1}. Biostimulant applications occurred twice during each crop cycle—on the 14th and the 21st of April for purslane and corn salad, and on the 3rd and 10th May for iceplant, with untreated plants serving as controls. The factorial combination of these two factors, N_r and biostimulant application, produced six distinct treatments: (1) high N_r—control; (2) high N_r—SW; (3) high N_r—PH; (4) low N_r—control; (5) low N_r—SW; and (6) low N_r—PH. Each treatment was replicated four times (n = 4), with each experimental unit consisting of one perlite bag containing five plants. This complete randomized design allowed for systematic evaluation of both individual and interactive effects of N_r and biostimulant application on plant performance.

2.3. Sampling and Laboratory Analysis

At harvest, the fresh weight (FW) of all the plants was recorded in order to calculate the yield (g $plant^{-1}$). One plant from each replication was collected, cleaned with distilled water, and dried to a constant weight at 65 °C to determine the mineral concentrations. The dried tissues were powdered using a blade mill and then subjected to the dry ashing procedure at 550 °C for 8 h. Subsequently, 10 mL of 0.25 M HCl solution was added to the ash of each sample to extract K^+, Ca^{2+}, Mg^{2+}, Na^+, Fe, Mn, and Zn [46]. The reduced N (N_{red}) content of the plant tissues was determined according to the Kjeldahl method, using a Labtec DT 220 digestion system coupled with a Tecator Kjeltec 8200 distillation unit (FOSS A/S, Hillerod, Denmark). To determine the mineral N (N_{min}) content, which corresponds to leaf nitrate (NO_3^-) concentration, the dried tissue samples were subjected to hot-water extraction in a water bath at 80 °C for 1 h, while the nitrate concentration in plant extracts was quantified photometrically at 410 nm via the salicylic acid method [47]. The total-N concentration of the dried plant tissues was calculated as the sum of N_{min} and N_{red} values, while the leaf NO_3^- concentrations were also given as mg of NO_3^- per kg of fresh weight (FW) (ppm).

2.4. Partial Budget Analysis

The partial budget analysis was performed to appraise the net economic profits that the growers may accrue by employing the SW or PH biostimulants. The economic method described by Giordano et al. [48] was used for the evaluation. To calculate the added net return sustained by both SW and PH, the subsequent formula was employed: Added net return = added gross return − added variable costs.

2.5. Statistical Analyses

The experiment was conducted with a completely randomized design with four replications per treatment. Treatment effects were analyzed using two-way factorial ANOVA to evaluate: (1) the main effects of nitrogen ratio (N_r) and biostimulant application, and (2) their interaction effects. When ANOVA indicated significant differences ($p < 0.05$), post hoc comparisons were performed using Duncan's multiple range test to identify specific treatment differences. The statistical analysis of the data, ANOVA and Duncan's multiple range test were performed using STATISTICA 12.5 for Windows (StatSoft Inc., Tulsa, Oklahoma, USA). Graphical representation of data was performed using the GraphPad Prism 8 software (GraphPad Software, La Jolla, CA, USA).

3. Results

3.1. Evolution of EC and pH in the Drainage Solution

Figure 1 shows the EC and pH evolution in the drainage solution throughout the cultivation period, depending on different NH_4^+/total-N ratio (N_r) treatments. In all three experiments, N_r had no significant impact on EC levels. For purslane and corn salad, EC ranged from 2.75 to 3.23 dS m^{-1} and 2.69 to 3.22 dS m^{-1}, respectively. For iceplant, EC values were between 2.80 and 3.23 dS m^{-1} for both N_r levels. Regarding pH, no significant differences were detected in the drainage solution for purslane and corn salad. During the cultivation period, pH decreased from 6.6 to 6.2, irrespective of the N_r treatment. In contrast, for iceplant, the pH values for the high-N_r treatment were consistently significantly lower, declining from 5.9 to approximately 5.1, compared to the low-N_r treatment, the pH values of which declined from 6.2 to approximately 5.5, indicating gradual acidification of the drainage solution in both cases.

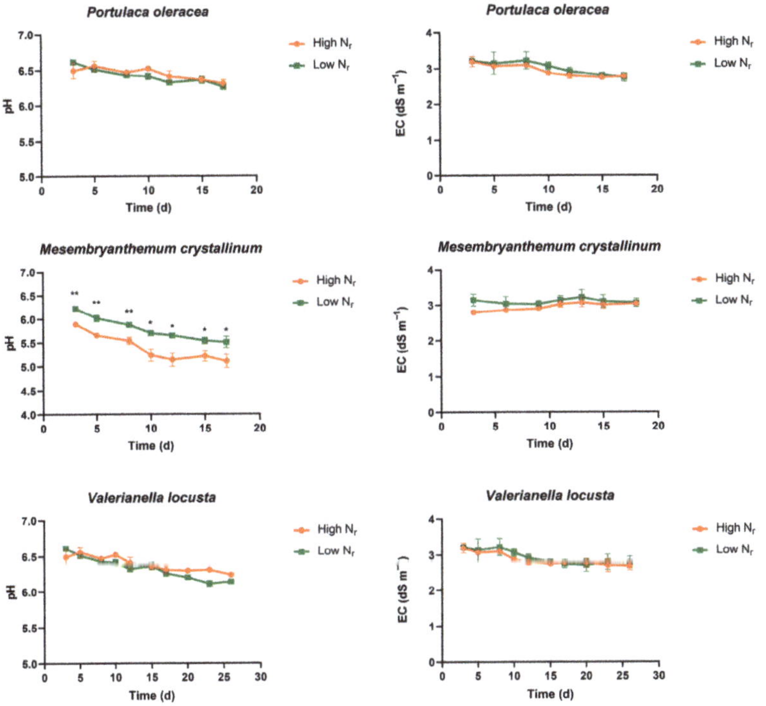

Figure 1. Evolution of EC (dS m^{-1}) and pH in the drainage solution of the two N_r treatments in the three experiments. Regarding the variance components, according to associated *p* values, ** ($p \leq 0.01$), and * ($p \leq 0.05$) represent statistical significance.

3.2. Yield and Dry Matter Content

The plant fresh weight (FW) of iceplant and corn salad was significantly affected by both the N_r and the biostimulant application, while interaction between the two factors was detected in both species. Specifically, under low-N_r conditions, the application of SW and PH significantly enhanced iceplant FW by 20.4% and 23.0%, respectively, compared to the corresponding control no biostimulant application. In contrast, under high N_r the FW of iceplant was significantly lower compared to low N_r, irrespective of the biostimulant application (Table 2).

Table 2. Impact of ammonium on N ratio (N_r) and biofertilizer application on fresh weight (FW) and dry matter content (DMC) (%) of purslane, iceplant and corn salad. SW: seaweed extract; PH: protein hydrolase.

N_r	Biostimulants	Purslane		Iceplant		Corn Salad	
		FW	DMC (%)	FW	DMC (%)	FW	DMC (%)
High		152.18	4.53	79.33 b	3.37 a	11.43 a	9.49 b
Low		147.29	4.37	110.72 a	2.76 b	9.34 b	9.98 a
	Control	154.58	4.43	89.15 c	3.16	10.23 b	9.72
	SW	143.38	4.43	99.05 a	3.05	10.11 b	9.81
	PH	151.13	4.47	97.56 ab	2.99	10.64 a	9.65
Interactions							
High	Control	160.93	4.53	77.96 c	3.54	11.48 a	9.40
	SW	145.44	4.49	80.49 c	3.34	11.47 a	9.54
	PH	150.84	4.57	79.45 c	3.26	11.32 a	9.50
Low	Control	149.35	4.35	97.65 b	2.85	8.8 c	10.03
	SW	141.57	4.38	117.61 a	2.75	8.83 c	10.08
	PH	151.43	4.36	120.17 a	2.66	10.22 b	9.79
Statistical significance							
N_r		NS	NS	***	***	***	***
Biostimulants		NS	NS	*	NS	*	NS
N_r X Biostimulants		NS	NS	*	NS	**	NS

In each column, means (n = 4) followed by different letters are significantly different according to Duncan's multiple range test ($p \leq 0.05$). Regarding the variance components, according to associated p values, *** ($p \leq 0.001$), ** ($p \leq 0.01$), and * ($p \leq 0.05$) represent statistical significance, while NS signifies non-significance.

For corn salad, PH application under low N_r significantly increased FW by 16.1%, compared to the control (no biostimulant application). When supplied with high N_r, corn salad exhibited a significantly higher FW than under low-N_r treatment. No significant differences were observed in the fresh yield of purslane, regardless of the treatments applied (Table 2).

Regarding dry matter content (DMC), significant differences were detected only in iceplant and corn salad. High N_r increased iceplant DMC by 22.1% relative to low N_r, whereas corn salad DMC increased by 5.3% under low N_r. In all three species, biostimulant applications had no significant effect on DMC (Table 2).

3.3. *Nitrogen and Leaf Nitrate Content*

The N nutrition of all three plants was significantly affected by both N_r and biostimulant application. Particularly, in purslane N, the reduced forms (N_{red}), mainly organic N, increased significantly by 15.3% under high N_r while the mineral form of N (N_{min}) significantly reduced by 21.3% under high N_r and by 15.9% with PH application. Consequently, total-N increased by 11.7% whereas NO_3^- content reduced by 19.3% under high N_r. Furthermore, the application of PH biostimulant also reduced the NO_3^- content by 14.5%, compared to the control (Table 3).

In iceplant, high N_r significantly decreased N_{red} by 4.2% and N_{min} by 17.5%, compared to low Nr, while PH increased N_{red} by 6.9% compared to the control (no biostimulant application). These changes led to a 5.1% decrease in total-N under high NH_4^+ concentration in the root zone (high N_r) compared to low N_r. Regarding NO_3^- content in fresh leaf tissues, only biostimulant application (SW: Algastar; PH: Tyson) had a significant effect, reducing it by 11.3% and 13.8%, respectively (Table 3).

Table 3. Impact of ammonium to N ratio (N_r) and biofertilizer application on reduced N (N_{red}), mineral N (N_{min}), total-N (N_{red} + N_{min}) content as percentage of dry matter (%), percentage of N assimilation (Nred/total-N %) and leaf nitrate (NO_3^-) content as mg of NO_3^- per Kg of fresh weight (FW) (ppm) of purslane, iceplant and corn salad. SW: seaweed extract; PH: protein hydrolase.

Main Effects N_r	**Biostimulants**	N_{red} (%)	N_{min} (%)	**Purslane** Total-N (%)	N_{ass} (%)	NO_3^- (ppm)
High		4.98 a	0.37 b	5.35 a	93.16 a	732 b
Low		4.32 b	0.47 a	4.79 b	90.15 b	907 a
	Control	4.64	0.44 a	5.07	91.35 b	855 a
	SW	4.60	0.45 a	5.04	91.01 b	873 a
	PH	4.75	0.37 b	5.12	92.56 a	731 b
		Statistical significance				
N_r		***	***	***	***	***
Biostimulants		NS	***	NS	**	***
N_r X Biostimulants		NS	NS	NS	NS	NS
N_r	**Biostimulants**			**Iceplant**		
		N_{red} (%)	N_{min} (%)	Total-N (%)	N_{ass} (%)	NO_3^- (ppm)
High		4.85 b	0.33 b	5.18 b	93.52 a	496
Low		5.06 a	0.40 a	5.46 a	92.69 b	481
	Control	4.90 b	0.38	5.28 b	92.78 b	530 a
	SW	4.72 b	0.36	5.08 b	92.91 b	470 b
	PH	5.24 a	0.36	5.60 a	93.64 a	457 b
		Statistical significance				
N_r		*	***	**	***	NS
Biostimulants		***	NS	***	**	**
N_r X Biostimulants		NS	NS	NS	NS	NS
N_r	**Biostimulants**			**Corn salad**		
		N_{red} (%)	N_{min} (%)	Total-N (%)	N_{ass} (%)	NO_3^- (ppm)
High		4.88	0.28 b	5.16	94.50 a	1191 b
Low		4.73	0.31 a	5.04	93.89 b	1358 a
	Control	4.83	0.30 a	5.13	94.15 ab	1295 a
	SW	4.77	0.31 a	5.08	93.92 b	1340 a
	PH	4.81	0.28 b	5.09	94.52 a	1194 b
		Statistical significance				
N_r		NS	***	NS	***	***
Biostimulants		NS	***	NS	**	***
N_r X Biostimulants		NS	NS	NS	NS	NS

In each column, means (n = 4) followed by different letters are significantly different according to Duncan's multiple range test ($p \leq 0.05$). Regarding the variance components, according to associated p values, *** ($p \leq 0.001$), ** ($p \leq 0.01$), and * ($p \leq 0.05$) represent statistical significance, while NS signifies non-significance.

For corn salad, neither N_r nor biostimulant application altered N_{red} and total-N. However, high N_r and PH application reduced N_{min} by 9.7% and 6.7%, respectively, compared to low N_r and no biostimulant application, which were also reflected in NO_3^- content, decreasing by 12.3% and 7.8%, respectively. Furthermore, N assimilation (N_{ass}), calculated as N_{red} to total-N, was significantly affected in all three species by both factors without any significant interactions between them. Specifically, N_{ass} was decreased under low N_r in both iceplant and corn salad, compared to low N, while the reverse was the case for purslane. As for the biostimulants, PH application consistently reduced NO_3^- and increased N_{ass} (Table 3).

3.4. Nutrient Concentrations in Plant Tissues

Regarding the K^+, Ca^{2+} and Mg^{2+} concentrations in purslane, only Ca^{2+} was significantly affected by N_r. Specifically, plants supplied with high N_r exhibited a 30.2% reduction in Ca^{2+} concentration compared to those grown under low N_r (Table 4). In contrast, K^+ content ranged from 63.69 to 67.00 mg g^{-1} and Mg^{2+} from 10.50 to 10.94 mg g^{-1} across all treatments, with no significant differences among either the N_r or the biostimulant application treatments (Table 4).

Table 4. Impact of ammonium to N ratio (N_r) and biofertilizer application on K, Ca, Mg and Na content as mg per g of dry matter of purslane, iceplant and corn salad. SW: seaweed extract; PH: protein hydrolase.

Main Effects		**Purslane**			
N_r	**Biostimulants**	**K (mg g^{-1})**	**Ca (mg g^{-1})**	**Mg (mg g^{-1})**	**Na (mg g^{-1})**
High		63.69	3.77 b	10.81	1.38
Low		67.00	5.40 a	10.74	1.32
	Control	64.89	4.91	10.94	1.29
	SW	64.29	4.64	10.91	1.47
	PH	66.44	4.13	10.50	1.32
		Statistical significance			
N_r		NS	***	NS	NS
Biostimulants		NS	NS	NS	NS
N_r X Biostimulants		NS	NS	NS	NS
N_r	**Biostimulants**	**Iceplant**			
		K (mg g^{-1})	Ca (mg g^{-1})	Mg (mg g^{-1})	Na (mg g^{-1})
High		20.69	0.72	4.74	10.23
Low		20.15	0.74	4.65	8.42
	Control	20.00	0.68	4.52	9.17
	SW	19.75	0.70	4.77	9.50
	PH	21.44	0.80	4.81	9.33
		Statistical significance			
N_r		NS	NS	NS	NS
Biostimulants		NS	NS	NS	NS
N_r X Biostimulants		NS	NS	NS	NS
N_r	**Biostimulants**	**Corn salad**			
		K (mg g^{-1})	Ca (mg g^{-1})	Mg (mg g^{-1})	Na (mg g^{-1})
High		57.80	0.58	4.07	0.44
Low		58.17	0.61	4.35	0.47
	Control	57.33	0.57	4.08	0.43
	SW	58.25	0.61	4.27	0.43
	PH	58.25	0.60	4.28	0.50
		Statistical significance			
N_r		NS	NS	NS	NS
Biostimulants		NS	NS	NS	NS
N_r X Biostimulants		NS	NS	NS	NS

In each column, means (n = 4) followed by different letters are significantly different according to Duncan's multiple range test ($p \leq 0.05$). Regarding the variance components, according to associated p values, *** ($p \leq 0.001$) represent statistical significance, while NS signifies non-significance.

Similarly, in iceplant and corn salad, macronutrient concentrations (K^+, Ca^{2+} and Mg^{2+}) were not significantly affected by the two factors examined (N_r and biostimulant) (Table 4). In iceplant, K^+ content ranged from 19.75 to 21.44 mg g^{-1}, Ca^{2+} from 0.68 to 0.80 mg g^{-1} and Mg from 4.52 to 4.81 mg g^{-1}. For corn salad, K^+ levels were 57.33–58.25 mg g^{-1}, Ca^{2+} 0.57–0.61 mg g^{-1} and Mg 4.07- 4.35 mg g^{-1}. Additionally, neither N_r nor biostimulants significantly impacted Na content, which ranged from 1.29 to 1.47 mg g^{-1} in purslane, 0.43 to 0.50 mg g^{-1} in corn salad, and 8.42 to 10.23 mg g^{-1} in iceplant.

In purslane leaf tissues, Fe (60.83–65.31 μg g^{-1}) and Mn (73.87 to 79.28 μg g^{-1}) concentrations remained stable and without any significant differences across treatments (Table 5). However, Zn concentration was significantly influenced by N_r, with high N_r increasing Zn content by 30.9% compared to low N_r.

Table 5. Impact of ammonium to N ratio (N_r) and biofertilizer application on Fe, Mn, and Zn content as μg per g of dry matter of purslane, iceplant and corn salad. SW: seaweed extract; PH: protein hydrolase.

Main Effects			Purslane	
N_r	**Biostimulants**	**Fe (μg g^{-1})**	**Mn (μg g^{-1})**	**Zn (μg g^{-1})**
High		64.30	75.61	49.60 a
Low		62.49	77.28	37.89 b
	Control	64.58	79.28	43.25
	SW	65.31	73.87	44.82
	PH	60.83	75.53	44.06
		Statistical significance		
N_r		NS	NS	***
Biostimulants		NS	NS	NS
N_r X Biostimulants		NS	NS	NS
N_r	**Biostimulants**		**Iceplant**	
		Fe (μg g^{-1})	Mn (μg g^{-1})	Zn (μg g^{-1})
High		170.63 a	70.38	83.04 a
Low		81.70 b	70.29	59.34 b
	Control	125.12	68.80	69.82
	SW	133.74	68.04	73.16
	PH	120.48	73.92	70.81
		Statistical significance		
N_r		***	NS	***
Biostimulants		NS	NS	NS
N_r X Biostimulants		NS	NS	NS
N_r	**Biostimulants**		**Corn salad**	
		Fe (μg g^{-1})	Mn (μg g^{-1})	Zn (μg g^{-1})
High		94.26 a	86.09 a	45.62 a
Low		87.54 b	78.52 b	38.59 b
	Control	91.73	84.62	44.30
	SW	87.70	79.8	38.75
	PH	92.64	82.12	42.94
		Statistical significance		
N_r		**	*	*
Biostimulants		NS	NS	NS
N_r X Biostimulants		NS	NS	NS

In each column, means (n = 4) followed by different letters are significantly different according to Duncan's multiple range test ($p \leq 0.05$). Regarding the variance components, according to associated p values, *** ($p \leq 0.001$), ** ($p \leq 0.01$), and * ($p \leq 0.05$) represent statistical significance, while NS signifies non-significance.

In iceplant, N_r significantly affected Fe and Zn concentrations, with Fe increasing by 109% and Zn by 40%, compared to low N_r (Table 5). No significant differences were observed in Mn concentration, which ranged from 68.04 to 73.92 $\mu g\ g^{-1}$ across all treatments.

In corn salad, high N_r led to significant increases in Fe, Mn and Zn concentrations of 7.7%, 9.6% and 18.2%, respectively. However, the application of biostimulants had no significant effect on Fe, Mn or Zn content in any of the tested plant species (purslane, iceplant, or corn salad).

3.5. Partial Budget Analysis

Table 6 shows the added net return obtained by the application of SW and PH on purslane, iceplant and corn salad.

Table 6. Added net return for purslane, iceplant and corn salad obtained by the application of seaweed extract (SW) or protein hydrolysate (PH).

Species	Biostimulant	Yield Increase (kg ha^{-1})	Price (EUR kg^{-1})	Added Gross Return (EUR ha^{-1})	Added Variable Cost (EUR ha^{-1})			Added Net Return (EUR ha^{-1})
					Biostimulant Treatment	Application	Total	
Purslane	SW	−2800.0	8.5	−23,800.0	81.0	200.0	281.0	−24,081.0
	PH	−862.5	8.5	−7331.3	51.0	200.0	251.0	−7582.3
Iceplant	SW	2475.0	41.0	101,475.0	81.0	200.0	281.0	101,194.0
	PH	2102.5	41.0	86,202.5	51.0	200.0	251.0	85,951.5
Corn salad	SW	−30.0	17.6	−528.0	81.0	200.0	281.0	−809.0
	PH	102.5	17.6	1804.0	51.0	200.0	251.0	1553.0

SW: 13.5 EUR/L; PH: 8.5 EUR/L.

The output underlined that biostimulants differently affected the added net return of the species. For purslane, negative values were found, independently of the biostimulant used. For iceplant, the application of both biostimulants gave positive outcomes in terms of added net return, with the peak in plants treated with SW (+101,194.0 EUR ha^{-1}). Lastly, corn salad showed different behavior depending on the biostimulant used with negative values when SW was applied and positive results when plants were treated with PH (+1553.0 EUR ha^{-1}).

4. Discussion

The integration of wild edible species and underutilized crops into highly intensive and productive agricultural systems, such as soilless culture, requires targeted research to understand and identify optimal agronomic practices for yield, quality and resource usage. To this end, effective and precise nutrition management is critical to optimize both yield production and product quality. For leafy greens, nutrient supply must be carefully regulated to minimize NO_3^- accumulation in edible tissues, ensuring compliance with the European Union's maximum permissible nitrate limits (Regulation No. 2023/915) and safeguarding human health [8–10,49].

In soilless cultivation systems, both NH_4^+ and NO_3^- forms are supplied to the plants. However, the ratio between the two nitrogen forms (N_r) is crucial for pH management in the root zone and N plant nutrition [50]. Plants preferentially take up NH_4^+ up to certain levels due to its lower energy requirements for N assimilation into carbon skeletons compared to NO_3^- [32]. However, excessive NH_4^+ can lead to ammonia toxicity, particularly under low-pH conditions in the root environment [51]. It has been observed that excessive consumption of sugars for ammonium assimilation caused by elevated NH_4^+ levels can result in carbohydrate limitation and elevated ammonia within the cytoplast [52]. This can reduce the availability of carbohydrates for photosynthesis and respiration, alter cytoplasmic pH, and induce oxidative stress [53].

As demonstrated in the present study, the response of plants to N_r levels depends on species and cultivation system. These species-specific differences likely reflect distinct capacities for ionic balance and tolerance to ammonium stress. The results of the fresh yield for the three examined leafy greens, purslane, iceplant and corn salad, varied with N_r levels. In the case of purslane, there were no significant differences in FW between low (0.04) and high N_r (0.12) levels, whereas iceplant was negatively affected by high N_r and corn salad demonstrated a positive response. Similar variability exists among other leafy vegetables. For instance, lettuce FW increased with N_r up to 0.30 [54]. Furthermore, Wenceslau et al. [52] recommended an N_r level of 0.23 for iceberg lettuce in a floating hydroponic system, while toxicity symptoms were observed at N_r levels higher than 0.5. Furthermore, the findings of Bulgari et al. [55] suggested that rocket yield remained unaffected by different N_r values from 0 to 0.50, consistent with the findings for *Cichorium spinosum* L. (stamnagathi) by Chatzigianni et al. [7]. On the other hand, sowthistle (*Sonchus oleraceus* L.), a wild edible leafy green, exhibited better performance in hydroponic culture with nutrient solutions of an N_r of 0.05 or less rather than higher values [56]. Regarding the effect of N_r on golden thistle (*Scolymus hispanicus* L.), another wild edible species, Papadimitriou et al. [57] suggested that the FW of substrate-grown plants was higher under standard N supply when 10% of the nitrogen was supplied in the form of NH_4^+ as compared to a supply of 20%.

In the present study, purslane FW and DMC remained unaffected by the two N_r treatments applied. Spyrou et al. [49] observed a 19.3% yield reduction in purslane grown in a floating hydroponic system when N_r increased from 0.07 to 0.14, while Chrysargyris et al. [58] found that purslane FW was reduced in an NFT hydroponic system when an N_r over 5% was supplied. The contradictory findings can mainly be attributed to the differences in the cultivation systems. These findings indicate that optimal N_r values are not universal but depend on the species and the cultivation system. In both studies, the plants were cultivated in water culture systems, i.e., hydroponics, whereas in the present study, a substrate-based system comprising perlite was employed. A key distinction is that in soilless substrate systems, the volume of the root solution is substantially less than in hydroponic (i.e., water culture) systems [59]. The higher solution volume per plant in hydroponic systems can sustain elevated NH_4^+ concentrations over extended periods. This phenomenon occurs despite the preferential uptake of NH_4^+ by the plants, as the total amount of N absorbed is considerably lower than the available NH_4^+ concentration of the root solution. Conversely, the limited root zone solution volume in substrate systems increases the likelihood of NH_4^+ depletion. Notably, the application of biostimulants had no effect on plant growth parameters, suggesting minimal stress impact under both N_r levels. These findings highlight the need to tailor fertigation strategies not only to plant species but also to the cultivation system.

Corn salad demonstrated a positive response to elevated N_r levels with increased NH_4^+ supply leading to enhanced plant FW. The slight increase in DMC under low N_r suggests that those plants can more efficiently utilize NH_4^+-N to cover their needs due to its lower energy demand for assimilation into organic compounds [53,60]. Both biostimulants further promoted plant growth under low NH_4^+ supply by increasing the FW. These findings align with Di Mola et al. [31], who suggested that plant-based protein hydrolysate biostimulant can effectively mitigate stress under reduced N availability in *V. locusta*. All the above highlight PH biostimulants' capacity to improve N nutritional status and stimulate growth under N-limited conditions in corn salad. Consequently, the results suggest that corn salad tolerated elevated NH_4^+ levels and benefited from biostimulant application, leading to enhanced growth.

In contrast, a clear negative effect of high NH_4^+ application has been observed in the case of iceplant. FW significantly decreased under high N_r while DMC increased, indicating that the plants were under stress. The pH levels in both N_r treatments showed a downward trend, a finding that is in accordance with the observations of Rincón-Cervera et al. [61] in soilless iceplant crops, with significantly lower values in high-N_r treatment due to high NH_4^+ supply [33]. However, pH was maintained at a level of over 5 for the entire growing cycle, which is the lower safety level for the plants [50]. It is thus evident that the observed growth restriction is predominantly ascribed to ammonia toxicity in the roots of iceplant [53]. However, both biostimulants enhanced plant growth, but only under optimal NH_4^+ supply. Choi et al. [62] recently reported that vegetal-derived PH biostimulants, such as 'Tyson' in the present study, can effectively protect hydroponically grown lettuce plants from yield loss caused by low NO_3:NH_4, supporting the role of biostimulants in counteracting ammonia toxicity due to high NH_4^+ supply.

Regarding N metabolism, the findings of the present study indicate that high N_r in the nutrient solution resulted in a promotion of N assimilation across all three plants. This phenomenon can be attributed to the N assimilation cycle, whereby NH_4^+-N is directly incorporated into glutamate via the glutamine synthetase pathway, whereas NO_3^- requires energy-intensive reduction to NH_4^+ via nitrate and nitrite reductases [63]. Hence, NH_4^+-N assimilation is a less energy-consuming process [32]. Therefore, under high N_r, NO_3^- concentration in the leaves of purslane and corn salad was restricted, primarily because higher NH_4^+ availability suppresses NO_3^- uptake and assimilation. Indeed, an increased supply of NH_4^+ is an effective practice for reducing the accumulation of NO_3^- in wild edible leafy greens, which are tolerant to elevated NH_4^+ concentrations in the root zone, such as *C. spinosum* [7] and *S. oleraceus* [56]. In the case of *M. crystallinum* plants, despite NH_4^+ stress, no differences in the leaf NO_3^- concentration were detected. This finding suggests that the assimilation of N was not affected while the differences in total-N were ascribed to reduced NO_3^- uptake. In addition, the cultivation of purslane plants in a water culture system, in conjunction with the imposition of elevated levels of NH_4^+ in the root zone, resulted in a decline in their FW, while the leaf NO_3^- concentration remained unaffected [49], mirroring iceplant responses in this study.

Elevated NO_3^- concentrations in food are associated with an increased risk of cancer [64,65]. As such, NO_3^- concentration in the edible parts of leafy vegetables is recognized as a pivotal quality index for ensuring human safety. The European Union has established regulatory measures for the maximum permissible limits of nitrate content for certain leafy greens, including lettuces, spinach and rucola, ranging from 2000 to 5000 mg kg^{-1} (Regulation No. 2023/915). Although there are no defined limits for the species examined in this study, the respective values for the other leafy greens serve as indicators. In this study, NO_3^- concentrations in corn salad remained below 2000 mg kg^{-1}, consistent with findings by Voutsinos-Frantzis et al. [10]. Manzocco et al. [66] reported that hydroponically grown crops with 18 mM N (NO_3^- + NH_4^+) in the nutrient solution can accumulate up to 4700 mg kg^{-1} of NO_3^-. Meanwhile, purslane and iceplant exhibited even lower NO_3^- levels (<1000 mg kg^{-1}), reflecting high product quality. Furthermore, Spyrou et al. [49] observed that hydroponic purslane grown in a 15 mM N solution can accumulate up to 2900 mg kg^{-1} of NO_3^-, suggesting that the present study's lower values may result from cultivation conditions. The low leaf NO_3^- concentration in all plant species may also be attributed to high solar radiation, given that the cultivation took place in spring under Mediterranean conditions [67]. An alternative approach to N nutrition management for soilless-grown leafy greens is to reduce N supply during the final days before harvest, which can limit NO_3^- accumulation while maintaining yield and enhancing the plants' antioxidant compound content [68–70].

The application of PH biostimulants, composed of peptides and amino acids, enhanced N assimilation in all three plant species and effectively reduced NO_3^- accumulation in their edible parts. It is evident that multiple physiological mechanisms are collectively involved in this plant response to PH biostimulants. Firstly, it is important to note that key enzymes involved in nitrogen metabolism, including nitrate reductase, glutamine synthetase, and glutamate synthase, have been shown to lead to increased conversion of NO_3^- into organic nitrogen compounds. This process is stimulated by PHs through enzymatic and molecular mechanisms, including the upregulation of related gene expression [71–73], both accelerating NO_3^- assimilation, thereby reducing its accumulation in leaves.

Additionally, PH biostimulants provide readily available organic nitrogen sources (amino acids and short peptides), which are absorbed by roots and leaves and directly used in protein synthesis, supplement N, and reduce the plant's dependence on the energy-demanding nitrate reduction pathway [74]. Furthermore, they contain bioactive peptides that act as signaling molecules, activating changes in gene expression related to N assimilation pathways, including transcriptional upregulation [75]. Amino acids also improve the efficiency of N uptake by regulating the activity of membrane transporters [76,77]. Collectively, these mechanisms enhance nitrogen use efficiency (NUE) and reduce NO_3^- levels in edible plant tissues, offering a promising strategy to improve both crop growth and product quality.

In purslane, Ca^{2+} concentration in the leaves significantly reduced under high N_r, while the Ca:Mg ratio also decreased. This suggests that the well-known competitive uptake mechanism between cations [53] was in favor of Mg^{2+} (NH_4^+-Mg^{2+}) rather than Ca^{2+} (NH_4^+-Ca^{2+}) [7]. The increased N_r in the nutrient solution also enhanced the uptake and accumulation of micronutrients such as Fe, Mn, and Zn in plant tissues of all three species. As demonstrated by White and Broadley [60], partial hydrolysis of metallic micronutrients, such as Fe, Mn, and Zn, has been observed in aqueous solutions, thereby rendering them unavailable to plants. This response is primarily attributed to the rhizosphere acidification induced by NH_4^+ uptake [78]. The higher availability of H^+ near the root cells, caused by their export from the cells in favor of NH_4^+ uptake, can increase the availability of these micronutrients. Higher leaf concentrations of Mn have also been reported in *C. spinosum* when the N_r increased from 0.05 to 0.25 [7]. Similarly, an increase in NH_4^+ supply in lettuce crops has elevated the concentration of Fe [54] and Zn [62].

Based on the results of this study, growers aiming to cultivate purslane, iceplant, and corn salad in substrate-based soilless systems should adjust N_r in the supplied solution and follow a biostimulant application strategy depending on each species' requirements. For purslane, the N_r can be maintained within the tested range (0.04–0.12) without compromising yield. To enhance product quality by reducing nitrate (NO_3^-) accumulation in edible tissues, growers may either use a high N_r (0.12) or apply protein hydrolysate (PH) biostimulants weekly. Corn salad benefits from a higher N_r (0.12), which increases FW while reducing NO_3^- concentrations. Supplementing with PH biostimulants further reduces leaf NO_3^- levels, enhancing food safety. Iceplant should be cultivated under low N_r (0.04) to prevent ammonium stress and yield losses, while applying either seaweed extracts or PH biostimulants can boost yield and further decrease leaf NO_3^- content.

The economic evaluation revealed that the added net return obtained by the application of biostimulants highly depends on the species and biostimulant used. Specifically, for purslane, the added net return was always negative, suggesting a non-economic convenience for both treatments. Contrariwise, for iceplant, both biostimulants gave positive added net return, with a peak obtained by SW treatments. Moreover, corn salad's added net return in response to biostimulant application depends on the biostimulant employed. Indeed, the application of SW determined negative added net return values, whereas the use of PH gave positive results. In general, the added variable costs for all species are very low; thus, the differences recorded can be attributable to the differences in yield performance and in price per kg, which in turn influence the added gross return. In fact, the negative values recorded for purslane are linked to the non-significant effect of biostimulants on yield and to the low price per kg, compared to the other two species. Furthermore, the peak recorded in iceplant can be attributed to the high increase in yield and to the very high price per kg (EUR 41.0).

5. Conclusions

This study demonstrates that optimizing the NH_4^+/total-N ratio (N_r) and applying biostimulants significantly influence yield performance, nitrate accumulation, and nutrient composition in underutilized leafy greens cultivated in soilless systems. The responses were highly species-dependent: corn salad exhibited improved fresh weight under high N_r (0.12), whereas iceplant suffered yield losses due to NH_4^+ toxicity, classifying it as an ammonium-sensitive species. In contrast, purslane maintained stable yields across N_r levels (0.04–0.12), reflecting its moderate NH_4^+ tolerance. Notably, protein hydrolysate (PH) biostimulants enhanced nitrogen assimilation efficiency and reduced nitrate accumulation in edible tissues, underscoring their potential to improve both crop productivity and food safety. Furthermore, elevated NH_4^+ supply enhanced the uptake of micronutrients (Fe, Mn, Zn), likely due to rhizosphere acidification favoring their solubility. These findings emphasize the need for species-specific nitrogen management strategies in soilless cultivation, integrating tailored N_r adjustments and biostimulant applications to optimize yield, nutritional quality, and resource-use efficiency. Moreover, the economic evaluation suggested that the application of both biostimulants is not convenient for purslane, but it is highly recommended for iceplant, and that—for corn salad—PH is the recommended biostimulant to increase the added net return. This study is preliminary and serves as a starting point for evaluating soilless cultivation and exploring the impact on the biochemical profiles and nutritional value of these three species. Additionally, future research should explore the molecular mechanisms underlying PH biostimulant efficacy and refine NH_4^+ thresholds for additional underutilized species to expand their sustainable production in controlled environments.

Supplementary Materials: The following supporting information can be downloaded at: https://www.mdpi.com/article/10.3390/horticulturae11091076/s1. Table S1: Composition and aminogram of Algastar® *Ascophyllum nodosum* seaweed extract; Table S2. Composition and aminogram of Tyson® protein hydrolysate.

Author Contributions: Conceptualization, G.N. and T.N.; data curation, T.N., I.K., G.P.S., E.G., S.N. and G.N.; formal analysis; T.N., G.P.S., S.M., T.K., S.N. and E.G.; investigation, T.N., I.K. and T.K.; methodology, T.N., I.K., S.M., E.G., G.P.S., T.K., S.N. and G.N.; Software, T.N., S.N., D.N. and E.G.; supervision, G.N.; validation, T.N., I.K., E.G., S.M., S.N., D.N. and G.N.; resourses, G.N.; project administration, G.N.; funding acquisition, G.N.; visualization, T.N., I.K. and S.N.; writing—original draft, T.N., I.K., E.G. and G.N.; writing—review and editing, T.N., I.K., E.G., G.P.S., T.K., S.M., S.N., D.N. and G.N. All authors have read and agreed to the published version of the manuscript.

Funding: This work was supported by the European Commission within the project "RADIANT: Realising Dynamic Value Chains for Underutilised Crops", which received funding from the European Union's Horizon 2020 research and innovation program under grant agreement no. 101000622.

Data Availability Statement: The data are included in the manuscript.

Conflicts of Interest: The authors declare no conflicts of interest.

References

1. European Commission. *EU Agricultural Outlook for Markets and Income, 2019–2030*; European Commission: Brussels, Belgium, 2019.
2. Pinto, E.; Ferreira, H.; Santos, C.S.; da Silva, M.N.; Styles, D.; Migliorini, P.; Ntatsi, G.; Karkanis, A.; Brémaud, M.F.; de Mey, Y.; et al. *Healthier and Sustainable Food Systems: Integrating Underutilised Crops in a 'Theory of Change Approach.' In Biodiversity, Functional Ecosystems and Sustainable Food Production*; Springer International Publishing: Berlin/Heidelberg, Germany, 2022; pp. 275–323, ISBN 9783031074349.
3. Lowe, N.M. The Global Challenge of Hidden Hunger: Perspectives from the Field. *Proc. Nutr. Soc.* **2021**, *80*, 283–289. [CrossRef]
4. Peduruhewa, P.S.; Jayathunge, K.G.L.R.; Liyanage, R. Potential of Underutilized Wild Edible Plants as the Food for the Future—A Review. *J. Food Secur.* **2021**, *9*, 136–147. [CrossRef]
5. Platis, D.P.; Papoui, E.; Bantis, F.; Katsiotis, A.; Koukounaras, A.; Mamolos, A.P.; Mattas, K. Underutilized Vegetable Crops in the Mediterranean Region: A Literature Review of Their Requirements and the Ecosystem Services Provided. *Sustainability* **2023**, *15*, 4921. [CrossRef]
6. Chatzigianni, M.; Savvas, D.; Papadopoulou, E.A.; Aliferis, K.A.; Ntatsi, G. Combined Effect of Salt Stress and Nitrogen Level on the Primary Metabolism of Two Contrasting Hydroponically Grown *Cichorium spinosum* L. Ecotypes. *Biomolecules* **2023**, *13*, 607. [CrossRef] [PubMed]
7. Chatzigianni, M.; Alkhaled, B.; Livieratos, I.; Stamatakis, A.; Ntatsi, G.; Savvas, D. Impact of Nitrogen Source and Supply Level on Growth, Yield and Nutritional Value of Two Contrasting Ecotypes of *Cichorium spinosum* L. Grown Hydroponically. *J. Sci. Food Agric.* **2018**, *98*, 1615–1624. [CrossRef]
8. Voutsinos-Frantzis, O.; Savvas, D.; Liakopoulos, G.; Karavidas, I.; Ntanasi, T.; Sabatino, L.; Marcelis, L.F.M.; Ntatsi, G. Optimizing Vertical Farm Cultivation of *Cichorium spinosum* L.: White Light's Influence and Nutrition Management. *Heliyon* **2024**, *10*, e37146. [CrossRef]
9. Voutsinos-Frantzis, O.; Karavidas, I.; Savvas, D.; Ntanasi, T.; Kaimpalis, V.; Consentino, B.B.; Aliferis, K.A.; Karkanis, A.; Sabatino, L.; Ntatsi, G. Impact of Nitrogen Limitation, Irrigation Levels, and Nitrogen-Rich Biostimulant Application on Agronomical and Chemical Traits of Hydroponically Grown *Cichorium spinosum* L. *Horticulturae* **2024**, *10*, 1063. [CrossRef]
10. Voutsinos-Frantzis, O.; Karavidas, I.; Petropoulos, D.; Zioviris, G.; Fortis, D.; Ntanasi, T.; Ropokis, A.; Karkanis, A.; Sabatino, L.; Savvas, D.; et al. Effects of NaCl and CaCl2 as Eustress Factors on Growth, Yield, and Mineral Composition of Hydroponically Grown *Valerianella locusta*. *Plants* **2023**, *12*, 1454. [CrossRef] [PubMed]
11. Uddin, K.; Quan, L.; Hasan, M.; Selamat Madom, M. Purslane: A Perspective Plant Source of Nutrition and Antioxidant. *Plant Arch.* **2020**, *20*, 1624–1630.
12. Kumar, A.; Sreedharan, S.; Kashyap, A.K.; Singh, P.; Ramchiary, N. A Review on Bioactive Phytochemicals and Ethnopharmacological Potential of Purslane (*Portulaca oleracea* L.). *Heliyon* **2022**, *8*, e08669. [CrossRef]
13. Uddin, M.K.; Juraimi, A.S.; Hossain, M.S.; Nahar, M.A.U.; Ali, M.E.; Rahman, M.M. Purslane Weed (*Portulaca oleracea*): A Prospective Plant Source of Nutrition, Omega-3 Fatty Acid, and Antioxidant Attributes. *Sci. World J.* **2014**, *2014*, 951019. [CrossRef] [PubMed]
14. Hosseinzadeh, M.H.; Ghalavand, A.; Mashhadi-Akbar-Boojar, M.; Modarres-Sanavy, S.A.M.; Mokhtassi-Bidgoli, A. Increased Medicinal Contents of Purslane by Nitrogen and Arbuscular Mycorrhiza under Drought Stress. *Commun. Soil Sci. Plant Anal.* **2019**, *51*, 118–135. [CrossRef]
15. Montoya-García, C.O.; Volke-Haller, V.H.; Trinidad-Santos, A.; Villanueva-Verduzco, C. Change in the Contents of Fatty Acids and Antioxidant Capacity of Purslane in Relation to Fertilization. *Sci. Hort.* **2018**, *234*, 152–159. [CrossRef]
16. Popescu, C.; Popescu, C.; Manea, S.; Vladut, V.; Caba, I.; Covaliu, I.C.; Abbas, H.; Dune, A.; Lupuleasa, D. Study on *Portulaca oleracea* Native Species as Vegetal Source of Omega-3 and Omega-6 Fatty Acids. *Rev. Chim.* **2018**, *69*, 2973–2980. [CrossRef]
17. Jin, R.; Wang, Y.; Liu, R.; Gou, J.; Chan, Z. Physiological and Metabolic Changes of Purslane (*Portulaca oleracea* L.) in Response to Drought, Heat, and Combined Stresses. *Front. Plant Sci.* **2016**, *6*, 1123. [CrossRef]
18. Xing, J.C.; Dong, J.; Wang, M.W.; Liu, C.; Zhao, B.Q.; Wen, Z.G.; Zhu, X.M.; Ding, H.R.; Zhao, X.H.; Hong, L.Z. Effects of NaCl Stress on Growth of *Portulaca oleracea* and Underlying Mechanisms. *Braz. J. Bot.* **2019**, *42*, 217–226. [CrossRef]
19. Chrysargyris, A.; Xylia, P.; Zengin, G.; Tzortzakis, N. Purslane (*Portulaca oleracea* L.) Growth, Nutritional, and Antioxidant Status under Different Nitrogen Levels in Hydroponics. *Horticulturae* **2024**, *10*, 1007. [CrossRef]
20. Loconsole, D.; Murillo-Amador, B.; Cristiano, G.; De Lucia, B. Halophyte Common Ice Plants: A Future Solution to Arable Land Salinization. *Sustainability* **2019**, *11*, 6076. [CrossRef]
21. Barkla, B.J.; Vera-Estrella, R. Single Cell-Type Comparative Metabolomics of Epidermal Bladder Cells from the Halophyte *Mesembryanthemum crystallinum*. *Front. Plant Sci.* **2015**, *6*, 435. [CrossRef] [PubMed]
22. Falleh, H.; Ksouri, R.; Medini, F.; Guyot, S.; Abdelly, C.; Magné, C. Antioxidant Activity and Phenolic Composition of the Medicinal and Edible Halophyte *Mesembryanthemum edule* L. *Ind. Crops Prod.* **2011**, *34*, 1066–1071. [CrossRef]
23. Eoh, G.; Kim, C.; Bae, J.; Park, J. Evaluation of Sodium Chloride Concentrations on Growth and Phytochemical Production of *Mesembryanthemum crystallinum* L. in a Hydroponic System. *Horticulturae* **2024**, *10*, 1304. [CrossRef]

24. Mndi, O.; Sogoni, A.; Jimoh, M.O.; Wilmot, C.M.; Rautenbach, F.; Laubscher, C.P. Interactive Effects of Salinity Stress and Irrigation Intervals on Plant Growth, Nutritional Value, and Phytochemical Content in *Mesembryanthemum crystallinum* L. *Agriculture* **2023**, *13*, 1026. [CrossRef]
25. Oliveira-Alves, S.C.; Andrade, F.; Sousa, J.; Bento-Silva, A.; Duarte, B.; Caçador, I.; Salazar, M.; Mecha, E.; Serra, A.T.; Bronze, M.R. Soilless Cultivated Halophyte Plants: Volatile, Nutritional, Phytochemical, and Biological Differences. *Antioxidants* **2023**, *12*, 1161. [CrossRef] [PubMed]
26. Ragaert, P.; Verbeke, W.; Devlieghere, F.; Debevere, J. Consumer Perception and Choice of Minimally Processed Vegetables and Packaged Fruits. *Food Qual. Prefer.* **2003**, *15*, 259–270. [CrossRef]
27. Fontana, E.; Nicola, S. Traditional and Soilless Culture Systems to Produce Corn Salad (*Valerianella olitoria* L.) and Rocket (*Eruca sativa* Mill.) with Low Nitrate Content. *J. Food Agric. Environ.* **2009**, *7*, 405–410.
28. Hernández, V.; Ángeles Botella, M.; Hellín, P.; Cava, J.; Fenoll, J.; Mestre, T.; Martínez, V.; Flores, P. Phenolic and Carotenoid Profile of Lamb's Lettuce and Improvement of the Bioactive Content by Preharvest Conditions. *Foods* **2021**, *10*, 188. [CrossRef]
29. Schmitzer, V.; Senica, M.; Slatnar, A.; Stampar, F.; Jakopic, J. Changes in Metabolite Patterns During Refrigerated Storage of Lamb's Lettuce (*Valerianella locusta* L. Betcke). *Front. Nutr.* **2021**, *8*, 731869. [CrossRef]
30. Ramos-Bueno, R.P.; Rincón-Cervera, M.A.; González-Fernández, M.J.; Guil-Guerrero, J.L. Phytochemical Composition and Antitumor Activities of New Salad Greens: Rucola (*Diplotaxis tenuifolia*) and Corn Salad (*Valerianella locusta*). *Plant Foods Hum. Nutr.* **2016**, *71*, 197–203. [CrossRef]
31. Di Mola, I.; Cozzolino, E.; Ottaiano, L.; Nocerino, S.; Rouphael, Y.; Colla, G.; El-Nakhel, C.; Mori, M. Nitrogen Use and Uptake Efficiency and Crop Performance of Baby Spinach (*Spinacia oleracea* L.) and Lamb's Lettuce (*Valerianella locusta* L.) Grown under Variable Sub-Optimal N Regimes Combined with Plant-Based Biostimulant Application. *Agronomy* **2020**, *10*, 278. [CrossRef]
32. Taiz, L.; Zeiger, E. *Plant Physiology*, 3rd ed.; Sinauer Associates: Sunderland, MA, USA, 2002; ISBN 0878938230.
33. Voogt, W.; Bar-Yosef, B. Water and Nutrient Management and Crops Response to Nutrient Solution Recycling in Soilless Growing Systems in Greenhouses. In *Soilless Culture: Theory and Practice*; Elsevier: Amsterdam, The Netherlands, 2019; pp. 425–507, ISBN 9780444636966.
34. Boschiero, B.N.; Mariano, E.; Azevedo, R.A.; Ocheuze Trivelin, P.C. Influence of Nitrate—Ammonium Ratio on the Growth, Nutrition, and Metabolism of Sugarcane. *Plant Physiol. Biochem.* **2019**, *139*, 246–255. [CrossRef] [PubMed]
35. Hawkesford, M.; Horst, W.; Kichey, T.; Lambers, H.; Schjoerring, J.; Møller, I.S.; White, P. Functions of Macronutrients. In *Marschner's Mineral Nutrition of Higher Plants*; Marschner, P., Ed.; Academic Press: London, UK, 2012; pp. 135–189.
36. Mansori, M.; Chernane, H.; Latique, S.; Benaliat, A.; Hsissou, D.; El Kaoua, M. Effect of Seaweed Extract (*Ulva rigida*) on the Water Deficit Tolerance of *Salvia officinalis* L. *J. Appl. Phycol.* **2016**, *28*, 1363–1370. [CrossRef]
37. Ibrahim, W.M.; Ali, R.M.; Hemida, K.A.; Sayed, M.A. Role of Ulva Lactuca Extract in Alleviation of Salinity Stress on Wheat Seedlings. *Sci. World J.* **2014**, *2014*, 847290. [CrossRef]
38. Di Stasio, E.; Van Oosten, M.J.; Silletti, S.; Raimondi, G.; dell'Aversana, E.; Carillo, P.; Maggio, A. Ascophyllum Nodosum-Based Algal Extracts Act as Enhancers of Growth, Fruit Quality, and Adaptation to Stress in Salinized Tomato Plants. *J. Appl. Phycol.* **2018**, *30*, 2675–2686. [CrossRef]
39. Hernández-Herrera, R.M.; Sánchez-Hernández, C.V.; Palmeros-Suárez, P.A.; Ocampo-Alvarez, H.; Santacruz-Ruvalcaba, F.; Meza-Canales, I.D.; Becerril-Espinosa, A. Seaweed Extract Improves Growth and Productivity of Tomato Plants under Salinity Stress. *Agronomy* **2022**, *12*, 2495. [CrossRef]
40. Sogoni, A.; Ngcobo, B.L.; Jimoh, M.O.; Kambizi, L.; Laubscher, C.P. Seaweed-Derived Bio-Stimulant (Kelpak®) Enhanced the Morphophysiological, Biochemical, and Nutritional Quality of Salt-Stressed Spinach (*Spinacia oleracea* L.). *Horticulturae* **2024**, *10*, 1340. [CrossRef]
41. Ntanasi, T.; Karavidas, I.; Zioviris, G.; Ziogas, I.; Karaolani, M.; Fortis, D.; Conesa, M.À.; Schubert, A.; Savvas, D.; Ntatsi, G. Assessment of Growth, Yield, and Nutrient Uptake of Mediterranean Tomato Landraces in Response to Salinity Stress. *Plants* **2023**, *12*, 3551. [CrossRef] [PubMed]
42. Colla, G.; Svecová, E.; Cardarelli, M.; Rouphael, Y.; Reynaud, H.; Canaguier, R.; Planques, B. Effectiveness of a Plant-Derived Protein Hydrolysate to Improve Crop Performances under Different Growing Conditions. *Acta Hortic.* **2013**, *1009*, 175–179. [CrossRef]
43. Lucini, L.; Rouphael, Y.; Cardarelli, M.; Canaguier, R.; Kumar, P.; Colla, G. The Effect of a Plant-Derived Biostimulant on Metabolic Profiling and Crop Performance of Lettuce Grown under Saline Conditions. *Sci. Hortic.* **2015**, *182*, 124–133. [CrossRef]
44. Zhou, W.; Zheng, W.; Wang, W.; Lv, H.; Liang, B.; Li, J. Exogenous Pig Blood-Derived Protein Hydrolysates as a Promising Method for Alleviation of Salt Stress in Tomato (*Solanum lycopersicum* L.). *Sci. Hortic.* **2022**, *294*, 110779. [CrossRef]
45. Sorrentino, M.; Panzarová, K.; Spyroglou, I.; Spíchal, L.; Buffagni, V.; Ganugi, P.; Rouphael, Y.; Colla, G.; Lucini, L.; De Diego, N. Integration of Phenomics and Metabolomics Datasets Reveals Different Mode of Action of Biostimulants Based on Protein Hydrolysates in *Lactuca sativa* L. and *Solanum lycopersicum* L. Under Salinity. *Front. Plant Sci.* **2022**, *12*, 808711. [CrossRef]

46. Campbell, C.R.; Plank, C.O. Preparation of Plant Tissue for Laboratory Analysis. In *Handbook of Reference Methods for Plant Analysis*; Kalra, Y., Ed.; CRC Press: Boca Raton, FL, USA, 1997; p. 320, ISBN 9780367802233.
47. Cataldo, D.A.; Maroon, M.; Schrader, L.E.; Youngs, V.L. Rapid Colorimetric Determination of Nitrate in Plant Tissue by Nitration of Salicylic Acid. *Soil Sci. Plant Anal.* **1975**, *6*, 71–80. [CrossRef]
48. Gurav, R.G.; Jadhav, J.P. A Novel Source of Biofertilizer from Feather Biomass for Banana Cultivation. *Environ. Sci. Pollut. Res.* **2013**, *20*, 4532–4539. [CrossRef]
49. Spyrou, G.P.; Karavidas, I.; Ntanasi, T.; Marka, S.; Giannothanasis, E.; Gohari, G.; Allevato, E.; Sabatino, L.; Savvas, D.; Ntatsi, G. Chloride as a Partial Nitrate Substitute in Hydroponics: Effects on Purslane Yield and Quality. *Plants* **2025**, *14*, 2160. [CrossRef]
50. Sonneveld, C.; Voogt, W. *Plant Nutrition of Greenhouse Crops*; Springer: Dordrecht, The Netherlands, 2009; ISBN 9789048125326.
51. Akl, I.A.; Savvas, D.; Papadantonakis, N.; Lydakis-Simantiris, N.; Kefalas, P. Influence of Ammonium to Total Nitrogen Supply Ratio on Growth, Yield and Fruit Quality of Tomato Grown in a Closed Hydroponic System. *Eur. J. Hortic. Sci.* **2003**, *68*, 204–211. [CrossRef]
52. Wenceslau, D.d.S.L.; Daniele, F.d.O.; Hudson, d.O.R.; Guilherme, F.F.; Luiz, A.A.G.; Erica, C.A.L.; Gustavo, C. Nitrate Concentration and Nitrate/Ammonium Ratio on Lettuce Grown in Hydroponics in Southern Amazon. *Afr. J. Agric. Res.* **2021**, *17*, 862–868. [CrossRef]
53. Marschner, P. *Marschner's Mineral Nutrition of Higher Plants*, 3rd ed.; Academic Press: London, UK, 2012; ISBN 978-0-12-384905-2.
54. Savvas, D.; Passam, H.C.; Olympios, C.; Nasi, E.; Moustaka, E.; Mantzos, N.; Barouchas, P. Effects of Ammonium Nitrogen on Lettuce Grown on Pumice in a Closed Hydroponic System. *HortScience* **2006**, *41*, 1667–1673. [CrossRef]
55. Bulgari, R.; Cocetta, G.; Prinsi, B.; Espen, L.; Ferrante, A. Influence of Different Ammonium and Nitrate Ratios on Quality of Rocket. *Acta Hortic.* **2021**, *1321*, 103–108. [CrossRef]
56. Chrysargyris, A.; Tzortzakis, N. Optimising Fertigation of Hydroponically Grown Sowthistle (*Sonchus oleraceus* L.): The Impact of the Nitrogen Source and Supply Concentration. *Agric. Water Manag.* **2023**, *289*, 108528. [CrossRef]
57. Papadimitriou, D.M.; Daliakopoulos, I.N.; Lydakis-Simantiris, N.; Cheiladaki, I.; Manios, T.; Savvas, D. Nitrogen Source and Supply Level Impact Water Uptake, Yield, and Nutrient Status of Golden Thistle in a Soilless Culture. *Sci. Hortic.* **2024**, *336*, 113384. [CrossRef]
58. Chrysargyris, A.; Hajisolomou, E.; Xylia, P.; Tzortzakis, N. Ammonium to Total Nitrogen Ratio Affects the Purslane (*Portulaca oleracea* L.) Growth, Nutritional, and Antioxidant Status. *Heliyon* **2023**, *9*, e21644. [CrossRef] [PubMed]
59. Mattson, N.; Lieth, J.H. Liquid Culture Hydroponic System Operation. In *Soilless Culture: Theory and Practice Theory and Practice*; Elsevier: Amsterdam, The Netherlands, 2019; pp. 567–585, ISBN 9780444636966.
60. White, P.J.; Broadley, M.R. Biofortification of Crops with Seven Mineral Elements Often Lacking in Human Diets—Iron, Zinc, Copper, Calcium, Magnesium, Selenium and Iodine. *New Phytol.* **2009**, *182*, 49–84. [CrossRef]
61. Rincón-Cervera, M.Á.; Pagan Loeiro da Cunha-Chiamolera, T.; Chileh-Chelh, T.; Carmona-Fernández, M.; Urrestarazu, M.; Guil-Guerrero, J.L. Growth Parameters, Phytochemicals, and Antitumor Activity of Wild and Cultivated Ice Plants (*Mesembryanthemum crystallinum* L.). *Food Sci. Nutr.* **2024**, *12*, 6548–6562. [CrossRef]
62. Choi, S.; Colla, G.; Cardarelli, M.; Kim, H. Effects of Vegetal Protein Hydrolysate Application Method, Nitrogen Level, and Nitrate-to-ammonium Ratio on Growth and Composition of Hydroponic Lettuce. *J. Sci. Food Agric.* **2025**, *105*, 599–610. [CrossRef]
63. Liu, X.; Hu, B.; Chu, C. Nitrogen Assimilation in Plants: Current Status and Future Prospects. *J. Genet. Genom.* **2022**, *49*, 394–404. [CrossRef]
64. Bryan, N.S.; Alexander, D.D.; Coughlin, J.R.; Milkowski, A.L.; Boffetta, P. Ingested Nitrate and Nitrite and Stomach Cancer Risk: An Updated Review. *Food Chem. Toxicol.* **2012**, *50*, 3646–3665. [CrossRef]
65. Zhang, F.-X.; Miao, Y.; Ruan, J.-G.; Meng, S.-P.; Dong, J.-D.; Yin, H.; Huang, Y.; Chen, F.-R.; Wang, Z.-C.; Lai, Y.-F. Association Between Nitrite and Nitrate Intake and Risk of Gastric Cancer: A Systematic Review and Meta-Analysis. *Med. Sci. Monit.* **2019**, *25*, 1788–1799. [CrossRef]
66. Manzocco, L.; Foschia, M.; Tomasi, N.; Maifreni, M.; Dalla Costa, L.; Marino, M.; Cortella, G.; Cesco, S. Influence of Hydroponic and Soil Cultivation on Quality and Shelf Life of Ready-to-Eat Lamb's Lettuce (*Valerianella locusta* L. Laterr). *J. Sci. Food Agric.* **2011**, *91*, 1373–1380. [CrossRef]
67. Agusta, H.; Kartika, J.G.; Sari, K.R. Nitrate Concentration and Accumulation on Vegetables Related to Altitude and Sunlight Intensity. *IOP Conf. Ser. Earth Environ. Sci.* **2021**, *896*, 012052. [CrossRef]
68. Zhou, W.; Chen, Y.; Xu, H.; Liang, X.; Hu, Y.; Jin, C.; Lu, L.; Lin, X. Short-Term Nitrate Limitation Prior to Harvest Improves Phenolic Compound Accumulation in Hydroponic-Cultivated Lettuce (*Lactuca sativa* L.) without Reducing Shoot Fresh Weight. *J. Agric. Food. Chem.* **2018**, *66*, 10353–10361. [CrossRef]
69. Zhou, W.; Liang, X.; Zhang, Y.; Li, K.; Jin, B.; Lu, L.; Jin, C.; Lin, X. Reduced Nitrogen Supply Enhances the Cellular Antioxidant Potential of Phenolic Extracts through Alteration of the Phenolic Composition in Lettuce (*Lactuca sativa* L.). *J. Sci. Food. Agric.* **2019**, *99*, 4761–4771. [CrossRef]

70. Voutsinos-Frantzis, O.; Savvas, D.; Antoniadou, N.; Karavidas, I.; Ntanasi, T.; Sabatino, L.; Ntatsi, G. Innovative Cultivation Practices for Reducing Nitrate Content in Baby Leaf Lettuce Grown in a Vertical Farm. *Horticulturae* **2024**, *10*, 375. [CrossRef]
71. Schiavon, M.; Ertani, A.; Nardi, S. Effects of an Alfalfa Protein Hydrolysate on the Gene Expression and Activity of Enzymes of the Tricarboxylic Acid (TCA) Cycle and Nitrogen Metabolism in *Zea mays* L. *J. Agric. Food Chem.* **2008**, *56*, 11800–11808. [CrossRef]
72. Ertani, A.; Schiavon, M.; Muscolo, A.; Nardi, S. Alfalfa Plant-Derived Biostimulant Stimulate Short-Term Growth of Salt Stressed *Zea mays* L. Plants. *Plant Soil* **2013**, *364*, 145–158. [CrossRef]
73. Ertani, A.; Cavani, L.; Pizzeghello, D.; Brandellero, E.; Altissimo, A.; Ciavatta, C.; Nardi, S. Biostimulant Activity of Two Protein Hydrolyzates in the Growth and Nitrogen Metabolism of Maize Seedlings. *J. Plant Nutr. Soil Sci.* **2009**, *172*, 237–244. [CrossRef]
74. Colla, G.; Nardi, S.; Cardarelli, M.; Ertani, A.; Lucini, L.; Canaguier, R.; Rouphael, Y. Protein Hydrolysates as Biostimulants in Horticulture. *Sci. Hortic.* **2015**, *196*, 28–38. [CrossRef]
75. Ryan, C.A.; Pearce, G.; Scheer, J.; Moura, D.S. Polypeptide Hormones. *Plant Cell* **2002**, *14*, S251–S264. [CrossRef]
76. Miller, A.J.; Fan, X.; Shen, Q.; Smith, S.J. Amino Acids and Nitrate as Signals for the Regulation of Nitrogen Acquisition. *J. Exp. Bot.* **2007**, *59*, 111–119. [CrossRef]
77. Fan, X.; Gordon-Weeks, R.; Shen, Q.; Miller, A.J. Glutamine Transport and Feedback Regulation of Nitrate Reductase Activity in Barley Roots Leads to Changes in Cytosolic Nitrate Pools. *J. Exp. Bot.* **2006**, *57*, 1333–1340. [CrossRef]
78. White, P.J. Ion Uptake Mechanisms of Individual Cells and Roots: Short-Distance Transport. In *Marschner's Mineral Nutrition of Higher Plants*; Marschner, P., Ed.; Academic Press: London, UK, 2012; pp. 7–48.

horticulturae

Article

Phytochemical Composition, Sensory Acceptance, and Cultivation Potential of *Sanguisorba verrucosa*, *Eruca vesicaria*, and *Scorzonera laciniata*

María Ángeles Botella [1], Pilar Hellín [2], Virginia Hernández [2], Mercedes Dabauza [2], Antonio Robledo [3], Alicia Sánchez [2], José Fenoll [2] and Pilar Flores [2,*]

[1] Department of Applied Biology, Polytechnic School of Orihuela (EPSO), Center for Agro-Food and Environmental Research and Innovation (CIAGRO), Miguel Hernández University, Carretera de Beniel km 3.2, 03312 Orihuela, Alicante, Spain; mangeles.botella@umh.es

[2] Murcian Institute of Agricultural and Environmental Research and Development (IMIDA), C/Mayor s/n, La Alberca, 30150 Murcia, Murcia, Spain; mariap.hellin@carm.es (P.H.); virginia.hernandez5@carm.es (V.H.); mercedes.dabauza@gmail.com (M.D.); alicia.sanchez15@carm.es (A.S.); jose.fenoll@carm.es (J.F.)

[3] ISLAYA Environmental Consulting, S.L., C/Ntra. Sra. de Fátima 34, 30151 Santo Ángel, Murcia, Spain; islaya@islaya.es

* Correspondence: mpilar.flores@carm.es

Abstract

Three wild edible plant species native to the Mediterranean region, *Sanguisorba verrucosa*, *Eruca vesicaria*, and *Scorzonera laciniata*, were investigated to evaluate their potential for cultivation and integration into human diets. All three species were positively rated in sensory evaluations and exhibited high levels of specific metabolites of nutritional and health-related interest. Moderate concentrations of β-carotene were found across all species. Notably, *S. verrucosa* and *E. vesicaria* contained appreciable amounts of vitamin C, and the phenolic content in *S. verrucosa* exceeded that of many commonly consumed vegetables. Each species also proved to be a rich source of distinct organic acids: *S. verrucosa* for fumaric acid, *E. vesicaria* for citric acid, and *S. laciniata* for quinic acid. Although domestication led to a reduction in several bioactive compounds, the nutritional value of these plants remains significant. The compositional and sensory profiles of these species highlight their promise as leafy vegetables for sustainable diets and as functional food ingredients. Furthermore, their cultivation could support biodiversity conservation efforts and reduce harvesting pressure on wild populations, contributing to more sustainable agricultural practices.

Keywords: secondary metabolites; traditional uses; agrobiodiversity; alternative crops; functional value; crop diversification

1. Introduction

The modern human diet is often based on the consumption of a few highly cultivated species, and diversifying dietary intake to include a broader array of plant species is thus a promising way to improve human welfare. Despite their global prevalence and cultural significance, wild edible plants (WEPs) are frequently underestimated as significant dietary components in developed countries [1]. However, they are a key component in people's diets worldwide [2], playing a crucial role in food security, especially in regions with limited agricultural resources or harsh climates [3,4]. The Food and Agriculture Organisation [5] reported that over 100 million people in the EU, 20% of the population, eat wild foods, and 65 million (14%) gather some wild food at least occasionally [6,7].

 https://doi.org/10.3390/horticulturae11091021

Moreover, there is currently a growing interdisciplinary interest in the study of wild plants as sources of both bioactive compounds and novel ingredients for haute cuisine. On the one hand, WEPs have rich nutritional profiles, providing essential micronutrients and macronutrients, including vitamins, proteins, minerals, and dietary fibers [8–10]. Furthermore, numerous studies have highlighted their therapeutic potential due to their antioxidant, antimicrobial, antiinflammatory, and antitumor properties [10–13], and they are natural antidiabetics as well [14]. On the other hand, the interest in haute cuisine is evidenced by chefs incorporating traditionally consumed wild species, particularly those from the Mediterranean diet, into their menus. Research by Łuczaj and Pieroni [15] highlighted the successful adoption of diverse local wild plants in avant-garde restaurants worldwide. Additionally, recent studies have explored the potential of wild plants to be marketed as novel functional foods, emphasizing innovative and health-promoting culinary applications [16].

For this study, three different species grown in the Mediterranean area were chosen: *Sanguisorba verrucosa* (*S. verrucosa*), *Eruca vesicaria* (*E. vesicaria*), and *Scorzonera laciniata* (*S. laciniata*). The genus *Sanguisorba* has proven to be a rich source of phytochemicals, with over 270 compounds identified to date. Among these, *Sanguisorba* officinalis stands out, with its extracts exhibiting a range of pharmacological activities, including notable anticancer, antiviral, and antiinflammatory effects [17]. More recently, *Sanguisorba minor* (*S. minor*), used as an edible plant in the Mediterranean diet, has been reported as a good source of bioactive compounds, particularly polyphenols, with a potential use in food supplements and pharmaceutical formulations [18]. Few studies have been found on *S. verrucosa*, such as some reports on the consumption of tender leaves and stems in salads in Spain [19]. Rivera et al. [20] documented the traditional medicinal use of *S. verrucosa* across various applications related to the circulatory, digestive, and integumentary systems. The genus *Eruca*, part of the Brassicaceae family, includes species commonly known as rocket salads, which hold an important place in the Mediterranean diet due to their rich phytochemical content [21,22]. In addition to their culinary relevance, traditional uses of these plants have been recorded in various ethnobotanical sources [23,24]. While *Eruca sativa* Mill. has been extensively studied, *E. vesicaria* has attracted comparatively limited scientific attention, despite its utilization in traditional dishes such as salads, omelets, and gazpachos [20]. Regarding *Scorzonera* species, some are utilized as vegetables or spices in parts of Asia and Europe [25], and a recent review encompassing 55 species highlighted their significant pharmacological potential [26]. Specifically, the young leaves of *S. laciniata* are occasionally consumed in salads [20], and evidence of their phytochemical richness and significant antioxidant activity suggests potential applications in dietary supplementation, as well as in microbiological and pharmacological research [27].

WEPs are predominantly harvested from their natural habitats. However, traditional foraging has several limitations, such as inconsistent supply, limited quality control, and the potential for ecological degradation due to overharvesting. In response to these challenges, recent research has increasingly turned to the cultivation of WEPs as a sustainable alternative [28–31]. Emerging evidence indicates that domestication and cultivation may offer significant advantages, such as enhanced yields and greater consistency in both nutritional and phytochemical profiles [32,33]. These findings underscore the potential of integrating WEP cultivation into rural Mediterranean agriculture, which may also contribute to the preservation of natural ecosystems by alleviating harvesting pressure on wild populations. Given the culinary and pharmaceutical potential of the three selected Mediterranean species and the limited available data on their organoleptic properties and cultivation performance, this study aims to investigate their sensory attributes and chemical composition. Special attention is given to the effects of cultivation on leaf composition, with a focus on evaluating the potential for large-scale commercial production and the corresponding implications for quality parameters.

2. Materials and Methods

The two-year study included both wild and cultivated harvesting.

2.1. Gathering of Wild Plants

Initial plant material was collected from various natural habitats, predominantly within the Region of Murcia. The taxonomic identification was performed by Antonio Robledo Miras (BSc Botany). The collection adhered to established protocols, i.e., recording parameters such as collection date, collector, specific locality, geographical coordinates, altitude, habitat description, land use, geological features, drainage characteristics, number of sampled plants, and estimated population size. Visual documentation, including images of whole plants and detailed leaf morphology, is provided for *S. oleraceus* and *S. tenerrimus* in Figure 1. Sampling was conducted over a two-month period at the beginning of the year (January to March). To ensure data representativeness, five distinct batches of each species were independently sampled from diverse locations (Figure 2) that comprised a heterogeneous mosaic of Quaternary, calcareous, siliceous, and mixed sedimentary substrates, including alluvial and colluvial deposits, marls, conglomerates, sandstones, schists, and other lithologies that generate diverse edaphic and hydrological conditions. Each batch was treated as an independent sample and contained at least 500 g of the edible portion from multiple individuals. These individuals were representative of the population at a mature harvest stage, characterized by fully expanded leaves and optimal leaf size. Furthermore, viable whole plants were transferred to IMIDA's facilities for cultivation, facilitating seed production for subsequent germination and cultivation trials. These seeds were also accessioned into the conservation collection of IMIDA's germplasm bank (BAGERIM). Seed cleaning was performed manually, with no disinfection procedures applied to preserve natural germination characteristics.

Figure 1. Plant and leaf detail of *S. verrucosa* (**A**,**D**), *E. vesicaria* (**B**,**E**), and *S. laciniata* (**C**,**F**).

Figure 2. Location of *S. verrucosa* (blue), *E. vesicaria* (green), and *S. laciniata* (orange) samples within the Murcia Region.

2.2. Germination Assay in Petri Dishes

One hundred seeds from each species were distributed across three Petri dishes (33, 33, and 34 seeds per dish) lined with filter paper. Deionized water was supplied on demand to maintain consistent moisture levels. The Petri dishes were taken to a controlled environment growth chamber with the following conditions: 16/8 h light/night photoperiod, at a 25 °C (day)/20 °C (night) temperature, 75% relative humidity, and 350 μmol m^{-2} s^{-1} flux density. Radicle emergence was established as the criterion for germination, at which point the germinated seeds were removed from the Petri dish. The duration of this germination trial was 15 days.

2.3. Field Germination Assays

During the spring season, an experiment was conducted using the obtained seeds at the IMIDA experimental facilities in La Alberca, Murcia, Spain. Prior to seed sowing, the plot was tilled, and compost was applied at a rate of 1 kg m^{-2}, and then thoroughly integrated into the soil via a second pass with a tractor. After that, seeds were sown with an optimum plant density established as 400 plants m^{-2}. The germination rates obtained from controlled Petri dish assays and seed weight (seeds per gram) were used to calculate the appropriate dose of each species.

Four different 1 m^2 plots (replications) were used for each species with a randomized block design. After the seeds were manually distributed into each plot, they were irrigated using an exudative pipe, and this continued until the end of the experiment. Twenty-four days after sowing, germination monitoring was carried out using a 60 cm × 60 cm template placed in the center of each of the 1 m^2 plots. Previously, in each plot, five 10 cm × 10 cm cells were randomly chosen.

2.4. Cultivation Trial

This experiment was carried out during the autumn–winter period. Seeds were initially germinated in Petri dishes in order to obtain better homogeneity. After radical emergence, the germinated seeds were transplanted to seedbeds containing commercial seedbed substrate. After that, the trays were taken to a growth chamber under controlled conditions with temperatures of 25 °C during the day and 20 °C during the night, 75% of humidity, and a photoperiod of 16 h day/8 h night. The seedlings were later transplanted into the experimental plots. A planting density of 400 plants m^{-2} was implemented, utilizing a 5 cm × 5 cm planting frame. The experimental design followed a randomized block design, with four blocks allocated to each species. Irrigation was delivered through a porous-pipe system and sustained for the duration of the experiment. Applications were timed at regular intervals and adjusted as necessary to maintain optimal soil moisture while preventing oversaturation. This regimen standardized water availability across individuals, resulting in uniform growth patterns. Plants were harvested during April–May when leaf length was between 5 and 10 cm and prior to flowering, corresponding to the mature harvest stage established for wild plants, characterized by fully expanded leaves and optimal leaf size. Samples consisted of the youngest aerial plant parts. Four replicates were collected for each species, with each replicate containing ten individual plants. These replicates were subsequently pooled and used for sensory attribute analysis on the same day.

2.5. Metabolite Analysis

2.5.1. Metabolite Extraction

At least 100 g from each replicate of both wild-gathered and cultivated plants were homogenized, flash-frozen in liquid nitrogen, and stored at −80 °C for subsequent metabolite analysis.

For the analysis of soluble sugars, organic acids, and total phenolic content, an extraction on 3 g of frozen material was carried out as described by López et al. [34], with 10 mL of ultrapure water (Millipore, Molsheim, France) in a Polytron (PT-MR 3100, Malters, Switzerland), and 25 mL of ethyl acetate (J.T. Baker, Deventer, Holland, The Netherlands) to remove pigments. The aqueous fraction was used for analyzing soluble sugars, organic acids, and total phenolic content.

The extraction of carotenoids and chlorophylls was performed according to Hernández et al. [35], by taking 1 g of frozen material. The extraction solvent consisted of 25 mL of a 1:1 (*v*/*v*) mixture of methanol and tetrahydrofuran (Scharlau, Sentmenat, Spain), supplemented with MgO (Merck, Darmstadt, Germany) and 0.1% (*w*/*v*) butylated hydroxytoluene (BHT) (Sigma-Aldrich, Saint Louis, MO, USA) to inhibit degradation. Post-extraction, the solvent was evaporated to achieve a final concentrated volume of 2 mL. β-apo-8′carotenal (Sigma-Aldrich, Saint Louis, MO, USA) was added to the extract to serve as an internal standard for accurate quantification.

For the extraction of vitamin C (ascorbic and dehydroascorbic), 3 g of fresh material were homogenized with EDTA, 0.05% (*w*/*v*), and dithiothreitol (Sigma, Steinheim, Germany).

2.5.2. Determination of Sugar and Organic Acids

The quantification of soluble sugars was performed using a Hewlett–Packard model 1100 High-Performance Liquid Chromatography (HPLC) system (Waldbronn, Germany) equipped with a Refractive Index (RI) detector. Chromatographic separation was achieved using a CARBOSep CHO-682 LEAD column (300 mm × 7.8 mm i.d., Transgenomic, Omaha, NE, USA). The isocratic mobile phase consisted of ultrapure water, delivered at a constant flow rate of 0.4 mL min^{-1}. To determine the linearity of the detector response and to establish the limits of detection for sucrose, glucose, and fructose, a series of standard solutions (≥99.5% purity, Sigma, Steinheim, Germany) was prepared and injected. These standards spanned a concentration range of 1–10 g·L^{-1}.

Organic acids were determined using a liquid chromatography system (Agilent 1200; Agilent Technologies, Santa Clara, CA, USA) coupled to a triple quadrupole tandem mass spectrometer (MS/MS). The analytical methodology was adapted from the protocol described by Flores et al. [36]. For calibration and to ascertain the linearity of the detector response, standard solutions were prepared from commercially sourced external standards (Sigma-Aldrich, St. Louis, MO, USA).

2.5.3. Chlorophyll and Carotenoid Analysis

The analysis of carotenoids and chlorophylls was performed according to Hernández et al. [35], utilizing an Agilent 1200 High-Performance Liquid Chromatography (HPLC) system (Agilent Technologies, Santa Clara, CA, USA) equipped with a photodiode array detector (PAD). Chromatographic separation was achieved using a binary solvent system consisting of methanol (solvent A) and methyl tert-butyl ether (solvent B) (Scharlau, Sentmenat, Spain). The mobile phase was delivered at a constant flow rate of 1.0 mL·min^{-1} with the following gradient program: an initial 15% solvent B for 20 min, followed by a linear gradient increasing to 30% solvent B over 20 min. This composition was then maintained for 10 min, culminating in an 80-min linear gradient to 90% solvent B. For quantification, carotenoids and chlorophylls were determined against commercially available external standards obtained from DHI LAB (Hoersholm, Denmark). Luteoxanthin was quantified relative to antheraxanthin, and the cis isomers of β-carotene were quantified with respect to their all-trans isomer.

2.5.4. Ascorbic Acid Analysis

The quantification of vitamin C (ascorbic acid and its derivative, dehydroascorbic acid) was carried out with an HPLC equipped with an MS/MS detector, according to Fenoll et al. [37]. The mobile phase utilized was a 0.2% (v/v) solution of formic acid (Scharlau, Sentmenat, Spain), delivered at a flow rate of 0.4 mL min^{-1}. For calibration, L-ascorbic acid standard solutions (Sigma-Aldrich, St. Louis, MO, USA) were prepared, with the addition of 0.05% (w/v) EDTA to ensure stability.

2.5.5. Total Phenolic Content Analysis

The total phenolic content was assessed using the Folin–Ciocâlteu method. A standard curve was generated from gallic acid (Fluka, Steinheim, Germany) solutions, with concentrations ranging from 50 to 400 mg·L^{-1}. Phenolic concentrations in samples were then calculated based on their absorbance at 765 nm, measured with a Shimadzu UV-2401PC spectrophotometer (Kyoto, Japan).

2.6. Sensory Assessment

A panel of fifteen trained assessors was used to assess the sensory profiles of the cultivated plants through the application of a descriptive analysis. Prior to evaluation, plants were prepared by thoroughly washing and then immersed in a solution of water containing food-grade hypochlorite. The leaves were subsequently rinsed with abundant drinking water, and excess moisture was removed using a kitchen centrifuge. The prepared leaves were then refrigerated until the sensory evaluation session. For the tasting session, the samples were presented raw and undressed in appropriately coded, food-grade plastic cups. To mitigate bias, the tasting order for each panelist was randomized. A separate sample, consisting of a single leaf from each species, was prepared for the assessment of leaf color and shape. All samples were coded with unique three-digit random numbers to prevent any potential influence that sequential numbering or alphabetical codes might exert on panelists' judgments. Sensory perception data were primarily captured numerically, based on an ideal reference pattern, a defined scale, or responses to specific questions, as described in reference [38]. Initially, trained tasters quantitatively evaluated five distinct sensory attributes: color, shape, smell, taste (flavor), and texture. Each attribute was rated using a nine-point hedonic scale, with "9" representing "I like it very much" (maximum value) and "1" signifying "I dislike it very much" (minimum value). Upon completion of the individual attribute evaluations, the panelists provided an overall assessment of general acceptability, integrating their perceptions of the aforementioned characteristics.

A subsequent questionnaire quantitatively assessed specific taste characteristics, including bitterness, sweetness, pungency (hotness), acidity, and texture (further detailed as crispness, stringiness, or fibrousness). These attributes were rated on a unipolar scale of ten values, ranging from "0" (minimum value) to "9" (maximum value). Assessors were also provided an open-ended section at the end of the questionnaire to record additional comments and relevant specific observations. The sensory evaluation sessions were conducted between 12:30 p.m. and 14:00 p.m. in a controlled environment. The tasting area was quiet, well-lit, and free from extraneous odors to prevent interference with the assessment process. Each panelist performed their evaluations individually, with no time constraints imposed. To ensure palate cleansing between samples, filtered water was provided to each assessor.

2.7. Statistical Analysis

The statistical analysis of the data was performed using IBM SPSS Advanced Statistics (version 25.0). Depending on the data distribution, either an Analysis of Variance (one-way ANOVA) or the Kruskal–Wallis test was applied.

3. Results and Discussion

3.1. Germination and Cultivation

The germination rates of wild species are generally lower than those of domesticated leafy vegetables [39,40]. Moreover, differences in germination percentages have been reported between direct sowing and pre-sowing on Petri dishes for wild species [33]. Therefore, assessing seed germination viability is essential for determining the most effective propagation strategy. Marked differences in germination performance were observed between controlled conditions in Petri dishes and open field conditions. In Petri dishes, germination was relatively high for all three species after 15 days, with values of 71.0 ± 11.4% for *S. laciniata*, 79.7 ± 17.6% for *S. verrucosa*, and 91.8 ± 7.3% for *E. vesicaria*. In contrast, field germination was lower, particularly for *S. verrucosa* (6.7 ± 2.9%) as compared to *E. vesicaria* (19.5 ± 8.9%) and *S. laciniata* (26.3 ± 7.2%). Based on the obtained results, and to ensure maximum uniformity in the cultivation trial, seeds were initially germinated in Petri dishes and then transferred to seedbeds.

3.2. Soluble Sugars

The concentrations of the three soluble sugars varied among the three studied species. In *E. vesicaria*, the sugars were present in decreasing order as glucose, sucrose, and fructose. In *L. laciniata*, the order was glucose, fructose, and sucrose, whereas in *S. verrucosa*, glucose was the most abundant, followed by sucrose and fructose (Table 1). This composition differs from the patterns observed in other *Sanguisorba* species. For instance, in *S. minor*, Karkanis et al. [41] reported fructose and glucose as the dominant sugars, while Ceccanti et al. [30] found sucrose to be most abundant, followed by glucose and fructose. These differences may be attributed to interspecific variation as well as environmental, genetic, or methodological factors influencing carbohydrate metabolism.

Among the species analysed, *S. verrucosa* showed the highest concentrations of all three sugars, with particularly elevated levels of glucose and sucrose as compared to *L. laciniata* and *E. vesicaria*. Additionally, *S. verrucosa* and *L. laciniata* had a markedly higher fructose content than other wild cultivated species, such as *Sonchus oleraceus* (*S. oleraceus*) and *Sonchus tenerrimus* (*S. tenerrimus*) [33]. When compared with commonly cultivated leafy vegetables, the studied species generally exhibited higher levels of soluble sugars.

Glucose content in *S. verrucosa* was higher than that reported for both lettuce and lamb's lettuce [42,43]. Similarly, fructose concentrations in *S. verrucosa* and *L. laciniata* exceeded those in lamb's lettuce, though they remained below the levels observed in lettuce. All three species showed higher sucrose levels than lamb's lettuce and baby lettuce [43,44]. As compared with lettuce, *S. verrucosa* had higher sucrose concentrations, *L. laciniata* had similar levels, and *E. vesicaria* exhibited lower values [42].

Table 1. Primary and secondary metabolite concentrations (μg gFW-1) in cultivated *S. verrucosa*, *E. vesicaria*, and *S. laciniata*. Values are mean ± SD (n = 4).

Metabolite	***S. verrucosa***	***E. vesicaria***	***S. laciniata***
Glucose	5275.0 ± 303.7	1877.3 ± 182.5	2527.3 ± 630.7
Sucrose	4699.7 ± 197.6	560.0 ± 29.2	1953.2 ± 217.8
Fructose	3837.1 ± 384.7	406.0 ± 21.1	2324.0 ± 541.5
Citric	762.4 ± 1.3	1235.9 ± 56.9	528.4 ± 80.7
Malic	445.4 ± 8.1	428.6 ± 18.6	1759.5 ± 72.3
Tartaric	2.3 ± 0.3	5.3 ± 0.2	7.8 ± 0.9
Fumaric	802.1 ± 38.7	54.2 ± 1.6	196.8 ± 34.8
Succinic	60.0 ± 4.6	30.8 ± 2.7	79.4 ± 9.0
Quinic	13.9 ± 0.7	9.6 ± 0.5	285.0 ± 5.3
Malonic	12.5 ± 1.3	9.9 ± 0.4	6.5 ± 1.0
Isocitric	21.2 ± 0.7	50.7 ± 3.1	11.7 ± 3.3
Ketoglutaric	2.8 ± 0.2	9.1 ± 0.6	16.7 ± 1.7
Glutamic	7.1 ± 0.3	7.4 ± 0.4	4.2 ± 0.4
Skimic	6.5 ± 0.0	0.9 ± 0.0	3.0 ± 0.3
Chlorophyll a	64.3 ± 1.2	96.4 ± 3.4	108.4 ± 3.4
Chlorophyll b	31.8 ± 0.9	45.4 ± 1.0	45.0 ± 1.5
All-trans-β-carotene	30.0 ± 1.0	16.0 ± 0.4	36.6 ± 0.0
Lutein	10.7 ± 0.3	13.0 ± 0.3	13.8 ± 0.2
All-trans violaxanthin	2.9 ± 0.0	3.6 ± 0.1	5.5 ± 0.2
9-cis-Neoxanthin	1.8 ± 0.1	2.6 ± 0.1	2.3 ± 0.0
9-cis-β-carotene	2.9 ± 0.1	18.3 ± 0.0	3.5 ± 0.0
Luteoxanthin	3.1 ± 0.2	6.4 ± 0.1	2.6 ± 0.1
13-cis-β-carotene	1.0 ± 0.1	1.7 ± 0.0	1.3 ± 0.0
Vitamin C	933.4 ± 204.3	1012.9 ± 128.5	530.8 ± 99.3
Total phenolics	3098.3 ± 93.6	246.7 ± 19.2	286.6 ± 11.8

3.3. *Organic Acids*

In the present study, the predominant organic acids identified across all three species were citric, malic, and fumaric acids (Table 1). However, the composition and concentration of the organic acids varied significantly among the species. In *S. verrucosa*, fumaric acid was the predominant organic acid, with levels exceeding those in *E. vesicaria* and *S. laciniata* by 15-fold and 4-fold, respectively. When compared to other wild species within the same genus, distinct profiles emerged. For instance, in *S. minor*, only a minor concentration of fumaric acid [30] or even trace amounts [8,41] were detected as compared to other organic acids. Regarding other cultivated wild and commonly consumed leafy vegetables, such as lettuce and lamb's lettuce, fumaric acid was found in higher concentrations in *S. verrucosa* [33,42,43]. The elevated fumaric acid content in *S. verrucosa* underscores the potential industrial relevance of this species, as fumaric acid is widely used as a food acidulant in beverages and baking powders [45]. Additionally, fumaric acid serves as a pharmaceutically active compound in the treatment of psoriasis and multiple sclerosis [46], highlighting its dual importance in both the food and pharmaceutical industries.

E. vesicaria was notable for its high content of citric acid and its positional isomer, isocitric acid (Table 1). This species also exhibited higher citric acid levels as compared to other cultivated wild species, such as *S. oleraceus* and *S. tenerrimus* [33] and similar concentration to lamb's lettuce [43]. Citric acid has been shown to protect both the liver and brain from oxidative damage and to support tissue regeneration, owing to its antioxidant properties [47]. Additionally, it contributes to reducing urinary acidity, alleviating stress, and slowing age-related processes. In the food industry, citric acid also plays a role in inhibiting lipid oxidation, thereby extending the shelf life of food products [48].

In *S. laciniata*, malic acid was the most abundant organic acid, with concentrations approximately four times higher than those observed in *S. verrucosa* and *E. vesicaria* (Table 1). This species also presented slightly higher malic acid levels as compared to other wild edible species, such as *S. oleraceus* and *S. tenerrimus* [33], but lower than commonly consumed vegetables such as lettuce and lamb's lettuce [42,43]. Malic acid has been reported to enhance the perception of sweetness when combined with sucrose [49]. In addition to malic acid, *S. laciniata* was distinguished by its elevated contents of quinic, fumaric, succinic, and ketoglutaric acids. Quinic acid, in particular, was found at levels 21 and 30 times higher than in *S. verrucosa* and *E. vesicaria*, respectively, while ketoglutaric acid concentrations were six and two times higher than in the same species. These results suggest that *S. laciniata* may serve as a valuable natural source of quinic acid, a compound associated with multiple biological activities, including antioxidant, antimicrobial, anticancer, antiviral, antidiabetic, antiageing, antinociceptive, and analgesic properties [50]. Finally, *S. laciniata* and *S. verrucosa* may serve as valuable sources of succinic acid, with *S. verrucosa* also being a notable source of fumaric acid when compared to commonly cultivated vegetables such as lettuce [33,42,43].

3.4. Carotenoids

Among the three species analyzed, *S. laciniata* showed the highest total carotenoid content (65.6 $\mu g\ g^{-1}$), followed by *E. vesicaria* (61.6 $\mu g\ g^{-1}$) and *S. verrucosa* (52.4 $\mu g\ g^{-1}$). The values observed were lower than those reported for other cultivated wild species, including *S. tenerrimus* and *S. oleraceus* [33], as well as *S. minor* [32]. In the case of lettuce, Kim et al. [51] documented comparable or higher concentrations, depending on the cultivar. Furthermore, in comparison to lamb's lettuce, the carotenoid content in the studied species was consistently lower across all cultivation conditions [35].

Regarding the carotenoid profile, carotenes represented a substantial proportion of the total carotenoid content, ranging from 58% to 65% in the studied species (Table 1). Whereas in *S. laciniata* and *S. verrucosa*, all-trans-β-carotene was identified as the predominant carotene, *E. vesicaria* exhibited higher levels of 9-cis-β-carotene, surpassing the concentration of the all-trans isomer. Interestingly, previous studies on *Eruca* sativa have identified zeaxanthin as the principal carotenoid, followed by *β*-carotene, cryptoxanthin, and lutein [52]. This discrepancy with our results may be attributable to interspecific variation within the genus *Eruca*, differences in environmental conditions, or methodological factors such as extraction and quantification techniques. Although total *β*-carotene concentrations in the three species were lower than those reported for vegetables such as chicory, dandelion, and both cultivated and wild forms of rocket [53], they may still be regarded as relatively good sources of this provitamin A compound, with levels generally exceeding those typically found in lettuce [51]. *β*-carotene is the major carotenoid present in the human diet [54], being the most potent precursor to vitamin A, and is naturally found as a mixture of its various cis and trans isomers. It has a potent antioxidant capacity and can function as a lipid scavenger and a singlet oxygen quencher [55]. Moreover, *β*-carotene has numerous associated health benefits, such as reducing the risk of cardiovascular diseases and specific cancers, enhancing the immune system, and offering protection against age-related macular degeneration, a primary cause of irreversible blindness in adults [56].

In vegetables, this carotenoid is predominantly found in the all-trans configuration, which can be converted into cis isomers during processing and intestinal digestion [57,58]. These cis forms differ in their biological functions, generally exhibiting a lower provitamin A activity and a reduced antioxidant capacity. Additionally, evidence from both in vitro and in vivo studies indicates that the all-trans isomer of β-carotene is more readily absorbed and utilized by the body than its cis counterparts [59,60]. According to our results, *S. verrucosa* and *S. laciniata* may represent more effective dietary sources of β-carotene than *E. vesicaria*, due to the greater proportion of the all-trans isomer present in these species, even though total β-carotene concentrations were similar.

Among the xanthophylls, lutein was the most abundant in all three species, constituting over 50% of the total xanthophyll fraction. This observation aligns with established findings that identify lutein as the predominant xanthophyll in the photosynthetic apparatus of higher plants [61]. Lutein and zeaxanthin are typically abundant in leafy vegetables such as kale, spinach, broccoli, and lettuce [62]. However, in the present analysis, zeaxanthin levels were undetectable in all species evaluated. The highest lutein concentration was observed in *S. laciniata* and *E. vesicaria*, and the lowest in *S. verrucosa*, being in all cases lower than those reported for certain cultivated *Sonchus* species [33] and for green vegetables such as parsley, spinach, and kale [63]. Luteoxanthin was the second most abundant xanthophyll, representing 11–25% of total xanthophylls. The highest concentration of luteoxanthin was observed in *E. vesicaria*, followed by *S. verrucosa* and *S. laciniata*, with all values falling within the range previously reported for *S. oleraceus* and *S. tenerrimus* [33]. Luteoxanthin has been reported to possess biological activities of pharmacological interest, including antimicrobial effects against *Helicobacter pylori* and anticancer properties. More recently, it has been shown to have an inhibitory activity against the angiotensin-converting enzyme 2 (ACE2) receptor, which is utilized by SARS-CoV-2 for cellular entry, suggesting a potential role in the development of antiviral agents [64–66]. According to previous findings, which identified lutein, β-carotene, violaxanthin, and neoxanthin as the predominant carotenoids present in the chloroplasts of terrestrial plants [67,68], violaxanthin and neoxanthin were the third and fourth most abundant xanthophylls in the species analysed. The highest concentrations of violaxanthin were observed in *S. laciniata*, with neoxanthin levels comparable to those found in *S. verrucosa* and *E. vesicaria*, and other cultivated and wild leafy vegetables [33,53]. Similar to the other carotenoids, neoxanthin and violaxanthin are recognized for their strong antioxidant potential [69].

3.5. Chlorophylls

Chlorophyll concentrations in all three species (Table 1) fell within the range previously reported for *S. minor* [32] and were slightly lower than those observed in cultivated wild species such as *S. oleraceus* and *S. tenerrimus* [33]. These values were considerably lower than those reported for lettuce [42], as well as for other cultivated leafy vegetables, including garden rocket and wild rocket (*Diplotaxis* spp.) [53], but were comparable to or higher than those found in lamb's lettuce [35]. Although less extensively studied than other plant-derived phytochemicals, chlorophylls and their derivatives have been suggested to exhibit various therapeutic properties. Notably, while the pharmacological effects of medicinal plants are typically attributed to their secondary metabolites (e.g., alkaloids, flavonoids, terpenes) [70,71], emerging evidence indicates that photosynthetic pigments, particularly chlorophylls, may also contribute significantly to their health-promoting properties [72,73]. In particular, these pigments may function as antioxidants, antimutagenic agents, and anticancer agents [74].

3.6. Vitamin C

The vitamin C content varied markedly among the species studied (Table 1). *S. verrucosa* and *E. vesicaria* were particularly rich in vitamin C, with levels sufficient to meet the recommended daily intake (RDI) for adult women (75 mg) with the consumption of just 80 g of *S. verrucosa* or 74 g of *E. vesicaria*. These findings indicate that both species may serve as effective natural sources of dietary vitamin C. When compared to spinach, which has been reported to contain up to 1500 μg g^{-1} of vitamin C in certain studies [75,76], *E. vesicaria* and *S. verrucosa* supplied approximately 64% and 59% of that amount, respectively. Additionally, all three species examined exhibited significantly higher vitamin C concentrations than those reported for lamb's lettuce grown under varying fertilisation conditions [35]. Relative to lettuce, *S. verrucosa* and *E. vesicaria* contained approximately five times more vitamin C, while *S. laciniata* contained around three times more [42]. Moreover, in comparison to other cultivated wild leafy vegetables, *E. vesicaria* and *S. verrucosa* showed vitamin C levels approximately twice as high as those found in *S. oleraceus* and *S. tenerrimus* [33], further highlighting their nutritional potential. These results are particularly relevant given the essential role of vitamin C in human health. As a key antioxidant, vitamin C supports numerous metabolic functions [77] and plays a central role in immune system function [78]. Since humans are incapable of synthesizing vitamin C endogenously, a regular intake from dietary sources is necessary to maintain adequate physiological levels [79]. In this context, the high concentrations found in *S. verrucosa* and *E. vesicaria* underscore their value as important contributors to dietary vitamin C intake.

3.7. Total Phenolic Content

The total phenolic content in the three species analyzed was notably high, with *S. verrucosa* exhibiting a particularly elevated concentration (Table 1). Leafy vegetables are generally rich in phenolic compounds, which contribute significantly to their antioxidant properties. *S. verrucosa* stands out as an excellent source of phenolic compounds, with a total phenolic content comparable to that reported for curry leaves (*Murraya koenigii*) (3469–5085 μg/100 g FW), which is considered among the vegetables with the highest phenolic content [80]. The content of *S. verrucosa* is generally much higher than the values reported for spinach, a leafy vegetable that is considered to be rich in phenolic compounds, with values ranging from 800 to 2088 μg/100 g FW [81–85]. Other vegetables, such as Chinese cabbage, broccoli, and lettuce, also show a high phenolic content, often above 1000 μg/g FW [86,87], values which were always lower than in *S. verrucosa*. When compared with cultivated *S. tenerrimus* and *S. oleraceus*, *S. verrucosa* showed markedly higher phenolic concentrations (up to 16 times more) [33]. This remarkable phenolic concentration is particularly relevant given the importance of phenolic compounds in *Sanguisorba* species due to their bioactive properties [41]. From a nutritional perspective, the inclusion of these wild species in the human diet could be highly beneficial.

3.8. Sensory Attributes

Median values were used to summarize the sensory evaluation parameters, as they provide a more reliable measure of central tendency in the presence of asymmetrical distributions or outliers. The qualitative satisfaction evaluation of the cultivated species showed a relatively high general acceptance in most parameters (Figure 3). In terms of color, all three species showed similar interquartile ranges, suggesting a generally consistent perception. However, *S. laciniata* had a slightly lower median, and *E. vesicaria* showed a greater variability, indicating a less favorable and more inconsistent evaluation, respectively. For shape, *S. verrucosa* and *E. vesicaria* received high median scores (7–8), suggesting a positive perception of leaf form. In contrast, *S. laciniata* had a lower median (5), indicating a less favorable assessment. Odor was the lowest-rated attribute overall, with a median score of 5 across all species. In *E. vesicaria*, despite a limited central dispersion, there was notable variability with extreme values, reflecting inconsistent evaluations. Regarding taste, *S. verrucosa* and *S. laciniata* received relatively high medians (7 and 6), suggesting good acceptance. *E. vesicaria* had a lower median (4), indicating a less favorable taste. Texture was generally rated positively, with medians of 7 for *S. verrucosa* and *E. vesicaria*, and 6 for *S. laciniata*. The overall scores varied: *S. verrucosa* received the highest median (7), followed by *S. lacinata* (5) and *E. vesicaria* (4). *E. vesicaria* also exhibited the greatest variability and most outliers. In summary, *S. verrucosa* was rated most favorably across attributes, with higher median scores and more consistent evaluations, indicating a strong overall preference among panelists.

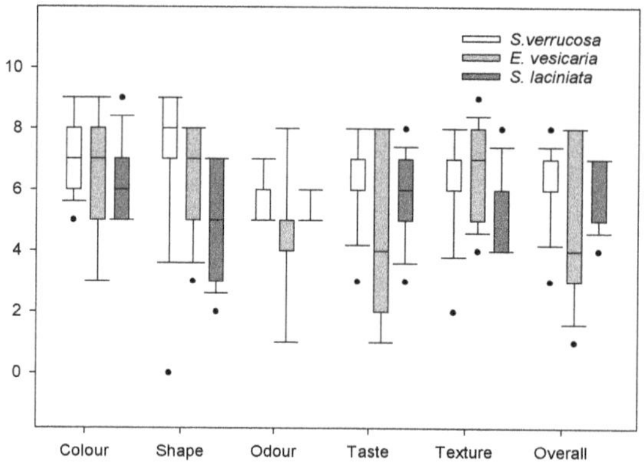

Figure 3. Box plot of the qualitative satisfaction evaluation for cultivated *S. verrucosa*, *E. vesicaria*, and *S. laciniata*. The boxes represent the interquartile range, the horizontal lines within the boxes indicate the median values, and the whiskers show the range of non-outlier data. Individual points beyond the whiskers denote outliers.

The sensory evaluation revealed notable differences among the three species across all attributes (Figure 4). *E. vesicaria* stood out with the highest median bitterness and the widest range, whereas *S. verrucosa* and *S. laciniata* were markedly less bitter, with median values of 2 and 1, respectively, and no values exceeding 5. For crunchiness, *E. vesicaria* also had the highest median (7), indicating that it was perceived as the crunchiest, although its lower whisker extended considerably, and thereby suggesting variability. The other two species had moderate median values (4) and broader distributions. All species showed a similar median in fibrosity (4–5), indicative of a moderate fibrous texture, with *E. vesicaria* showing the greatest variability within the interquartile range. Sweetness was generally low across species; *E. vesicaria* had the lowest median (2), while *S. verrucosa* and *S. laciniata* had higher, though still modest, median values (4). *E. vesicaria* also exhibited a much higher pungency than the other two species, which had very low median values (0–1) and narrow ranges, indicating a general lack of pungency. Lastly, acidity was perceived as low in all species (medians 1–2), although *E. vesicaria* and *S. laciniata* showed broader ranges, suggesting occasional samples with perceptible acidity, unlike the more uniformly mild *S. verrucosa*. The results highlight distinct sensory differences among the three species evaluated. *E. vesicaria* was characterized by higher levels of bitterness, crunchiness, and pungency, as well as a greater variability in several attributes. In contrast, *S. verrucosa* and *S. laciniata* showed milder profiles, with consistently lower scores for bitterness and pungency and slightly higher sweetness. Overall, these differences suggest that each species offers a unique sensory profile, which may inform on their potential applications in culinary use or selection in breeding programs aimed at specific flavor qualities.

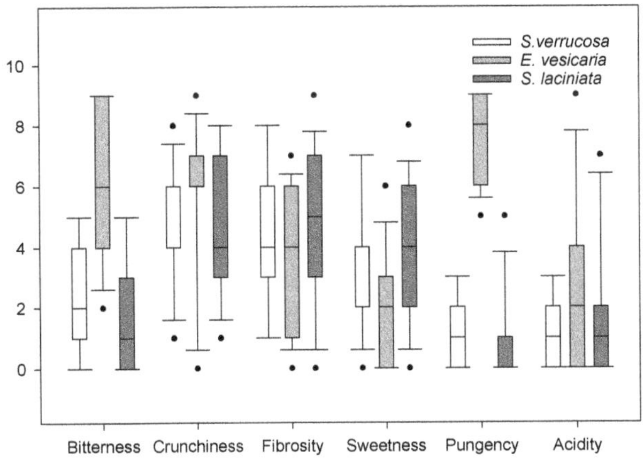

Figure 4. Box plot of the organoleptic descriptive tests for cultivated *S. verrucosa*, *E. vesicaria*, and *S. laciniata*. The boxes represent the interquartile range, the horizontal lines within the boxes indicate the median values, and the whiskers show the range of non-outlier data. Individual points beyond the whiskers denote outliers.

3.9. Wild-Gathered vs. Cultivated Plants

The changes in the metabolite profiles associated with the organoleptic and functional quality following cultivation are shown in Figure 5, expressed as percentages relative to their levels in wild-collected plants. Supplementary Table S1 provides the detailed compositions of primary and secondary metabolites in wild specimens. The domestication of the three studied species resulted in a consistent decline across all analyzed metabolites. Concerning specific metabolite families, *E. vesicaria* exhibited the greatest reductions in total sugar and organic acid contents after cultivation. The most substantial decrease in carotenoids was observed in *S. verrucosa*, while the largest reductions in chlorophyll concentrations occurred in both *S. verrucosa* and *E. vesicaria*. The standard deviations associated with vitamin C levels reflect the high variability among wild populations from different geographical locations. Total phenolic compounds decreased by approximately 60% across all species following domestication.

Numerous studies have consistently demonstrated that wild edible plants tend to contain higher concentrations of phytochemicals as compared to their domesticated counterparts [88,89]. In the case of *Sanguisorba minor*, domestication has not been shown to significantly alter free sugar content, although wild populations exhibited slightly higher levels of fructose and glucose [30]. Conversely, an increase in organic acid concentrations has been observed in cultivated plants [30,41]. Regarding pigment content, in a previous study evaluating the effect of open-field, pot, and soilless cultivation as compared to wild plants, carotenoid and chlorophyll contents showed strong species- and system-dependent variation, with open-field cultivation generally resulting in higher carotenoid levels than other cultivation methods [90]. A general decline in total phenolic compounds following domestication has been reported in several species, including *S. minor* [90], *Portulaca oleracea*, and *Porophyllum ruderale* [91]. However, it has also been shown that the phenolic content, and consequently the antimicrobial and antitumor potential, of *S. minor* can be enhanced under specific substrate and fertilization regimes [41]. Closely related species, such as *S. oleraceus* and *S. tenerrimus*, exhibit distinct responses to cultivation. Notably, *S. tenerrimus* showed significant increases in total soluble sugars, organic acids, β-carotene, and vitamin

C, whereas *S. oleraceus* experienced a decrease in phenolic content [33]. These interspecific differences are likely attributable to both species-specific physiological responses and the substantial influence of environmental factors. Abiotic conditions, including light intensity, temperature, water availability, and soil nutrient levels, significantly affect the biochemical composition of both wild and cultivated plants [92,93]. Cultivated plants were grown with organic amendments and regular irrigation, whereas wild plants were collected from habitats mostly characterized by low rainfall and poor soil fertility. Elevated levels of sugars, organic acids, and bioactive compounds in wild populations may reflect adaptive responses to environmental stressors such as drought, nutrient limitations, and poor soil conditions. Under such conditions, wild plants accumulate soluble sugars to maintain osmotic balance and cellular integrity [94,95], while organic acids such as malate, citrate, and succinate contribute to stress tolerance by enhancing resistance to water deficits, metal toxicity, and hypoxic conditions [96]. Additionally, exposure to environmental stress can induce the synthesis of secondary metabolites, particularly phenolic compounds, as part of the plant's defense mechanisms [97]. Similarly, vitamin C functions as an antioxidant, and its production is enhanced under stress to counteract reactive oxygen species [98]. Consequently, wild-collected plants tend to have higher vitamin C levels, reflecting their exposure to greater environmental stress compared to cultivated plants.

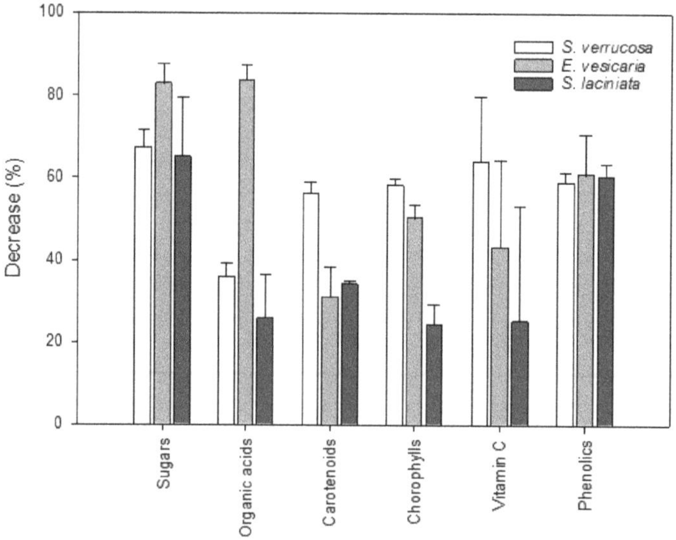

Figure 5. Decreased percentage of metabolite content in cultivated plants as compared to wild plants.

4. Conclusions

The integration of WEPs into both agricultural systems and human diets offers numerous advantages. WEPs are well adapted to Mediterranean climates, and their cultivation contributes to biodiversity conservation efforts by alleviating harvesting pressure on wild populations. From a nutritional perspective, WEPs contribute to dietary diversification and are rich in bioactive compounds, offering potential health benefits while also opening venues for rural development and local economic growth. Although the domestication process led to some decline in phytochemical content in the three species studied, they continue to exhibit a significant nutritional value. After cultivation, all three species retained sugar levels higher than those found in many commonly consumed vegetables and remained valuable sources of specific organic acids: *S. verrucosa* for fumaric acid, *E. vesicaria* for citric acid, and *S. laciniata* for quinic acid. Additionally, moderate levels of β-carotene were observed, exceeding those in lamb's lettuce, and both *S. verrucosa* and *E. vesicaria* contained appreciable amounts of vitamin C. Notably, *S. verrucosa* demonstrated phenolic concentrations surpassing those of most traditional vegetables. These findings indicate that targeted fertilization and cultivation strategies could further enhance the nutritional quality of these species, underscoring their relevance in the development of sustainable and health-oriented food systems.

Supplementary Materials: The following supporting information can be downloaded at: https://www.mdpi.com/article/10.3390/horticulturae11091021/s1, Table S1: Primary and secondary metabolite concentrations (μg gFW^{-1}) in wild-gathered *S. verrucosa, E. vesicaria,* and *S. laciniata.* Values are mean ± SD (n = 5).

Author Contributions: Conceptualization, M.D., P.F., A.R., M.Á.B. and P.H.; methodology, M.D., A.R., P.F. and P.H.; software, P.F. and A.R.; validation, M.Á.B., V.H., A.S. and P.F.; formal analysis, A.R., M.D., P.H., P.F., V.H. and A.S.; investigation, M.D., P.F., P.H., A.R. and A.S.; resources, M.D., P.F. and P.H.; data curation, J.F. and P.F.; writing—original draft preparation, M.Á.B. and P.F.; writing—review and editing, M.Á.B. and P.F.; visualization, M.Á.B., V.H., A.S. and J.F.; project administration, M.D.; funding acquisition, M.D., P.F. and P.H. All authors have read and agreed to the published version of the manuscript.

Funding: This research was co-funded (80%) by the European Union through the European Regional Development Fund (ERDF) (project PO07-050).

Data Availability Statement: The original contributions presented in this study are included in the article and supplementary material. Further inquiries can be directed to the corresponding author.

Acknowledgments: The authors are grateful to Pascual Romero, Juana Cava Artero, Inmaculada Garrido González, María V. Molina Menor, Elia Molina Menor, Carlos Colomer, and Ángeles Jiménez for technical assistance.

Conflicts of Interest: Author Antonio Robledo was employed by the company ISLAYA Environmental Consulting. The remaining authors declare that the research was conducted in the absence of any commercial or financial relationships that could be construed as a potential conflict of interest.

References

1. Bacchetta, L.; Visioli, F.; Cappelli, G.; Caruso, E.; Martin, G.; Nemeth, E.; Bacchetta, G.; Bedini, G.; Wezel, A.; van Asseldonk, T.; et al. A manifesto for the valorization of wild edible plants. *J. Ethnopharmacol.* **2016**, *191*, 180–187. [CrossRef]
2. Sánchez-Mata, M.C.; Tardio, J. (Eds.) *Mediterranean Wild Edible Plants. Ethnobotany and Food Composition Tables*; Springer: New York, NY, USA, 2016. [CrossRef]
3. Nirmala, C.; Shahar, B.; Dolma, N.; Santosh, O. Promising underutilized wild plants of cold desert Ladakh, India for nutritional security and health benefits. *Appl. Food Res.* **2022**, *2*, 100145. [CrossRef]
4. Saha, S.; Saha, S.; Mandal, S.; Rahaman, C. Unco nventional but valuable phytoresources: Exploring the nutritional benefits of 18 wild edible Asteraceae from West Bengal, India. *Genet. Resour. Crop Evol.* **2023**, *70*, 2161–2192. [CrossRef]
5. FAO. Biodiversity and the Ecosystem Approach in Agriculture, Forestry and Fisheries. In Proceedings of the Ninth Regular Session of the Commission on Genetic Resources for Food and Agriculture, Rome, Italy, 12–13 October 2002; Inter-Departmental Working Group on Biological Diversity for Food and Agriculture: Rome, Italy, 2002.
6. Luczaj, L.; Pieroni, A.; Tardío, J.; Pardo-de-Santayana, M.; Soukand, R.; Svanberg, I.; Kalle, R. Wild food plant use in 21st century Europe: The disappearance of old traditions and the search for new cuisines involving wild edibles. *Acta Soc. Bot. Pol.* **2012**, *81*, 359–370. [CrossRef]
7. Schulp, C.; Thuiller, W.; Verburg, P. Wild food in Europe: A synthesis of knowledge and data of terrestrial wild food as an ecosystem service. *Ecol. Econ.* **2014**, *105*, 292–305. [CrossRef]
8. Romojaro, A.; Botella, M.; Obón, C.; Pretel, M. Nutritional and antioxidant properties of wild edible plants and their use as potential ingredients in the modern diet. *Int. J. Food Sci. Nutr.* **2013**, *64*, 944–952. [CrossRef]

9. de Medeiros, P.; Figueiredo, K.; Gonçalves, P.; Caetano, R.; Santos, É.; dos Santos, G.; Barbosa, D.; de Paula, M.; Mapeli, A. Wild plants and the food-medicine continuum-an ethnobotanical survey in Chapada Diamantina (Northeastern Brazil). *J. Ethnobiol. Ethnomed.* **2021**, *17*, 37. [CrossRef] [PubMed]
10. Clemente-Villalba, J.; Burló, F.; Hernández, F.; Carbonell-Barrachina, A. Valorization of wild edible plants as food ingredients and their economic value. *Foods* **2023**, *12*, 1012. [CrossRef] [PubMed]
11. Alu'datt, M.; Rababah, T.; Alhamad, M.; Gammoh, S.; Al-Mahasneh, M.; Tranchant, C.; Rawshdeh, M. Pharmaceutical, nutraceutical and therapeutic properties of selected wild medicinal plants: Thyme, spearmint, and rosemary. In *Therapeutic, Probiotic, and Unconventional Foods*; Mihai, A., Holban, A., Eds.; Academic Press: Cambridge, MA, USA, 2018; pp. 275–290.
12. Fukalova, T.; Martínez, M.; Raigón, M. Five undervalued edible species inherent to autumn-winter season: Nutritional composition, bioactive constituents and volatiles profile. *PeerJ* **2021**, *9*, e12488. [CrossRef] [PubMed]
13. Huyan, T.; Li, Q.; Wang, Y.; Li, J.; Zhang, J.; Liu, Y.; Shahid, M.; Yang, H.; Li, H. Anti-tumor effect of hot aqueous extracts from *Sonchus oleraceus* (L.) L. and *Juniperus sabina* L—Two traditional medicinal plants in China. *J. Ethnopharmacol.* **2016**, *185*, 289–299. [CrossRef]
14. Ajayi, A.F.; Akhigbe, R.E.; Adewumi, O.M.; Okeleji, L.O.; Mujaidu, K.B.; Olaleye, S.B. Effect of ethanolic extract of *Cryptolepis sanguinolenta* stem on in vivo and in vitro glucose absorption and transport: Mechanism of its antidiabetic activity. *Indian J. Endocrinol. Metab.* **2012**, *16*, S91–S96. [CrossRef] [PubMed]
15. Łuczaj, Ł.; Pieroni, A. Nutritional ethnobotany in Europe: From emergency foods to healthy folk cuisines and contemporary foraging trends. In *Mediterranean Wild Edible Plants: Ethnobotany and Food Composition Tables*; Sánchez-Mata, M.C., Tardío, J., Eds.; Springer: New York, NY, USA, 2016; pp. 75–90. [CrossRef]
16. Pereira, A.; Fraga-Corral, M.; García-Oliveira, P.; Jimenez-Lopez, C.; Lourenço-Lopes, C.; Carpena, M.; Otero, P.; Gullón, P.; Prieto, M.; Simal-Gandara, J. Culinary and nutritional value of edible wild plants from northern Spain rich in phenolic compounds with potential health benefits. *Food Funct.* **2020**, *11*, 8493–8515. [CrossRef]
17. Zhou, P.; Li, J.; Chen, Q.; Wang, L.; Yang, J.; Wu, A.; Jiang, N.; Liu, Y.; Chen, J.; Zou, W.; et al. A comprehensive review of genus sanguisorba: Traditional uses, chemical constituents and medical applications. *Front. Pharmacol.* **2021**, *12*, 750165. [CrossRef] [PubMed]
18. Tocai, A.; Kokeric, T.; Tripon, S.; Barbu-Tudoran, L.; Barjaktarevic, A.; Cupara, S.; Vicas, S. *Sanguisorba minor* Scop.: An overview of its phytochemistry and biological effects. *Plants* **2023**, *12*, 2128. [CrossRef]
19. Tardio, J.; Pardo-De-Santayana, M.; Morales, R. Ethnobotanical review of wild edible plants in Spain. *Bot. J. Linn. Soc.* **2006**, *152*, 27–71. [CrossRef]
20. Rivera, D.; Alcaraz, F.J.; Verde, A.; Fajardo, J.; Obón, C. *Las Plantas en la Cultura Popular. Enciclopedia Divulgativa de la Historia Natural de Jumilla-Yecla*; Fundación Cajamurcia: Murcia, Spain, 2008.
21. Schaffer, S.; Schmitt-Schillig, S.; Muller, W.E.; Eckert, G.P. Antioxidant properties of Mediterranean food plant extracts: Geographical differences. *J. Physiol. Pharmacol.* **2005**, *56* (Suppl. S1), 115–124.
22. Bogani, P.; Visioli, F. Antioxidants in the Mediterranean diets: An update. *World Rev. Nutr. Diet.* **2007**, *97*, 162–179. [CrossRef]
23. Perry, L.M.; Metzger, J. *Medicinal Plants of East and Southeast Asia: Attributed Properties and Uses*; MIT Press: Cambridge, MA, USA, 1978.
24. Yaniv, Z.; Schafferman, D.; Amar, Z. Tradition, uses and biodiversity of rocket (*Eruca sativa, Brassicaceae*) in Israel. *Econ. Bot.* **1998**, *52*, 394–400. [CrossRef]
25. Zidorn, C.; Ellmerer-Müller, E.; Stuppner, H. Sesquiterpenoids from *Scorzonera hispanica* L. *Pharmazie* **2000**, *55*, 550–551.
26. Gong, Y.; Shi, Z.; Yu, J.; He, X.; Meng, X.; Wu, Q.; Zhu, Y. The genus *Scorzonera* L. (Asteraceae): A comprehensive review on traditional uses, phytochemistry, pharmacology, toxicology, chemotaxonomy, and other applications. *J. Ethnopharmacol.* **2024**, *320*, 116787. [CrossRef] [PubMed]
27. Erden, Y.; Kırbağ, S. Chemical and biological activities of some *Scorzonera* species: An in vitro study. *Proc. Natl. Acad. Sci. India Sect. B Biol. Sci.* **2015**, *85*, 319–326. [CrossRef]
28. Puccinelli, M.; Carmassi, G.; Pardossi, A.; Incrocci, L. Wild edible plant species grown hydroponically with crop drainage water in a Mediterranean climate: Crop yield, leaf quality, and use of water and nutrients. *Agric. Water Manag.* **2023**, *282*, 108275. [CrossRef]
29. Misra, S. Survey of edible plants for human consumption in South Odisha, India. *Int. J. Emerg. Technol. Innov. Res.* **2020**, *7*, 277–296. Available online: http://www.jetir.org/papers/JETIR2012040.pdf (accessed on 25 June 2025).
30. Ceccanti, C.; Finimundy, T.; Barros, L. Nutritional value of wild and domesticated *Sanguisorba minor* Scop. *Plant. Hortic.* **2023**, *9*, 560. [CrossRef]
31. Chrysargyris, A.; Baldi, A.; Lenzi, A.; Bulgari, R. Wild plant species as potential horticultural crops: An opportunity for farmers and consumers. *Horticulturae* **2023**, *9*, 1193. [CrossRef]
32. Lenzi, A.; Orlandini, A.; Bulgari, R.; Ferrante, A.; Bruschi, P. Antioxidant and mineral composition of three wild leafy species: A comparison between microgreens and baby greens. *Foods* **2019**, *8*, 487. [CrossRef]

33. Botella, M.; Hellín, P.; Hernández, V.; Dabauza, M.; Robledo, A.; Sánchez, A.; Fenoll, J.; Flores, P. Chemical composition of wild collected and cultivated edible plants (*Sonchus oleraceus* L. and *Sonchus tenerrimus* L.). *Plants* **2024**, *13*, 269. [CrossRef] [PubMed]
34. López, A.; Javier, G.; Fenoll, J.; Hellín, P.; Flores, P. Chemical composition and antioxidant capacity of lettuce: Comparative study of regular-sized (Romaine) and baby-sized (Little Gem and Mini Romaine) types. *J. Food Compos.* **2014**, *33*, 39–48. [CrossRef]
35. Hernández, V.; Botella, M.; Hellín, P.; Cava, J.; Fenoll, J.; Mestre, T.; Martínez, V.; Flores, P. Phenolic and carotenoid profile of lamb's lettuce and improvement of the bioactive content by preharvest conditions. *Foods* **2021**, *10*, 188. [CrossRef]
36. Flores, P.; Hellín, P.; Fenoll, J. Determination of organic acids in fruits and vegetables by liquid chromatography with tandem-mass spectrometry. *Food Chem.* **2012**, *132*, 1049–1054. [CrossRef]
37. Fenoll, J.; Martinez, A.; Hellín, P.; Flores, P. Simultaneous determination of ascorbic and dehydroascorbic acids in vegetables and fruits by liquid chromatography with tandem-mass spectrometry. *Food Chem.* **2011**, *127*, 340–344. [CrossRef]
38. Sancho, J.; Bota, E.; Castro, J. *Introducción al Análisis Sensorial de los Alimentos*; Editorial Alfaomega: Barcelona, Spain, 1999.
39. Ceccanti, C.; Landi, M.; Benvenuti, S.; Pardossi, A.; Guidi, L. Mediterranean wild edible plants: Weeds or "new functional crops"? *Molecules* **2018**, *23*, 2299. [CrossRef] [PubMed]
40. Maleki, K.; Soltani, E.; Seal, C.E.; Pritchard, H.W.; Lamichhane, J.R. The seed germination spectrum of 528 plant species: A global meta-regression in relation to temperature and water potential. *bioRxiv* **2022**. [CrossRef]
41. Karkanis, A.; Fernandes, Å.; Vaz, J.; Petropoulos, S.; Georgiou, E.; Ciric, A.; Sokovic, M.; Oludemi, T.; Barros, L.; Ferreira, I. Chemical composition and bioactive properties of *Sanguisorba minor* Scop. under Mediterranean growing conditions. *Food Funct.* **2019**, *10*, 1340–1351. [CrossRef] [PubMed]
42. Flores, P.; López, A.; Fenoll, J.; Hellín, P.; Kelly, S. Classification of organic and conventional sweet peppers and lettuce using a combination of isotopic and bio-markers with multivariate analysis. *J. Food Compos. Anal.* **2013**, *31*, 217–225. [CrossRef]
43. Schmitzer, V.; Senica, M.; Slatnar, A.; Stampar, F.; Jakopic, J. Changes in metabolite patterns during refrigerated storage of lamb's lettuce (*Valerianella locusta* L. Betcke). *Front. Nutr.* **2021**, *8*, 731869. [CrossRef]
44. Spinardi, A.; Ferrante, A. Effect of storage temperature on quality changes of minimally processed baby lettuce. *J. Food Agric. Environ.* **2012**, *10*, 38–42.
45. Yadav, P.; Chauhan, A.K.; Singh, R.B.; Khan, S.; Halabi, G. Organic acids: Microbial sources, production, and applications. In *Functional Foods and Nutraceuticals in Metabolic and Non-Communicable Diseases*; Singh, R.B., Watanabe, S., Isaza, A.A., Eds.; Academic Press: Cambridge, MA, USA, 2022; pp. 325–337. [CrossRef]
46. Gold, R.; Linker, R.; Stangel, M. Fumaric acid and its esters: An emerging treatment for multiple sclerosis with antioxidative mechanism of action. *Clin. Immunol.* **2012**, *142*, 44–48. [CrossRef]
47. Singh, S.K.; Kaldate, R.; Bisht, A. Citric acid, antioxidant effects in health. In *Antioxidants Effects in Health*; Nabavi, S.M., Silva, A.S., Eds.; Elsevier: Amsterdam, The Netherlands, 2022; pp. 309–322. [CrossRef]
48. Gordon, M.H. The mechanism of antioxidant action in vitro. In *Food Antioxidants*; Hudson, B.J.F., Ed.; Elsevier: New York, NY, USA, 1990; pp. 1–18.
49. Fabian, F.; Blum, H. Relative taste potency of some basic food constituents and their competitive and compensatory action. *J. Food Sci.* **2006**, *8*, 179–193. [CrossRef]
50. Benali, T.; Bakrim, S.; Ghchime, R.; Benkhaira, N.; El Omari, N.; Balahbib, A.; Taha, D.; Zengin, G.; Hasan, M.; Bibi, S.; et al. Pharmacological insights into the multifaceted biological properties of quinic acid. *Biotechnol. Genet. Eng. Rev.* **2024**, *40*, 3408–3437. [CrossRef]
51. Kim, D.; Shang, X.; Assefa, A.; Keum, Y.; Saini, R. Metabolite profiling of green, green/red, and red lettuce cultivars: Variation in health beneficial compounds and antioxidant potential. *Food Res. Int.* **2018**, *105*, 361–370. [CrossRef]
52. Komeroski, M.; Portal, K.; Comiotto, J.; Klug, T.; Flores, S.; Rios, A. Nutritional quality and bioactive compounds of arugula (*Eruca sativa* L.)sprouts and microgreens. *Int. J. Food Sci. Technol.* **2023**, *58*, 5089–5096. [CrossRef]
53. Znidarcic, D.; Ban, D.; Sircelj, H. Carotenoid and chlorophyll composition of commonly consumed leafy vegetables in Mediterranean countries. *Food Chem.* **2011**, *129*, 1164–1168. [CrossRef]
54. Johnson, E.J. The role of carotenoids in human health. *Nutr. Clin. Care* **2002**, *5*, 56–65. [CrossRef]
55. Grune, T.; Lietz, G.; Palou, A.; Ross, A.; Stahl, W.; Tang, G.; Thurnham, D.; Yin, S.; Biesalski, H. β-Carotene is an important vitamin a source for human. *J. Nutr.* **2010**, *140*, 2268S–2285S. [CrossRef] [PubMed]
56. Gul, K.; Tak, A.; Singh, A.K.; Singh, P.; Yousuf, B.; Wani, A.A. Chemistry, encapsulation, and health benefits of β-carotene—A review. *Cogent Food Agric.* **2015**, *1*, 1018696. [CrossRef]
57. Schieber, A.; Carle, R. Occurrence of carotenoid cis-isomers in food: Technological, analytical, and nutritional implications. *Trends Food Sci. Technol.* **2005**, *16*, 416–422. [CrossRef]
58. Tao, W.; Ye, X.; Cao, Y. Isomerization and degradation of *all-trans-*β-carotene during in-vitro digestion. *Food Sci. Hum. Wellness* **2021**, *10*, 370–374. [CrossRef]
59. Deming, D.; Teixeira, S.; Erdman, J. All-*trans* β-carotene appears to be more bioavailable than 9-*cis* or 13-*cis* β-carotene in gerbils given single oral doses of each isomer. *J. Nutr.* **2002**, *132*, 2700–2708. [CrossRef]

60. Erdman, J.; Thatcher, A.; Hofmann, N.; Lederman, J.; Block, S.; Lee, C.; Mokady, S. All-trans β-carotene is absorbed preferentially to 9-*cis* β-carotene, but the latter accumulates in the tissues of domestic ferrets (*Mustela putorius puro*). *J. Nutr.* **1998**, *128*, 2009–2013. [CrossRef]
61. Dall'Osto, L.; Lico, C.; Alric, J.; Giuliano, G.; Havaux, M.; Bassi, R. Lutein is needed for efficient chlorophyll triplet quenching in the major LHCII antenna complex of higher plants and effective photoprotection in vivo under strong light. *BMC Plant Biol.* **2006**, *6*, 32. [CrossRef]
62. Perry, A.; Rasmussen, H.; Johnson, E.J. Xanthophyll (lutein, zeaxanthin) content in fruits, vegetables and corn and egg products. *J. Food Comp. Anal.* **2009**, *22*, 9–15. [CrossRef]
63. Maiani, G.; Castón, M.; Catasta, G.; Toti, E.; Cambrodon, I.; Bysted, A.; Granado-Lorencio, F.; Olmedilla-Alonso, B.; Knuthsen, P.; Valoti, M.; et al. Carotenoids: Actual knowledge on food sources, intakes, stability and bioavailability and their protective role in humans. *Mol. Nutr. Food Res.* **2009**, *53*, S194–S218. [CrossRef] [PubMed]
64. Molnár, P.; Deli, J.; Tanaka, T.; Kann, Y.; Tani, S.; Gyémánt, N.; Molnár, J.; Kawase, M. Carotenoids with anti-Helicobacter pylori activity from Golden Delicious apple. *Phytother. Res.* **2010**, *24*, 644–648. [CrossRef]
65. Joshi, B.C.; Mukhija, M.; Kalia, A.N. Pharmacognostical review of *Urtica dioica* L. *Int. J. Green Pharm.* **2014**, *8*, 201–209. [CrossRef]
66. Upreti, S.; Prusty, J.; Pandey, S.; Kumar, A.; Samant, M. Identification of novel inhibitors of angiotensin-converting enzyme 2 (ACE-2) receptor from *Urtica dioica* to combat coronavirus disease 2019 (COVID-19). *Mol. Divers.* **2021**, *25*, 1795–1809. [CrossRef]
67. Al-Babili, S.; Bouwmeester, H.J. Strigolactones, a novel carotenoid-derived plant hormone. *Annu. Rev. Plant Biol.* **2015**, *66*, 161–186. [CrossRef] [PubMed]
68. Ruiz-Sola, M.A.; Rodríguez-Concepción, M. Carotenoid biosynthesis in *Arabidopsis*: A colorful pathway. *Arab. Book* **2012**, *10*, e0158. [CrossRef]
69. Rumengan, A.P.; Mandiangan, E.S.; Tanod, W.A. Identification of pigment profiles and antioxidant activity of *Rhizophora mucronata* mangrove leaves originating from Lembeh, North Sulawesi, Indonesia. *Biodiversitas* **2021**, *22*, 2805–2816. [CrossRef]
70. Chassagne, F.; Huang, X.; Lyles, J.; Quave, C. Validation of a 16th century traditional Chinese medicine use of *Ginkgo biloba* as a topical antimicrobial. *Front. Microbiol.* **2019**, *10*, 775. [CrossRef]
71. Ibrahim, M.; Jaafar, H.; Karimi, E.; Ghasemzadeh, A. Primary, secondary metabolites, photosynthetic capacity and antioxidant activity of the Malaysian herb Kacip Fatimah (*Labisia Pumila Benth*) exposed to potassium fertilization under greenhouse conditions. *Int. J. Mol. Sci.* **2012**, *13*, 15321–15342. [CrossRef] [PubMed]
72. Roca, M.; Chen, K.; Pérez-Gálvez, A. 6-Chlorophylls. In *Handbook on Natural Pigments in Food and Beverages*; Carle, R., Schweiggert, R.M., Eds.; Woodhead Publishing: Cambridge, UK, 2016; pp. 125–158.
73. Zepka, L.; Jacob-Lopes, E.; Roca, M. Catabolism and bioactive properties of chlorophylls. *Curr. Opin. Food Sci.* **2019**, *26*, 94–100. [CrossRef]
74. Martins, T.; Barros, A.; Rosa, E.; Antunes, L. Enhancing health benefits through chlorophylls and chlorophyll-rich agro-food: A comprehensive review. *Molecules* **2023**, *28*, 5344. [CrossRef] [PubMed]
75. Sandrika, A.; Munir, M.A.; Aprilia, V.; Emelda, E. Determination of vitamin C in spinach (*Amaranthus* sp.) using titration method. *J. Pena Sains* **2023**, *10*, 73–79. [CrossRef]
76. Ibrahim, I.M.; Oluchukwu, N.V.; Salisu, A.; Nkemakonam, O.M. Nutritional and phytochemical analysis of spinach leaf aqueous extract: A comprehensive study on proximate composition, minerals, vitamins, and antioxidant activity. *Eurasian J. Sci. Technol.* **2025**, *5*, 302–311. [CrossRef]
77. Hasanah, U. Penentuan kadar vitamin menggunakan metode iodometri. *J. Kel. Sehat Sejah.* **2018**, *16*, 90–95. [CrossRef]
78. Evana, E.; Barek, M.S. Determination of vitamin C (ascorbic acid) contents in two varieties of melon fruits (*Cucumis melo* L.) by iodometric titration. *Fuller. J. Chem.* **2021**, *6*, 143–147.
79. Rahayuningsih, J.; Sisca, V.; Angasa, E.E. Analisis vitamin C pada buah jeruk Pasaman untuk meningkatkan kekebalan tubuh pada masa pandemi COVID-19. *J. Rekayasa Chem.* **2022**, *4*, 29–33. [CrossRef]
80. Mohankumar, J.B.; Uthira, L.; Maheswari, S.U. Total phenolic content of organic and conventional green leafy vegetables. *J. Nutr. Hum. Health* **2018**, *2*, 1–6. [CrossRef]
81. Chu, Y.; Sun, J.; Wu, X.; Liu, R. Antioxidant and antiproliferative activities of common vegetables. *J. Agric. Food Chem.* **2002**, *50*, 6910–6916. [CrossRef]
82. Bunea, A.; Andjelkovic, M.; Socaciu, C.; Bobis, O.; Neacsu, M.; Verhe, R.; Van Camp, J. Total and individual carotenoids and phenolic acids content in fresh, refrigerated and processed spinach (*Spinacia oleracea* L.). *Food Chem.* **2008**, *108*, 649–656. [CrossRef] [PubMed]
83. Bayili, R.; Abdoul-Latif, F.; Kone, O.; Diao, M.; Bassole, I.; Dicko, M. Phenolic compounds and antioxidant activities in some fruits and vegetables from Burkina Faso. *Afr. J. Biotechnol.* **2011**, *10*, 13543–13547. [CrossRef]
84. Mazzucotelli, C.; González-Aguilar, G.; Villegas-Ochoa, M.; Domínguez-Avila, A.; Ansorena, M.; Di Scala, K. Chemical characterization and functional properties of selected leafy vegetables for innovative mixed salads. *J. Food Biochem.* **2017**, *42*, e12461. [CrossRef]

85. Machado, R.M.A.; Alves-Pereira, I.; Ferreira, R.M.A. Plant growth, phytochemical accumulation and antioxidant activity of substrate-grown spinach. *Heliyon* **2018**, *4*, e00751. [CrossRef]
86. Bahorun, T.; Luximon-Ramma, A.; Crozier, A.; Aruoma, O. Total phenol, flavonoid, proanthocyanidin and vitamin C levels and antioxidant activities of Mauritian vegetables. *J. Sci. Food Agric.* **2004**, *84*, 1553–1561. [CrossRef]
87. Mattila, P.; Hellström, J. Phenolic acids in potatoes, vegetables, and some of their products. *J. Food Compos. Anal.* **2007**, *20*, 152–160. [CrossRef]
88. Davis, D. Declining fruit and vegetable nutrient composition: What is the evidence? *HortScience* **2009**, *44*, 15–19. [CrossRef]
89. Di Gioia, F.; Avato, P.; Serio, F.; Argentieri, M. Glucosinolate profile of *Eruca sativa*, *Diplotaxis tenuifolia* and *Diplotaxis erucoides* grown in soil and soilless systems. *J. Food Compos. Anal.* **2018**, *69*, 197–204. [CrossRef]
90. Ceccanti, C.; Landi, M.; Incrocci, L.; Pardossi, A.; Venturi, F.; Taglieri, I.; Ferroni, G.; Guidi, L. Comparison of three domestications and wild-harvested plants for nutraceutical properties and sensory profiles in five wild edible herbs: Is. domestication possible? *Foods* **2020**, *9*, 1065. [CrossRef]
91. Fukalova, T.; Garcia-Martinez, M.; Raigon, M. Nutritional composition, bioactive compounds, and volatiles profile characterization of two edible undervalued plants: *Portulaca oleracea* L. and *Porophyllum ruderale* (Jacq.) *Cass. Plants* **2022**, *11*, 377. [CrossRef]
92. Riquelme, J.; Olaeta, J.; Gálvez, L.; Undurraga, P.; Fuentealba, C.; Osses, A.; Orellana, J.; Gallardo, J.; Pedreschi, R. Nutritional and functional characterization of wild and cultivated *Sarcocornia neei* grown in Chile. *Cienc. Investig. Agrar.* **2016**, *43*, 283–293. [CrossRef]
93. Paschoalinotto, B.; Polyzos, N.; Compocholi, M.; Rouphael, Y.; Alexopoulos, A.; Dias, M.; Barros, L.; Petropoulos, S. Domestication of wild edible species: The response of *Scolymus hispanicus* plants to different fertigation regimes. *Horticulturae* **2023**, *9*, 103. [CrossRef]
94. Yamada, K.; Osakabe, Y. Sugar compartmentation as an environmental stress adaptation strategy in plants. *Semin. Cell Dev. Biol.* **2018**, *83*, 106–114. [CrossRef]
95. Gurrieri, L.; Merico, M.; Trost, P.; Forlani, G.; Sparla, F. Impact of drought on soluble sugars and free proline content in selected arabidopsis mutants. *Biology* **2020**, *9*, 367. [CrossRef]
96. de Oliveira, H.; Siqueira, J.; Medeiros, D.; Fernie, A.; Nunes-Nesi, A.; Araújo, W. Harnessing the dynamics of plant organic acids metabolism following abiotic stresses. *Plant Physiol. Biochem.* **2025**, *220*, 109465. [CrossRef] [PubMed]
97. Kumar, K.; Debnath, P.; Singh, S.; Kumar, N. An overview of plant phenolics and their involvement in abiotic stress tolerance. *Stresses* **2023**, *3*, 570–585. [CrossRef]
98. Paciolla, C.; Fortunato, S.; Dipierro, N.; Paradiso, A.; De Leonardis, S.; Mastropasqua, L.; De Pinto, M. Vitamin C in plants: From functions to biofortification. *Antioxidants* **2019**, *8*, 519. [CrossRef] [PubMed]

horticulturae

MDPI

Article

Foliar Application of Iron and Zinc Affected Aromatic Plants Grown Under Conventional and Organic Agriculture Differently

Nikolaos Tzortzakis [1,*], Efraimia Hajisolomou [1], Nikoletta Zaravelli [1,2] and Antonios Chrysargyris [1,*]

1 Department of Agricultural Sciences, Biotechnology and Food Science, Cyprus University of Technology, 3603 Limassol, Cyprus

2 School of Forestry and Natural Environment, Aristotle University of Thessaloniki, 54124 Thessaloniki, Greece

* Correspondence: nikolaos.tzortzakis@cut.ac.cy (N.T.); a.chrysargyris@cut.ac.cy (A.C.)

Abstract

The utilization of organic fertilizers for the cultivation of wild edible and medicinal plants offers agronomic and ecological benefits, given their suitability to low-input and sustainable production systems. Under such conditions, these species may also benefit from targeted foliar applications of micronutrients to enhance their nutritional quality. This study examined the effects of a vinasse-based organic fertilizer and conventional fertilization regime, in combination with foliar applications of iron (Fe) and zinc (Zn), on the biomass, leaf photochemistry, and plant stress-related responses of *Sideritis cypria* and *Origanum dubium*. In *S. cypria*, organic fertilization resulted in a similar yield compared to conventional fertilization, while *O. dubium* showed a significant decrease in yield when using organic fertilizers. The impact of spraying with Zn on *S. cypria* dry matter content was related to the availability of nutrients, particularly nitrogen, while in *O. dubium* Zn spraying induced a decrease in dry matter. The total phenols content and antioxidant activity of *S. cypria* were elevated by conventional fertilization and foliar application of Fe, while the combination of organic fertilization and foliar application of Fe and Zn reduced lipid peroxidation. In *O. dubium*, foliar application of Fe and Zn led to a reduction in total phenols content, antioxidant capacity, and hydrogen peroxide content under adequate nutrition. In general, foliar spraying with Zn tended to improve water use efficiency under specific fertilization practices on both species, while the positive effect of conventional fertilization on nutrient use efficiency still requires further validation. Ultimately, the efficiency of organic fertilization was related to the examined species, inducing variations in leaf chlorophyll content. In addition, foliar application of Fe and Zn affected the antioxidant capacity and mineral content of the examined species. Thus, appropriate fertilization management is vital to fully realize the specific benefits of foliar micronutrient addition.

Keywords: medicinal plants; oregano; *Origanum dubium*; sideritis; *Sideritis cypria*; organic farming; sustainable fertilization; antioxidants; water and nutrient use efficacy

1. Introduction

Today's agriculture faces enormous challenges in ensuring safe and healthy food production in regard to a constantly expanding population [1,2]. In parallel, the agriculture sector is one of the main key players in environmental constraints, natural resource deterioration, and decreased biodiversity of living organisms [3]. In order to preserve the

https://doi.org/10.3390/horticulturae11080967

availability of food, this concern necessitates not only a boost in worldwide crop production but also the implementation of strategies that guarantee food safety and nutritional value [4,5]. Therefore, crop yields may increase under intensive cultivation systems [6], which primarily depend on the use of overfertilization practices, mostly with inorganic fertilizers [7]. The literature has extensively documented the impact of cultivation methods on crops, since choosing the right ones enhances crop yield and quality [8–10]. However, the use of an inappropriate practice can lower crop yields, raise soil mineral content to harmful levels, lose minerals through leaching or mobilization, pollute water supplies, and deteriorate soil quality [11,12]. Farmers frequently choose intensive agricultural schemes, such as traditional crop production with high fertilizer, phytochemical, and water needs that result in large yields but not necessarily high-quality products. Nevertheless, this strategy has a number of negative effects on human health and the environment, such as soil deterioration, nutrient loss through leaching, and a rise in greenhouse gas emissions, as well as ultimately air, water, and soil contamination [4,11].

As part of sustainability, the preservation of biodiversity and natural resources and organic methods restrict the usage of inorganic fertilizers and instead employ environmentally friendly products like organic manure, natural-based products, and soil restoration practices [4,13,14]. A potentially environmentally friendly method of producing crops, organic agriculture strategies can yield high-quality products of nutritional value and qualities equivalent to those produced conventionally [15,16]. Organic farming is an integrated system under ecological and restoration principles. Since wild plant species have innate native adaptations to adverse soil and climate conditions, they have substantial ecological significance in this context, especially when it comes to soil deterioration [17]. Furthermore, organic farming has the ability to retain a smaller environmental impact while producing plant products that are healthier and of greater quality [7,18].

This is especially crucial for medicinal and aromatic plants (MAPs) since their quality and advantageous molecules are what really make them valuable [4,19]. Since these species are defined by minimal input needs (e.g., fertilizers, agrochemicals, watering), the incorporation of MAPs in organic agricultural production falls within its purview [20]. Although MAPs have comparatively poor yields, consumers' interest in organically grown products and herbs has been steadily rising at the same time [16,21]. However, there is not much research that compares the conventional and organic methods of growing MAPs [9,22]. Prior reports indicated that although there were no appreciable changes in overall water usage, organic MAPs cultivation had better energy use and carbon footprints than conventional ones [16].

Awareness in using organic farming methods has grown as a result of the environmental impact of fertilizer and agrochemical inputs used in the production of MAPs, as well as a shortage of approved pesticides and herbicides for different MAPs [23,24]. Since MAPs have minimal nutrient requirements and are resilient to biotic and abiotic stresses, low-input cultivation techniques are appropriate for MAPs production. Aside from the environmental concerns, growing MAPs organically is linked to higher-quality final products than conventional methods, and using fewer chemical pesticides and fertilizers guarantees the safety of the final products because there are fewer heavy metals and chemical residues [22,25]. A further consideration to take into account is that commercial cultivation of MAPs improved access to the market year-round and allowed for the standardization of the final product while adhering to certain safety protocols. Moreover, MAPs cultivation is also lowering the possibility of genetic degradation brought on by irrational wild plant harvesting [24,26].

In agriculture, foliar mineral applications—especially with micronutrients such as zinc (Zn), manganese (Mn), and copper (Cu)—are one method of regulating mineral intake and needs through leaves in parallel with root absorption [27]. Foliar application may help to reduce oxidative stress or nutrient deficits caused by several abiotic and biotic causes [28–30] and act as a biofortification strategy to improve micronutrient uptake in plants [31]. Among the micronutrients, both iron (Fe) and Zn are essential components for plant growth and participate in several metabolic processes either as metal components of various enzymes or as structural, functional, or regulatory cofactors [32,33]. Iron and Zn appear to have a beneficial effect on plant development, as they stimulate chlorophyll content and photosynthetic rates, induce indole-3-acetic acid (IAA) biosynthesis (by Zn), contribute to protein biosynthesis and membrane stability, and enhance oxidation and reduction processes [34]. Previous research examined the impact of foliar administration of micronutrients on MAPs development and their essential oil production, including chamomile (*Matricaria chamomilla* L.) [35], sweet basil (*Ocimum basilicum* L.) [28], coriander (*Coriandrum sativum* L.) [36], and rosemary (*Rosmarinus officinalis* L.) [34].

Research into lesser-known local MAPs and their biological attributes is becoming increasingly popular [37,38]. With over 150 species, the *Sideritis* L. genus (Lamiaceae) is mostly distributed in the Mediterranean region [39]. One of them is the native Cypriot *Sideritis cypria* Post. [39], which has a phytochemical profile rich in secondary metabolites with potent cytotoxic and antioxidant properties linked to apigenin and apigenin derivatives [40] and antimicrobial properties towards *Candida albicans* and Gram-positive bacteria [39,41]. The principal components of the essential oil profile of *S. cypria* were β-phellandrene (25.11%), β-caryophyllene (22.52%), and α-pinene (11.92%) in leaves, which were linked with its biocidal activities [39].

Another important MAP species is known as Cypriot oregano, *Origanum dubium* Boiss., which is a perennial plant of the *Origanum* genus from the Lamiaceae family that thrives across the Mediterranean region and is well appreciated in Cyprus and in the eastern Mediterranean area. The plant favors great antioxidant and antiseptic/antimicrobial properties [42,43], is used against migraines, toothaches, and respiratory problems [44], and has been extensively utilized since ancient times. The principal components of the essential oil profile of *O. dubium* were carvacrol (44.59–68.67%), trans-sabinene hydrate (12.71–23.10%), *p*-cymene (2.97–9.01%), and γ-terpinene (3.53–8.22%) [45] with well-documented biocidal properties of the main components [43].

In the framework of putting sustainable agricultural techniques into reality, it is crucial to guarantee appropriate yields and improved quality variables, given the possibility for *S. cypria* and *O. dubium* to be commercial crop options, with the former being almost unexploited on a commercial scale. Recently their precise mineral needs in hydroponics were documented for the first time, with nitrogen-N, phosphorus-P and potassium-K of 150, 50, and 350 mg L^{-1}, respectively, for *S. cypria* [46] and with N, P, and K of 150, 75, and 350 mg L^{-1}, respectively, for *O. dubium* [45]. Moreover, the phytochemical profile of the selected species was varied, with *S. cypria* posing increased cytotoxic activities against breast cancer cells [47]. However, little knowledge exists on the cultivation practices of the above-mentioned species. Due to the growing demand for certification of the organic products, these approaches will assist in lessening the negative environmental effects of the present agriculture industry [48]. Consequently, the present research investigated the effects of a commonly accessible organic fertilizer on the development and quality characteristics of *S. cypria* and *O. dubium* plants cultivated in field conditions, compared to traditional inorganic fertilization. Furthermore, the impacts of micronutrient foliar application were also examined to further support the plants biomass production, mineral content, and antioxidant capacity.

2. Materials and Methods

2.1. Plant Species, Soil Characteristics, and Experimental Setup

S. cypria and *O. dubium* seedlings were purchased from the Cypriot National Centre of Medicinal/Aromatic Plants, Nicosia, in plastic trays, at the growth stage of 3–4 leaves and 5–6 cm height. Plants were transplanted in soil in a field of 1000 m^2 during spring of 2024 in a commercial organic farm, Anogyra, Limassol, Cyprus (34°44′16.53″ N; 32°44′40.86″ E, 428 m above sea level). The climate of the region during the experimental study had average minimum and maximum temperatures of 13.6 and 33.4 °C, respectively, air humidity ranged from 64 to 73%, and precipitation ranged from 0.5 to 34 mm.

Soil samples, collected from four points, were air-dried in an oven. Particle size distribution was determined by dry sieving with mesh sizes of 4.75, 2.00, 0.850, 0.425, 0.250, 0.150, and 0.075 mm. Additionally, a soil subsample was hand-sieved to pass a 2 mm mesh, and several soil physicochemical properties were determined as described previously [49]. Soil type was determined by the hydrometer method with the Bouyoucos scale. Calcium carbonate ($CaCO_3$) was determined using the Bernard calcimetry method, and results were expressed as a percentage. Organic matter was determined using the Walkley-Black method, and results were expressed as a percentage. Electrical conductivity (EC) and pH were determined using the 1:5 dilution method. Following extraction, P was determined using the Olsen sodium bicarbonate method spectrophotometrically, and K and sodium (Na) content were quantified using flame photometry. Nitrogen was determined by the Kjeldahl method (BUCHI, Digest automat K-439 and Distillation Kjeldahl K-360, Flawil, Switzerland). Magnesium (Mg) and calcium (Ca) were determined by an atomic absorption spectrophotometer (PG Instruments AA500FG, Leicestershire, UK) [9]. Mineral content in soil was expressed as grams per kilogram of dry weight (g kg^{-1} DW).

Plants were arranged in single rows (inter-row of 1.5 m and intra-row of 0.33 m), resulting in a plant density of 20,368 plants ha^{-1}. Plants were grown for about six months in soil. The experimental farm was divided into three main plots with three planting rows each: (i) No fertilizers (NoFert), (ii) Organic fertilization (OrFert), and (iii) Conventional fertilization (CoFert). Each main plot was divided into three subplots, where the foliar applications took place as follows: (A) foliar with water (H_2O), (B) foliar with iron (Fe), and (C) foliar with zinc (Zn). Therefore, the experimental setup had nine treatments, and each treatment had 12 plants in each row (36 plants per treatment) for each of the examined species (324 plants for *S. cypria* and 324 plants for *O. dubium*).

The plants were irrigated every day for 10 min or according to the plant's needs. The application of synthetic fertilizers [20-20-20 and 0-0-22+22S+11Mg (potassium magnesium sulfate-$K_2SO_4 \cdot MgSO_4$)] and the organic fertilizer (Phenix, Hello NATURE) took place three times (24 April 2024, 15 May 2024, and 28 May 2024). The fertilizers were applied on soil and directly incorporated in the soil to diminish any losses. Organic fertilization was carried out using Phenix (Hello Nature, Italy), a vinasse-based fertilizer with 6% N, 8% phosphorus pentoxide (P_2O_5), 15% potassium oxide (K_2O), and 3% magnesium oxide (MgO), and 50% organic matter content. Considering the chemical composition of the tested fertilizers, the amount of organic fertilizers was almost 1.5 times higher compared to the synthetic fertilizers to maintain a balanced mineral application. The cumulative amounts of N, P, K, and Mg were 75.21, 37.61, 138.95, and 22.49 kg ha^{-1}, respectively. The foliar application took place twice with 1.79 mM Fe and 1.74 mM Zn by using iron (II) sulfate ($FeSO_4 \cdot H_2O$) and zinc sulfate ($ZnSO_4 \cdot H_2O$) chemical fertilizers, respectively. No pesticides were used in the present study.

2.2. Plant Biomass and Photochemistry

Various measurements were implemented at the end of the experiment. Twelve replicated plants were harvested at 1 cm above soil, the upper fresh weight was weighed (g), dried at 42 °C to constant weight, and dry weight was recorded [45]. Dry matter content (%) was calculated.

Leaf photochemistry was also assessed by measuring the relative chlorophyll content by employing an optical chlorophyll meter (SPAD-502, Minolta, Osaka, Japan) [49]. Additionally, fresh plant tissue (six replications/treatment; each replication was a pool of two plant tissues; 0.1 g) was used for chlorophyll extraction [50]. The absorbance was then measured at 470, 653, and 666 nm (Multiskan GO, Thermo Fischer Scientific Oy, Vantaa, Finland). Photosynthetic leaf pigments, chlorophyll a (Chl a), chlorophyll b (Chl b), total chlorophyll (total Chl), and carotenoid (Car) content were then calculated (mg g^{-1} fresh weight). The ratio of chlorophyll a to chlorophyll b (Chl a/Chl b) and of carotenoids to total chlorophylls (Car/Total Chls) were also calculated.

2.3. Total Phenols, Total Flavonoids, and Antioxidant Capacity

Total phenols content was measured on the methanolic extracts from the different samples by using the Folin-Ciocalteu method; the samples were optometrically measured at 755 nm, and results were expressed as gallic acid equivalents (mg GAE g^{-1} of fresh weight) as described previously [49]. Total flavonoid content was determined from the same extracts according to the aluminium chloride colorimetric method [51]; the absorbance of the solution was measured at 510 nm (Multiskan GO, Thermo Fischer Scientific Oy, Vantaa, Finland) and expressed as rutin equivalents (mg Rutin g^{-1} of fresh weight).

The antioxidant activity of the methanolic plant extracts was determined by using three assays of 2,2-diphenyl-1-picrylhydrazyl (DPPH) and ferric reducing antioxidant power (FRAP), as previously described [49], as well as the 2,2′-azino-bis(3-ethylbenzothiazoline-6-sulphonic acid) (ABTS) assay, as described previously [52]. To give a more thorough and accurate evaluation of the plant's antioxidant capacity, especially when various fertilization techniques are used, several radical scavenging tests were employed. It aids in taking into consideration the complexity of plant chemistry and the impact of various farming techniques; in fact, using at least two diverse approaches is advised [53]. The three spectrophotometric techniques that were used evaluated the capacity to decrease ferric ions (FRAP) and the free radical scavenging activity (DPPH, ABTS) [54]. Results were expressed as trolox (($\pm$)-6-Hydroxy-2,5,7,8-tetramethylchromane-2-carboxylic acid) equivalent (mg Trolox g^{-1} of fresh weight).

2.4. Lipid Peroxidation and Hydrogen Peroxide Content

The content of hydrogen peroxide (H_2O_2) was assessed according to Loreto and Velikova [55], and lipid peroxidation was assessed according to De Azevedo Neto et al. [56]. The results were expressed as micromoles of H_2O_2 per gram of fresh weight, while lipid peroxidation was measured in terms of malondialdehyde (MDA) content and expressed as nanomoles of MDA per gram of fresh weight.

2.5. Nutrient Content in Plants

At the end of the experiment, samples from the upper parts of the plant (leaves and stems) were used to determine the mineral content in six replications per treatment (three pooled plants per replication). Plant tissue was dried to constant weight (at 42 °C), then was ash-burned at 450 °C for 6 h and acid-digested (2 M HCl) [45]. The nutrient content was determined, with K and Na measured photometrically (Flame photometer, Lasany Model 1832, Lasany International, Panchkula, India), and P measured spectrophotometrically

(Multiskan GO, Thermo Fischer Scientific, Waltham, MA, USA). Nitrogen was determined by the Kjeldahl method (BUCHI, Digest automat K-439 and Distillation Kjeldahl K-360, Flawil, Switzerland). Data were expressed in g kg^{-1} of dry weight [45].

2.6. Statistical Analysis

Statistical analysis was performed using IBM SPSS version 22. Data means were compared with one-way and two-way analysis of variance (ANOVA) and Duncan's multiple range tests for comparisons of treatment means at $p < 0.05$. Measurements were done in four to six biological replications/treatments (each replication consisted of a pool of three individual measures/samples).

3. Results

3.1. Soil Analysis

Following analysis, the soil was characterized by a clay-loam texture, with almost 80% of the particles being lower than 4.75 mm. Additionally, the soil had 4.61% organic matter, available $CaCO_3$ of 60.67%, a pH of 7.64, and an EC of 1.08 mS cm^{-1}. Finally, the soil contained N 2.431 g kg^{-1}, P 0.054 g kg^{-1}, K 0.358 g kg^{-1}, Ca 6.298 g kg^{-1}, Mg 0.598 g kg^{-1}, and Na 0.126 g kg^{-1}.

3.2. Two-Way Anova Analysis

Table 1 exhibits the results yielded by the two-way ANOVA analysis on the main impacts of the fertilization application and foliar application, as well as their interaction, on *S cypria* and *O. dubium* plants. In the case of *S cypria*, the different fertilization schemes affected the chlorophyll a:chlorophyll b ratio at $p < 0.05$; plant fresh weight, dry matter content, macronutrient content (N, P, K, and Na), total phenolics content, antioxidant capacity (DPPH, FRAP, and ABTS), flavonoids, and stress indicators (MDA and H_2O_2) at $p < 0.001$. The different foliar applications mainly affected dry matter content, ABTS, and flavonoids at $p < 0.05$; FRAP at $p < 0.01$; and the macronutrient content, total phenolics, DPPH, and stress indicators at $p < 0.001$. Finally, their interaction mainly affected the plants' fresh weight at $p < 0.05$; plants' macronutrient content, total phenolics, antioxidant capacity, flavonoids, and stress indicators at $p < 0.001$.

Table 1. Effect of different fertilization applications, foliar applications, and their interaction on *Sideritis cypria* and *Origanum dubium* plant biomass, physiology, nutrient content, antioxidant capacity, and stress response.

	Sideritis cypria			*Origanum dubium*		
	Fertilization	**Foliar**	**Fertilization × Foliar**	**Fertilization**	**Foliar**	**Fertilization × Foliar**
Plant FW	***	ns	*	***	ns	*
Plant DW	ns	ns	ns	ns	ns	ns
Plant DM	***	*	ns	***	*	ns
SPAD	ns	ns	ns	ns	ns	ns
Chlorophyll a	ns	ns	ns	**	**	***
Chlorophyll b	ns	ns	ns	ns	*	***
Total chlorophylls	ns	ns	ns	ns	**	***
Carotenoids	ns	ns	ns	***	ns	***
Chl a/Chl b	*	ns	ns	**	ns	***

Table 1. *Cont.*

	Sideritis cypria			*Origanum dubium*		
	Fertilization	**Foliar**	**Fertilization × Foliar**	**Fertilization**	**Foliar**	**Fertilization × Foliar**
Car/Total Chls	ns	ns	ns	**	ns	***
N	***	***	***	***	***	***
P	***	***	***	***	***	***
K	***	***	***	***	***	***
Na	***	***	***	***	***	***
Phenols	***	***	***	*	***	***
DPPH	***	***	***	*	***	***
FRAP	***	**	***	**	***	**
ABTS	***	*	***	***	***	***
Flavonoids	***	*	***	***	***	***
MDA	***	***	***	***	***	***
H_2O_2	***	***	***	***	***	***

*** Significant at $p < 0.001$; ** significant at $p < 0.01$; * significant at $p < 0.05$; ns: not significant, according to two-way ANOVA. Fresh weight (FW); dry weight (DW); dry matter (DM); chlorophyll a (Chl a); chlorophyll b (Chl b); total chlorophylls (Total Chl); total chlorophylls (Total Chl); carotenoids (Car); nitrogen (N); phosphorus (P); potassium (K); sodium (Na); 2,2-diphenyl-1-picrylhydrazyl (DPPH); ferric reducing antioxidant power (FRAP); 2,20-azino-bis(3-ethylbenzothiazoline-6-sulphonic acid) (ABTS); hydrogen peroxide (H_2O_2); malondialdehyde (MDA).

In the case of *O. dubium*, the different fertilization applications affected the total phenolics and DPPH at $p < 0.05$; chlorophyll a, the chlorophyll a to chlorophyll b ratio, the carotenoids to total chlorophylls ratio FRAP at $p < 0.01$; SPAD, dry matter content, macronutrient content (N, P, K, and Na), carotenoids, ABTS, flavonoids, and stress indicators (MDA, H_2O_2) at $p < 0.001$. The different foliar applications mainly affected dry matter content and chlorophyll b at $p < 0.05$; total chlorophylls at $p < 0.01$; macronutrient content, total phenolics, the antioxidant capacity (assayed by DPPH, FRAP, and ABTS), flavonoids, and stress indicators at $p < 0.001$. Finally, their interaction mainly affected the plants' fresh weight at $p < 0.05$; FRAP at $p < 0.01$; chlorophylls (a, b, and total), macronutrient content, total phenolics, the antioxidant capacity (DPPH and ABTS), flavonoids, and stress indicators at $p < 0.001$.

3.3. *Plant Biomass, Photochemistry, and Mineral Content*

The impacts of fertilization and foliar application of Fe and Zn on *S. cypria* plants are shown in Table 2. The highest fresh weight was recorded for plants grown with OrFert and foliar Zn, and the lowest for plants grown with NoFert when sprayed with H_2O. Interestingly, in NoFert, the foliar application of both Fe and Zn increased the plants biomass, while in OrFert, only the application of Zn increased the plants biomass, compared to the foliar application of H_2O. In contrast, in CoFert, a significant reduction in fresh weight was evidenced after Zn spraying, compared to foliar application of H_2O. Furthermore, the highest and lowest dry weights were found in the NoFert treatment, with the Zn spraying and without spraying, respectively. Finally, the use of OrFert and CoFert, in combination with Zn spraying, resulted in the lowest and highest dry matter content, respectively.

In the case of *O. dubium* plants, the highest fresh weight was marked in plants grown with CoFert and foliar applications of H_2O, while no significant changes were evidenced with the foliar application of Fe and Zn in the same fertilization application (Table 2). Furthermore, a significant decrease was evidenced in the dry matter content of plants grown with CoFert, regardless of foliar application, in comparison to the rest of the treatments. Finally, the plants' dry weight was unchanged by the tested applications.

Table 2. Effect of different fertilization applications [No fertilization (NoFert), Organic fertilization (OrFert), Conventional fertilization (CoFert)] and foliar applications (water-H_2O, iron-Fe, zinc-Zn) on *Sideritis cypria* and *Origanum dubium* plant biomass fresh weight (FW; g plant^{-1}), dry weight (DW; g plant^{-1}), and dry matter content (DM; %).

	Fertilization	**Foliar**	**Plant FW**	**Plant DW**	**Plant DM**
Sideritis cypria	NoFert	H_2O	193.37 ± 13.84 cd	83.54 ± 7.64 d	43.60 ± 3.89 cd
		Fe	260.43 ± 36.26 abc	111.08 ± 12.58 bcd	43.59 ± 1.83 cd
		Zn	305.58 ± 21.65 ab	150.09 ± 8.30 a	49.78 ± 2.66 bc
	OrFert	H_2O	243.12 ± 23.29 bc	121.30 ± 11.56 abc	50.06 ± 1.26 bc
		Fe	234.87 ± 22.56 bcd	126.81 ± 10.40 ab	54.48 ± 1.57 ab
		Zn	335.65 ± 37.20 a	140.88 ± 12.56 ab	42.58 ± 1.54 d
	CoFert	H_2O	254.80 ± 26.39 abc	111.34 ± 7.00 bcd	44.83 ± 2.68 cd
		Fe	224.48 ± 32.52 bcd	109.50 ± 15.18 bcd	48.73 ± 2.02 bcd
		Zn	155.45 ± 15.36 d	88.63 ± 8.40c	57.12 ± 1.25 a
	Fertilization	**Foliar**	**Plant FW**	**Plant DW**	**Plant DM**
Origanum dubium	NoFert	H_2O	164.90 ± 31.92 c	66.03 ± 13.42	40.49 ± 2.48 ab
		Fe	209.62 ± 29.20 bc	84.20 ± 10.19	40.60 ± 0.92 ab
		Zn	172.43 ± 16.10 c	69.42 ± 6.21	40.32 ± 0.83 ab
	OrFert	H_2O	140.42 ± 15.20 c	61.98 ± 6.81	42.73 ± 1.34 a
		Fe	141.53 ± 19.16 c	58.82 ± 10.11	40.73 ± 1.54 ab
		Zn	234.68 ± 40.20 bc	74.09 ± 11.39	33.74 ± 3.69 b
	CoFert	H_2O	350.77 ± 34.28 a	93.14 ± 14.24	25.73 ± 1.73 c
		Fe	283.75 ± 24.20 ab	70.95 ± 10.38	24.09 ± 2.34 c
		Zn	287.35 ± 40.72 ab	70.06 ± 16.53	22.18 ± 3.18 c

Significant differences ($p < 0.05$) among applications are indicated by different letters. Values are means (±SE) of six replicates for each treatment.

The photochemistry and therefore the photosynthetic attributes in plants were affected by the treatments examined. In *S. cypria* leaves, SPAD was significantly increased with the use of NoFert with Fe spraying and CoFert with Zn spraying, compared to the use of OrFert with Zn spraying (Table 3). Furthermore, chlorophyll a was significantly increased with the application of NoFert without foliar sprays and with foliar Zn. Chlorophyll b was significantly increased with the application of NoFert with foliar Zn. Total chlorophylls were significantly increased with the use of NoFert without foliar sprays and with Zn spraying, as well as OrFert with Zn spraying, compared to the lowest values of chlorophylls (a, b, and total) achieved by CoFert with foliar Fe. Furthermore, the chlorophyll a to chlorophyll b ratio was significantly increased with the use of CoFert with Fe spraying, compared to NoFert with foliar Zn and OrFert with foliar Fe and Zn.

In the case of *O. dubium* leaves, chlorophyll a was the highest with the use of CoFert and Zn spraying (Table 3). Interestingly, while the treatment of NoFert and Fe spraying did not differ significantly from the previously mentioned treatment (CoFert + Zn), the treatment of NoFert and Zn spraying produced plants with the lowest chlorophyll a content. Similarly, the NoFert and Zn spraying treatment resulted in the lowest total chlorophyll content, while in the same treatment, the CoFert with foliar application of H_2O resulted in the lowest chlorophyll b content. In contrast, the application of NoFert and Fe spraying produced plants with the highest content of chlorophyll b and total chlorophylls, while the same treatment recorded the lowest content of carotenoids, as well as the lowest ratios of chlorophyll a to chlorophyll b and carotenoids to chlorophylls. Regarding carotenoids, the highest content was recorded in plants grown with the CoFert and Fe spraying treatment. The highest chlorophyll a to chlorophyll b ratios were recorded for NoFert and Zn spraying, OrFert and Fe spraying, and CoFert and foliar application of H_2O. Additionally, the highest carotenoids to chlorophylls ratio was recorded for the CoFert and foliar application of H_2O.

Table 3. Effect of different fertilization applications [No fertilization (NoFert), Organic fertilization (OrFert), Conventional fertilization (CoFert)] and foliar applications (water-H_2O, iron-Fe, zinc-Zn) on *Sideritis cypria* and *Origanum dubium* SPAD value, chlorophyll content (Chlorophyll a: Chl a; Chlorophyll b: Chl b; Total chlorophylls: Total chls; mg g^{-1} FW), carotenoid content (Car: mg g^{-1} FW), chlorophyll a to chlorophyll b and carotenoids to Total chlorophylls ratios.

	Fertilization	Foliar	SPAD	Chl a	Chl b	Total Chls	Car	Chl a/Chl b	Car/Total Chl
Sideritis cypria	NoFert	H_2O	56.10 ± 1.65 ab	0.76 ± 0.04 a	0.38 ± 0.03 ab	1.13 ± 0.07 a	0.118 ± 0.003	2.05 ± 0.07 ab	0.105 ± 0.005
		Fe	67.45 ± 5.21 a	0.70 ± 0.03 ab	0.35 ± 0.02 ab	1.04 ± 0.05 ab	0.115 ± 0.005	2.04 ± 0.05 ab	0.108 ± 0.003
		Zn	49.35 ± 3.69 b	0.81 ± 0.03 a	0.42 ± 0.03 a	1.23 ± 0.05 a	0.120 ± 0.007	1.93 ± 0.09 b	0.098 ± 0.006
	OrFert	H_2O	57.98 ± 5.56 ab	0.71 ± 0.04 ab	0.35 ± 0.03 ab	1.06 ± 0.07 ab	0.110 ± 0.009	2.02 ± 0.04 ab	0.105 ± 0.005
		Fe	56.75 ± 2.33 ab	0.72 ± 0.05 ab	0.37 ± 0.05 ab	1.10 ± 0.10 ab	0.110 ± 0.004	1.98 ± 0.12 b	0.103 ± 0.009
		Zn	47.30 ± 3.79 b	0.74 ± 0.02 ab	0.38 ± 0.01 ab	1.12 ± 0.03 a	0.110 ± 0.007	1.96 ± 0.03 b	0.103 ± 0.006
	CoFert	H_2O	53.68 ± 3.35 ab	0.72 ± 0.02 ab	0.34 ± 0.01 ab	1.06 ± 0.03 ab	0.115 ± 0.003	2.11 ± 0.02 ab	0.110 ± 0.004
		Fe	57.45 ± 6.83 ab	0.62 ± 0.04 b	0.28 ± 0.02 b	0.89 ± 0.06 b	0.103 ± 0.005	2.22 ± 0.03 a	0.115 ± 0.003
		Zn	66.73 ± 3.76 a	0.71 ± 0.06 ab	0.35 ± 0.05 ab	1.07 ± 0.10 ab	0.115 ± 0.003	2.06 ± 0.12 ab	0.113 ± 0.013
	Fertilization	**Foliar**	**SPAD**	**Chl a**	**Chl b**	**Total chls**	**Car**	**Chl a/Chl b**	**Car/Total chl**
Origanum dubium	NoFert	H_2O	36.93 ± 2.96	0.76 ± 0.02 de	0.46 ± 0.01 cd	1.22 ± 0.03 de	0.10 ± 0.00 abc	1.63 ± 0.03 bc	0.08 ± 0.00 bcd
		Fe	40.55 ± 2.49	0.84 ± 0.02 ab	0.61 ± 0.03 a	1.45 ± 0.04 a	0.06 ± 0.01 d	1.39 ± 0.07 d	0.04 ± 0.01 e
		Zn	40.23 ± 3.96	0.69 ± 0.01 f	0.37 ± 0.01 e	1.06 ± 0.02 f	0.09 ± 0.00 bc	1.84 ± 0.03 a	0.08 ± 0.00 abc
	OrFert	H_2O	36.95 ± 3.60	0.78 ± 0.03 cde	0.49 ± 0.04 bcd	1.26 ± 0.06 cd	0.09 ± 0.01 c	1.63 ± 0.08 bc	0.07 ± 0.01 cd
		Fe	43.95 ± 3.21	0.80 ± 0.01 bcd	0.42 ± 0.01 de	1.21 ± 0.02 de	0.10 ± 0.00 abc	1.91 ± 0.02 a	0.08 ± 0.00 abc
		Zn	41.28 ± 2.24	0.81 ± 0.01 bc	0.53 ± 0.03 bc	1.34 ± 0.04 bc	0.09 ± 0.01 c	1.54 ± 0.06 cd	0.06 ± 0.01 d
	CoFert	H_2O	37.78 ± 3.10	0.75 ± 0.01 e	0.38 ± 0.01 e	1.13 ± 0.02 ef	0.11 ± 0.00 ab	1.98 ± 0.04 a	0.10 ± 0.00 a
		Fe	38.93 ± 3.22	0.80 ± 0.01 bcde	0.45 ± 0.02 de	1.24 ± 0.03 cd	0.11 ± 0.00 a	1.80 ± 0.09 ab	0.09 ± 0.00 ab
		Zn	44.45 ± 2.45	0.86 ± 0.00 a	0.54 ± 0.03 b	1.40 ± 0.02 ab	0.09 ± 0.01 abc	1.62 ± 0.09 bc	0.07 ± 0.01 cd

Significant differences ($p < 0.05$) among applications are indicated by different letters. Values are means (±SE) of six replicates for each treatment.

Results regarding the macronutrient content of *S. cypria* plant leaves are presented in Figure 1, where fluctuations were recorded for the examined fertilization and foliar applications. Particularly, plants treated with CoFert and Zn spraying recorded the highest content of N, P, K, and Na (16.56, 20.72, 3.92, and 1.98 g kg^{-1}, respectively) (Figure 1(A1–A4)), while plants treated with NoFert and Zn spraying recorded the lowest content of N, showing a decrease of 33.9% over the former treatment. Additionally, the use of NoFert with foliar application of H_2O resulted in the lowest P and K contents in leaves (Figure 1(A2,A3)), showing a decrease of 36.4% and 67.3%, compared to the CoFert and Zn spraying treatment, respectively. Finally, the lowest overall Na was evidenced in plants treated with NoFert and foliar Fe, noting a 77.3% decrease compared to plants treated with CoFert and Zn spraying (Figure 1(A4)).

In the case of the *O. dubium*, at the NoFert treatments, the foliar application of H_2O resulted in the lowest content of N, while a significant increase was observed with the foliar application of both Fe and Zn over the rest of the treatments (Figure 1(B1)). Similarly, the Fe spraying increased the N content of plants grown with OrFert, an effect that was also exhibited in both P and K contents (Figure 1(B2,B3)). For K particularly, the highest and lowest K content was found in plants grown with OrFert, with Fe spraying and without foliar applications, respectively. Finally, the lowest Na content was marked for plants treated with NoFert, with Fe or without foliar applications, while a significant increase was evidenced with Zn spraying over the abovementioned treatments (Figure 1(B4)). Similarly, foliar applications in the OrFert treatment increased the content of Na over foliar applications of H_2O, with the OrFert and Zn treatment recording the highest Na content.

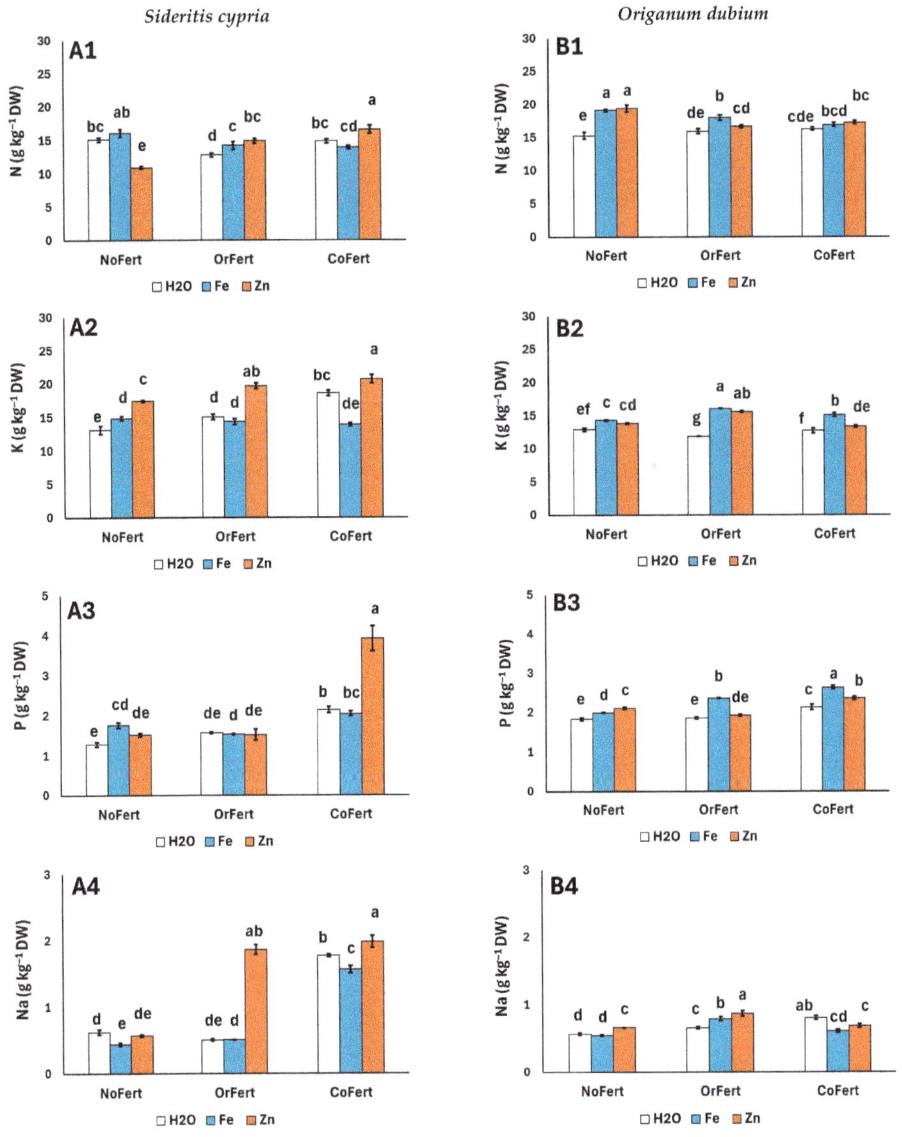

Figure 1. Effect of different fertilization applications [No fertilization (NoFert), Organic fertilization (OrFert), Conventional fertilization (CoFert)] and foliar applications (water-H_2O, iron-Fe, zinc-Zn) on *Sideritis cypria* (**A1–A4**) and *Origanum dubium* (**B1–B4**) macronutrient content (g kg^{-1}). Nitrogen-N (**A1**,**B1**), potassium-K (**A2**,**B2**), phosphorus-P (**A3**,**B3**), and sodium-Na (**A4**,**B4**). Significant differences ($p < 0.05$) among applications are indicated by different letters. Values are means (±SE) of six replicates for each treatment. DW: dry weight.

Figure 2 presents the results regarding total phenolics, flavonoids, and antioxidant capacity of *S. cypria* (Figure 2(A1–A5)) and *O. dubium* (Figure 2(B1–B5)) plants grown under varying fertilization and foliar applications. In the case of *S. cypria*, total phenolics were the highest when plants were grown with CoFert, using foliar application of H_2O or of Fe, while the use of Zn spraying caused a significant decrease over the previously mentioned treatments (Figure 2(A1)). Furthermore, total phenolics were the lowest when plants were grown with NoFert, using foliar application of H_2O, while Fe and Zn spraying increased total phenolics. The plants that received CoFert and Fe spraying had the highest flavonoids content, while the ones that received OrFert and Fe spraying had the lowest flavonoids content (Figure 2(A2)). The highest DPPH was recorded in plants treated with CoFert with foliar application of H_2O (Figure 2(A3)), the highest FRAP was recorded in plants treated with CoFert and Fe spraying (Figure 2(A4)), while the highest ABTS was recorded in both abovementioned treatments (Figure 2(A5)). In contrast, the antioxidant capacity (assayed by FRAP and ABTS) was the lowest in plants treated with OrFert and Fe spraying (Figure 2(A4,A5)).

In the case of the *O. dubium*, total phenolics were increased with foliar application of H_2O under all fertilizer applications, while in the case of NoFert, an increase was observed with Zn spraying (Figure 2(B1)). In the case of NoFert, the application of Fe decreased the content of total phenolics, while in the case of OrFert and CoFert, both Fe and Zn alleviated the production of total phenolics. Flavonoids followed the trend of the type of fertilization applied, as the highest contents were observed with no fertilizer application, while a decrease was found with the use of OrFert and NoFert (Figure 2(B2)). The antioxidant capacity (after DPPH and FRAP assays) of *O. dubium* plants increased without the foliar applications under all fertilizer applications, while a significant decrease was noted with the Fe and Zn spraying (Figure 2(B3,B4)). Furthermore, the highest ABTS content was observed in plants grown with NoFert or OrFert, treated with H_2O foliar applications and foliar Zn, and plants grown with CoFert, without foliar applications (Figure 2(B5)). Generally, the lowest antioxidant capacity (after DPPH, FRAP, and ABTS assays) was evidenced in plants treated with CoFert and Zn spraying.

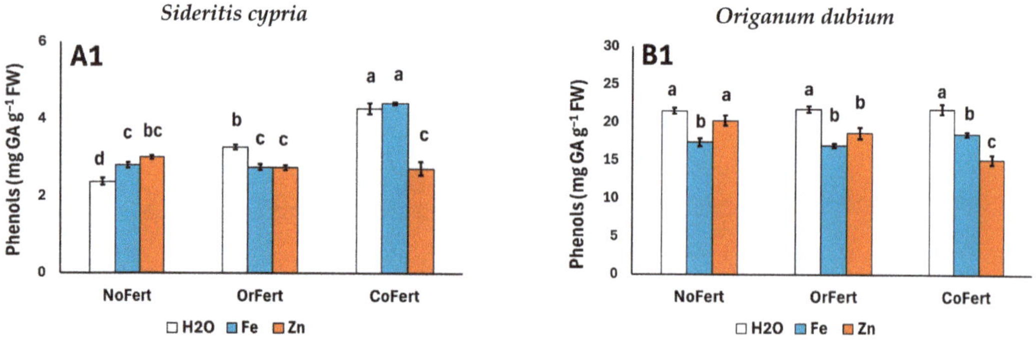

Figure 2. *Cont.*

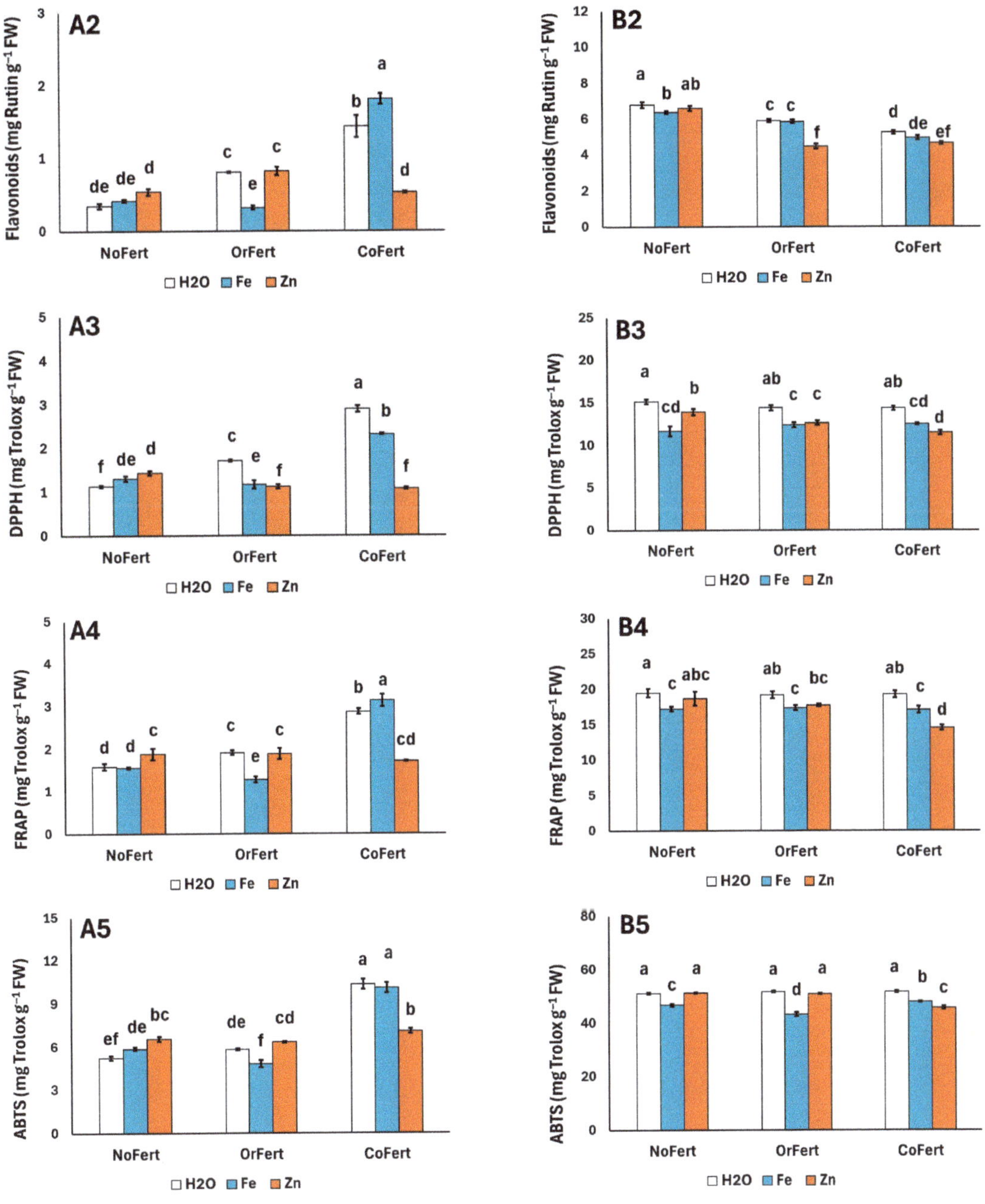

Figure 2. Effect of different fertilization applications [No fertilization (NoFert), Organic fertilization (OrFert), Conventional fertilization (CoFert)] and foliar applications (water-H_2O, iron-Fe, and zinc-Zn) on *Sideritis cypria* (**A1–A5**) and *Origanum dubium* (**B1–B5**) total phenols (**A1**,**B1**; mg GA g^{-1} FW), flavonoids (**A2**,**B2**; mg Rutin g^{-1} FW), and antioxidant activity (**A3**,**B3** for DPPH, **A4**,**B4** for FRAP, and **A5**,**B5** for ABTS; mg Trolox g^{-1} FW). Significant differences ($p < 0.05$) among applications are indicated by different letters. Values are means (±SE) of six replicates for each treatment. FW: Fresh weight; GA: gallic acid.

The stress response of *S. cypria* plants, as assayed by MDA content and H_2O_2 production, was presented in Figure 3(A1,A2). Plants that received conventional fertilization, in conjunction with Fe spraying, had the highest content of MDA and H_2O_2. Furthermore, in the case of MDA, the lowest content was observed in plants grown with OrFert and supplied with Fe and Zn spraying (Figure 3(A1)). Finally, the lowest H_2O_2 was observed in plants treated with NoFert and Fe spraying, exhibiting a significant decrease compared to the other foliar treatments of NoFert (Figure 3(A2)).

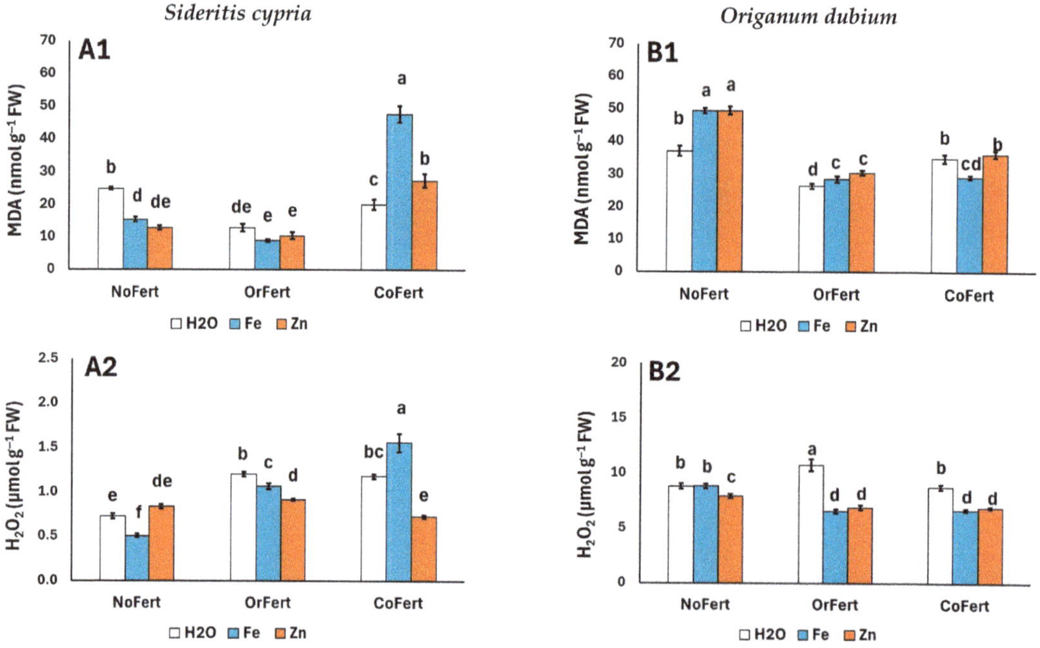

Figure 3. Effect of different fertilization applications [No fertilization (NoFert), Organic fertilization (OrFert), Conventional fertilization (CoFert)] and foliar applications (water-H_2O, iron-Fe, zinc-Zn) on *Sideritis cypria* (**A1**,**A2**) and *Origanum dubium* (**B1**,**B2**) lipid peroxidation-malondialdehyde (MDA) (**A1**,**B1**; nmol g^{-1} FW) and hydrogen peroxide-H_2O_2 (**A2**,**B2**; μmol g^{-1} FW). Significant differences ($p < 0.05$) among applications are indicated by different letters. Values are means (±SE) of six replicates for each treatment. FW: Fresh weight.

Regarding the effects of the examined applications on the stress response of *O. dubium* plants, the highest lipid peroxidation, through the MDA content, was found in plants grown using no fertilization and either foliar application of Fe or Zn, while lower values were observed without foliar applications (Figure 3(B1)). A significant decrease in MDA was found in plants treated with both OrFert and CoFert. In the case of OrFert, MDA increased with the Fe and Zn spraying, compared to H_2O spraying, while in the case of CoFert, MDA was decreased with the Fe spraying, compared to Zn and H_2O spraying. The accumulation of H_2O_2 was also influenced by the treatments examined (Figure 3(B2)). The highest H_2O_2 content was observed in plants grown with OrFert without foliar application. The Fe or Zn spraying at OrFert and CoFert applications caused a significant decrease in H_2O_2 when compared to foliar application of H_2O.

In *S. cypria*, the WUE tended to increase only after Zn spraying for both the NoFert and OrFert treatments, while the opposite results were evidenced after Zn spraying at the CoFert treatment (Figure 4(A1)). The NUE after fertilization was obviously increased after OrFert and CoFert applications, while NUE was decreased after Zn spraying at the CoFert treatment compared to the H_2O or Fe spraying at the same treatment (i.e., CoFert) (Figure 4(A2)). The foliar efficacy, after comparing the Fe or Zn spraying with the H_2O spraying, revealed a similar trend to the WUE, as the foliar efficacy was increased after Zn spraying at the NoFert and OrFert treatments and decreased after Zn spraying at the CoFert treatment (Figure 4(A3)).

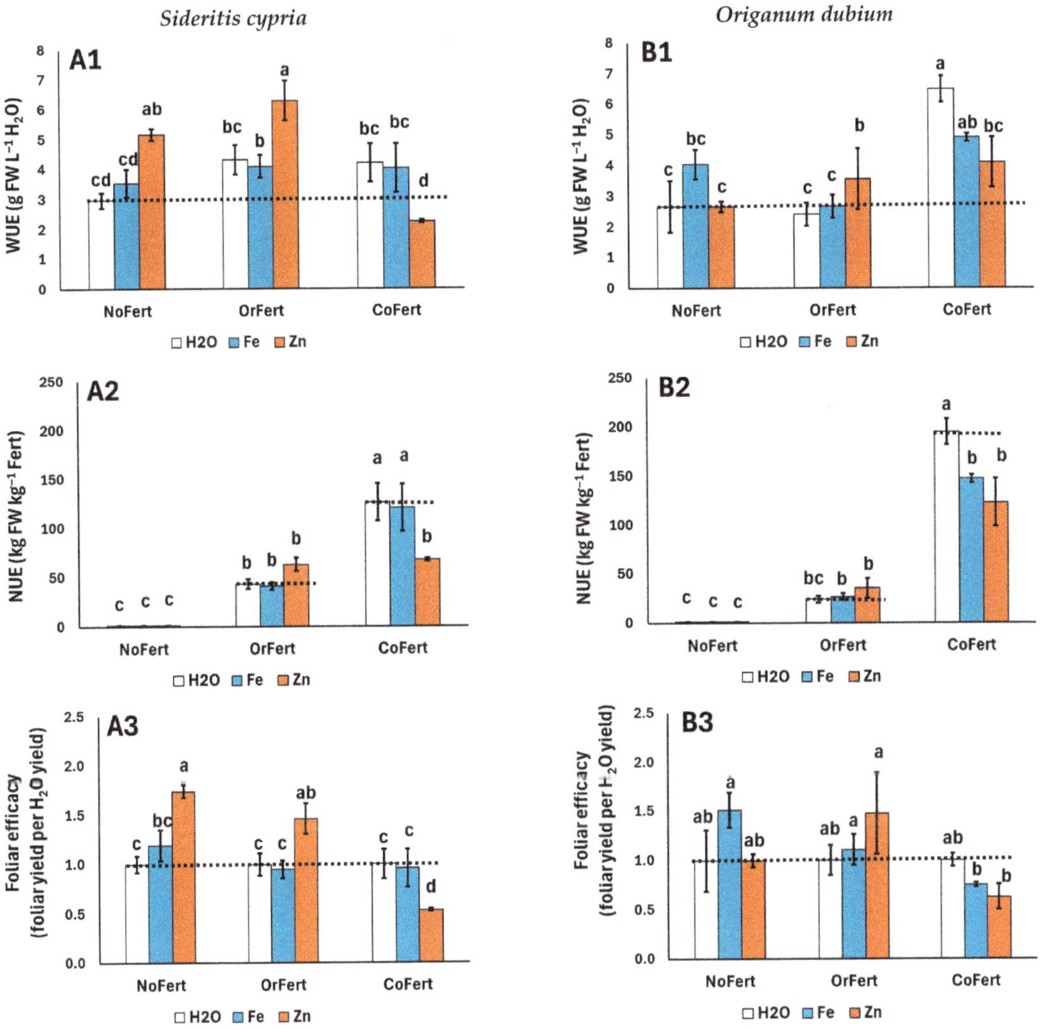

Figure 4. Effect of different fertilization applications [No fertilization (NoFert), Organic fertilization (OrFert), Conventional fertilization (CoFert)] and foliar applications (water-H_2O, iron-Fe, zinc-Zn) on *Sideritis cypria* (**A1–A3**) and *Origanum dubium* (**B1–B3**) water use efficiency-WUE (**A1**,**B1**; g FW L^{-1} H_2O), nutrient use efficiency-NUE (**A2**,**B2**; kg FW kg^{-1} Fertilizer), and foliar application efficacy (**A3**,**B3**; yield after foliar application per yield after H_2O application). Significant differences ($p < 0.05$) among applications are indicated by different letters. Values are means (±SE) of six replicates for each treatment. Dotted line indicates values at the NoFert and H_2O applications. FW: Fresh weight.

In the case of *O. dubium*, WUE tended to increase at OrFert but decreased at CoFert treatments after Zn spraying, whereas Fe spraying did not have any effects (Figure 4(B1)). The NUE was negatively affected by the Fe and Zn spraying only at the CoFert treatment (Figure 4(B2)).

The foliar application efficacy in *S. cypria* was evidenced with the Zn spraying at the NoFert and OrFert treatments but not in the case of the CoFert treatment, where the foliar efficacy decreased compared to the H_2O or Fe spraying (Figure 4(A3)). In the case of *O. dubium*, the foliar efficacy was not evidenced, as foliar application seems to not alter the fresh biomass produced (Figure 4(B3)).

4. Discussion

Sustainability and preservation of biodiversity are the guiding principles of organic farming and the utilization of organic fertilizers [4,13]. Since MAPs and other underutilized crops are adapted to both abiotic and biotic stressors and require minimal inputs, they are well-suited for integration into organic farming systems [20,57]. *S. cypria* and *O. dubium* are two promising MAPs that were used in the current research to investigate and encourage the adoption of sustainable farming techniques. Maintaining a proper balance of inputs is essential in the context of ecologically sound and organic agriculture, given that fertilization has a critical role in crop development and quality and ultimately in crop output [17,58]. Application rates and fertilizer type determine how organic fertilization and the combination of chemical and organic fertilizers affect crop development and soil fertility. Organic fertilizer application rates are generally determined by crop N requirements and the predicted organic fertilizer N supply rates; however, P and K content are not considered in general. Optimal fertilization techniques are crucial in boosting the plant growth and quality-related characteristics of MAPs [59]. On the other hand, conventional fertilization might raise the MAP yield but not necessarily mirror advances in the quality and safety of the consumed/used products, because the level of bioactive substances and, thus, the plants nutritional value are decreased by mineral fertilizers [58].

Prior studies reported that conventional fertilization resulted in similar *Sideritis perfoliata* yield compared to organic fertilization [9], which is in agreement with the *S. cypria* yield for the present study. However, *O. dubium* had 2.5 times higher yield in conventional fertilization compared to the organic scheme, indicating the various responses of plant metabolism to fertilization and different plants mineral needs. Such variation in mineral needs is important at a farm level in order to optimize the crop selection based on mineral needs and soil fertility. Zinc spraying increased *S. cypria* fresh weight in the NoFert and OrFert treatments, but Fe spraying did not have such effects.

According to other studies, organic medicinal plants had a much higher dry matter content and a lower water content than conventionally produced plants [58], which aligns only with the findings of the *O. dubium* plants, since *S. cypria* plants had the same dry matter content in all fertilized or non-fertilized plants. Previous reports indicated similar dry matter content in organically vs. conventionally grown peppermint (*Mentha piperita* L.) and rosemary (*R. officinalis* L.) [58]. Moreover, irrigation is also an important cultivation practice to be considered, as the MAPs under drought stress activate metabolic processes towards water stress, with reduced cell multiplication and expansion, which negatively affects plant growth with decreased yield, increased dry matter content, and the activation of various antioxidant mechanisms towards the water shortage challenge [9,60]. Interestingly, Zn spraying exerted fertilizer-specific effects on *S. cypria* dry matter content, decreasing it under the OrFert treatment but increasing it under the CoFert treatment. The observed variation in plant responses as a function of base fertilization is related to the differential availability of nutrients, particularly N, which has a significant role in the efficacy of utilization of exogenously applied Zn [61]. In addition, the dry matter content of *O. dubium* plants with OrFert application was negatively impacted by the Zn spraying, compared to H_2O spraying. Similarly, Adamczyk-Szabela and Wolf [62] demonstrated that Zn supplementation had a differential impact on the dry biomass. It was mentioned that dry biomass was decreased with Zn supplementation in common nettle (*Urtica dioica* L.), increased in borage (*Borago officinalis* L.), or unaffected in basil (*Ocimum basilicum* L.) and peppermint (*M. piperita* L.) [62].

Fertilization constitutes a critical factor in the synthesis of chlorophylls by supplying key macronutrients such as N. Variations in the availability and mineralization of nutrients between organic and conventional fertilization can significantly affect the production of chlorophylls, thus influencing photosynthesis and overall plant development [54]. Chrysargyris et al. [18] observed that organic fertilization decreased the relative chlorophyll content of spearmint (*M. spicata* L.) leaves, as indicated by lower SPAD values, compared to conventional fertilization. In contrast, Hajisolomou et al. [54] reported that *Portulaca oleracea* L. and *Sonchus oleraceus* L. plants subjected to organic fertilization had SPAD values and chlorophyll contents comparable to those observed under conventional fertilization. Similarly, in the current study, the fertilization type did not significantly affect SPAD or total chlorophyll content in the leaves of *S. cypria* or *O. dubium*. However, plant genotype responses depend on variable soil fertility [63]. For instance, in *O. dubium*, the interaction of fertilization and foliar applications influenced the content of chlorophylls. More specifically, total chlorophyll contents were significantly elevated under NoFert with Fe spraying and CoFert with Zn spraying. This may be related to the critical roles of Fe and Zn in the biosynthesis of chlorophyll, as Fe is involved in the formation of chlorophyll precursors, while Zn influences enzyme activities, the metabolism of photosynthetic proteins, and N uptake [64–66]. In addition, N fertilization has a synergistic effect on Zn accumulation, especially when Zn supply is not limiting, thus influencing chlorophyll content [67]. In accordance with these results, Fe and Zn spraying have been associated with greater chlorophyll accumulation in pepper (*Capsicum annum* L.) and wheat (*Triticum aestivum* L.), respectively [64,68].

The mineral content of MAPs is a significant factor influencing their medicinal efficacy and nutritional status [45]. For optimal growth, plants require an adequate supply of nutrients, tailored to their specific needs and developmental stage. Organic farming generally consists of lower nutrient addition with organic fertilizers compared to conventional ones, resulting in a decreased mineral content in soils [9]. In addition, nutrient availability is harder to predict, as it depends on the mineralization of organic matter, which is influenced by factors such as temperature, available microorganisms, pH, and soil moisture [48]. Furthermore, to enhance the nutritional value of MAPs, agronomic biofortification applications have been gaining interest [57]. In the current study, both examined plants exhibited varying responses to basic fertilization and foliar applications. Nitrogen content in *S. cypria* leaves was increased by Zn spraying under both CoFert and OrFert, likely due to the positive effect of Zn supply on N absorption. However, this effect is contingent upon N availability and form [67]. Zn is an important component of enzymes that regulate key metabolic processes in plants, while also having a critical role in auxin and protein synthesis that contribute to enhanced N uptake [69]. This is consistent with previous research. Tzortzakis et al. [31] reported that under sufficient N levels in hydroponically grown *S. cypria*, Zn spraying mediated N uptake. In addition, Alinejad Elahshah et al. [70] demonstrated that Zn spraying significantly enhanced N absorption by strawberry (*Fragaria* × *ananassa* Duch.) plants. In *O. dubium*, however, N content was significantly increased by both Fe and Zn spraying, but only under NoFert, suggesting that under nutrient limitation, foliar micronutrients boost N assimilation. Phosphorus content in *S. cypria* was also improved under Zn spraying with CoFert, possibly due to Zn stimulating the activation of P-solubilizing enzymes such as phosphatase and phytase [71]. In *O. dubium*, Fe spraying was linked to increased P acquisition, particularly under OrFert and CoFert. Similarly, Neofytou et al. [72] reported elevated P accumulation in *S. cypria* leaves after application of Fe spraying under increased P availability. Potassium uptake was enhanced with Zn spraying under fertilized *S. cypria* treatments. This result aligns with the findings of Alinejad Elahshah et al. [70], as Zn spraying led to increased K uptake in strawberry (*F.* × *ananassa* Duch.) plant leaves. Interestingly, *O. dubium* was also found to have increased K levels under Zn spraying, especially under CoFert, while Fe spraying had a more pronounced effect on leaf K content, being in accordance with previous reports in peppermint (*M. piperita* L.) [73].

The evaluated species exhibited distinct variations in their phenolic profile and antioxidant capacity in response to the applied fertilization schemes and/or the foliar applications. Total phenols, flavonoids, and antioxidant capacity of *S. cypria* were triggered by CoFert following foliar with H_2O or Fe, underscoring the critical role of Fe in this species in its metabolic and physiological processes. For instance, Fe spraying to *S. cypria* plants induced non-enzymatic antioxidant mechanisms and flavonoid contents, contributing to enhanced antioxidant capacity [33]. This effect was limited to CoFert, where nutrient availability, particularly N, was more immediate and sufficient, especially during periods of rapid plant growth. In contrast, organic fertilization, which releases nutrients more gradually, may not adequately meet the demands of the plants [74]. This may have limited the plants' metabolic response to Fe spraying, leading to decreased phenolics, flavonoids, and antioxidant capacity of *S. cypria*. Nonetheless, Fe and Zn spraying modulated the production of phenolic compounds and antioxidants, causing a reduction in lipid peroxidation [72]. Although organic fertilization is often associated with enhanced produce quality, particularly in terms of phytochemical richness, this is not consistently observed [74]. For instance, Lv et al. [25] found no changes on total phenols but only on individual phenolic components on peppermint (*M. piperita* L.) and cinnamon (*Cinnamomum verum* L.) that were grown in organic and conventional farms. In contrast, organically grown *S. perfoliata* [9], *R. officinallis*, and *Melissa officinalis* [58] revealed higher total phenolics than the conventional one.

In the case of the *O. dubium*, total phenolics, flavonoids, and antioxidants were stimulated with the foliar application of H_2O, while both Fe and Zn applications resulted in plants with decreased levels of phenolics, flavonoids, and antioxidants, especially under fertilization schemes (CoFert and OrFert). In contrast, Fe and Zn nanoparticle applications in *M. officinalis* L. reportedly improved biochemical attributes, including total phenolic and flavonoid content [75]. This discrepancy may be related to the increased availability of nutrients, and especially N, as it influences nitrogen-containing metabolites, including free amino acids, proteins, and alkaloids. This in turn causes a decline in phenolics, flavonoids, and vitamin C content [58]. In addition, the carbon:nitrogen (C:N) ratio influences the synthesis of key plant-defense components [76]. In contrast, N shortage stimulates the production of defense components [18]. This is caused by the reallocation of available carbon, with plants diversifying and increasing secondary metabolism, which includes phenolics, flavonoids, and vitamin C increases [77].

Application of organic fertilizers has been associated with increased stress tolerance by improving water and nutrient uptake, antioxidant activity, and gene expression [78]. In addition, Fe and Zn have direct roles in mitigating oxidative damage by activating several antioxidant enzymes, which decrease reactive oxygen species (ROS) [79]. Zinc is involved in several physiological processes that influence plants' stress response, including the protection and maintenance of the cell membrane's stability and being a cofactor of enzymes such as superoxide dismutase, peroxidase, and catalase [79,80]. For instance, Zn had a significant effect on the resistance of rice plants to low temperatures, influencing the activity of antioxidant enzymes and subsequently decreasing the plants' MDA content [81]. Likewise, Fe is associated with antioxidant enzymes, being present in the active site of enzymes involved in ROS scavenging and thus facilitating tolerance to various abiotic stressors [65,82]. For *S. cypria*, the lowest MDA content was observed in plants grown with OrFert and treated with Fe and Zn spraying. This indicates the influence of organic fertilization in conjunction with foliar micronutrient applications in attenuating oxidative damage markers and enhancing antioxidant activities [83]. In the case of *O. dubium*, the highest lipid peroxidation was observed under the NoFert treatment, while a mediation effect was observed following fertilization. In general, nutrient deprivation and deficiencies affect metabolic processes, which potentially result in a stress response [84]. Interestingly, Fe and Zn spraying modulated the production of H_2O_2, especially in fertilized treatments. This is attributable to the enhancement of antioxidant enzyme activities that scavenge ROS, thus mitigating oxidative stress [85].

The current study demonstrated that Zn spraying tended to increase WUE of *S. cypria* plants under NoFert and OrFert treatments and of *O. dubium* plants under OrFert. Zinc has been found to have an influence on WUE by increasing osmolytes, improving stomatal conductance, and maintaining membrane integrity, supporting these findings [86]. In contrast, the opposite occurred under CoFert for both *S. cypria* and *O. dubium*. This may be related to nutrient imbalances, particularly the elevated uptake and accumulation of P observed in *S. cypria*, which could have interfered with Zn utilization [87]. In addition, both plant species exhibited increased NUE under CoFert application. Unlike organic fertilizers, where nutrient release depends on mineralization, conventional fertilizers immediately supply nutrients under high plant demand. The result indicates that the steady but delayed nutrient supply of organic amendments may not have been fully realized within the duration of the experiment [88]. In contrast, under CoFert, decreases occurred when Zn spraying was applied, mainly due to the decrease in biomass production.

5. Conclusions

This study examined the use of organic and conventional fertilizers in the field cultivation of *Sideritis cypria* and *Origanum dubium*, in conjunction with foliar applications of iron (Fe) and zinc (Zn). In terms of yield, *S. cypria* presented minute differences among organic fertilization (OrFert) and conventional fertilization (CoFert), but also with no fertilization. In contrast, *O. dubium* revealed a 2.5 times increase in fresh weight under conventional fertilization relative to organic fertilization, indicating a species-specific response to fertilization method. In terms of foliar applications, Zn produced fertilizer-specific effects on *S. cypria* dry matter content, related to the availability of nutrients, particularly N, which influences the efficacy of Zn. In contrast, the dry matter content of *O. dubium* under OrFert application was decreased under Zn spraying, demonstrating application responses that are specific to each species. *S. cypria* showed distinct phenolic and antioxidant responses to fertilization and foliar treatments, mostly benefiting from conventional fertilization and Fe spraying, while organic fertilization showed variable effects, likely due to slower nutrient release. In contrast, the lowest malondialdehyde content was observed in *S. cypria* grown with OrFert and treated with Fe and Zn spraying, with their combination indicating an amelioration effect. In *O. dubium*, the phenolics content, antioxidant activity, and hydrogen peroxide production were decreased by Fe and Zn spraying under adequate nutrition. Finally, Zn spraying resulted in improvements of water use efficiency on both species, contingent on fertilization application, while conventional cultivation led to an improvement on nutrient use efficiency, provided by the immediate supply of nutrients. In conclusion, the utilization of organic fertilizers in *S. cypria* and *O. dubium* cultivation, although inducing variations in terms of leaf chlorophyll content, exhibited adequate fertilizer efficiency. Additionally, Fe and Zn spraying had a significant role in the antioxidant capacity and mineral uptake of the examined species. However, appropriate fertilization management is needed to fully utilize the benefits of foliar applications, especially in the context of organic practices.

Author Contributions: Conceptualization, N.T.; methodology, A.C. and N.T.; software, A.C. and N.Z.; validation, E.H., and A.C.; formal analysis, E.H., N.Z. and A.C.; investigation, E.H., N.Z. and A.C.; resources, N.T.; data curation, N.T.; writing—original draft preparation, A.C., and N.T.; writing—review and editing, A.C. and N.T.; visualization, A.C.; supervision, N.T.; project administration, N.T.; funding acquisition, N.T. All authors have read and agreed to the published version of the manuscript.

Funding: This research was funded by the Project "Opti-AromaQ" EXCELLENCE/0421/0299 which is co-financed by the European Union and the Republic of Cyprus through the Research and Innovation Foundation.

Data Availability Statement: Data are contained within the article.

Conflicts of Interest: The authors declare no conflicts of interest.

References

1. Ventura, Y.; Eshel, A.; Pasternak, D.; Sagi, M. The development of halophyte-based agriculture: Past and present. *Ann. Bot.* **2015**, *115*, 529–540. [CrossRef]
2. Łuczaj, Ł.; Pieroni, A.; Tardío, J.; Pardo-De-Santayana, M.; Sõukand, R.; Svanberg, I.; Kalle, R. Wild food plant use in 21st century Europe: The disappearance of old traditions and the search for new cuisines involving wild edibles. *Acta Soc. Bot. Pol.* **2012**, *81*, 359–370. [CrossRef]
3. Feng, T.; Xiong, R.; Huan, P. Productive use of natural resources in agriculture: The main policy lessons. *Resour. Policy* **2023**, *85*, 103793. [CrossRef]
4. Rostaei, M.; Fallah, S.; Carrubba, A.; Lorigooini, Z. Organic manures enhance biomass and improve content, chemical compounds of essential oil and antioxidant capacity of medicinal plants: A review. *Heliyon* **2024**, *10*, e36693. [CrossRef]
5. Seppelt, R.; Klotz, S.; Peiter, E.; Volk, M. Agriculture and food security under a changing climate: An underestimated challenge. *iScience* **2022**, *25*, 105551. [CrossRef] [PubMed]
6. Mardanluo, S.; Souri, M.K.; Ahmadi, M. Plant growth and fruit quality of two pepper cultivars under different potassium levels of nutrient solutions. *J. Plant Nutr.* **2018**, *41*, 1604–1614. [CrossRef]
7. Du, J.; Yu, Y.; Tang, C.; Zong, K.; Zhang, S.; Zhang, Q.; Fang, L.; Li, Y. Organic fertilizers increase the proportion of saprotrophs favoring soil nitrification under medicinal plants *Fritillaria thunbergii*. *Ind. Crops Prod.* **2024**, *219*, 119129. [CrossRef]
8. Järvan, M.; Edesi, L. The effect of cultivation methods on the yield and biological quality of potato. *Agron. Res.* **2009**, *7*, 289–299.
9. Chrysargyris, A.; Kloukina, C.; Vassiliou, R.; Tomou, E.-M.; Skaltsa, H.; Tzortzakis, N. Cultivation strategy to improve chemical profile and anti-oxidant activity of *Sideritis perfoliata* L. subsp. perfoliata. *Ind. Crops Prod.* **2019**, *140*, 111694. [CrossRef]
10. Manik, S.M.N.; Pengilley, G.; Dean, G.; Field, B.; Shabala, S.; Zhou, M. Soil and crop management practices to minimize the impact of waterlogging on crop productivity. *Front. Plant Sci.* **2019**, *10*, 140. [CrossRef]
11. Kopittke, P.M.; Menzies, N.W.; Wang, P.; McKenna, B.A.; Lombi, E. Soil and the intensification of agriculture for global food security. *Environ. Int.* **2019**, *132*, 105078. [CrossRef]
12. Dekemati, I.; Simon, B.; Vinogradov, S.; Birkás, M. The effects of various tillage treatments on soil physical properties, earthworm abundance and crop yield in Hungary. *Soil Tillage Res.* **2019**, *194*, 104334. [CrossRef]
13. Tscharntke, T.; Grass, I.; Wanger, T.C.; Westphal, C.; Batáry, P. Beyond organic farming–harnessing biodiversity-friendly landscapes. *Trends Ecol. Evol.* **2021**, *36*, 919–930. [CrossRef]
14. Fernández, J.A.; Ayastuy, M.E.; Belladonna, D.P.; Comezaña, M.M.; Contreras, J.; de Maria Mourão, I.; Orden, L.; Rodríguez, R.A. Current trends in organic vegetable crop production: Practices and techniques. *Horticulturae* **2022**, *8*, 893. [CrossRef]
15. Barbieri, P.; Pellerin, S.; Nesme, T. Comparing crop rotations between organic and conventional farming. *Sci. Rep.* **2017**, *7*, 13761. [CrossRef] [PubMed]
16. Litskas, V.; Chrysargyris, A.; Stavrinides, M.; Tzortzakis, N. Water-energy-food nexus: A case study on medicinal and aromatic plants. *J. Clean. Prod.* **2019**, *233*, 1334–1343. [CrossRef]
17. Carrascosa, Á.; Pascual, J.A.; Ros, M.; Petropoulos, S.; del Mar Alguacil, M. The effect of fertilization regime on growth parameters of *Sonchus oleraceus* and two genotypes of *Portulaca oleracea*. *Biol. Life Sci. Forum* **2022**, *16*, 7. [CrossRef]
18. Chrysargyris, A.; Koutsoumpeli, E.; Xylia, P.; Fytrou, A.; Konstantopoulou, M.; Tzortzakis, N. Organic cultivation and deficit irrigation practices to improve chemical and biological activity of *Mentha spicata* plants. *Agronomy* **2021**, *11*, 599. [CrossRef]

19. Molina, M.; Tardío, J.; Aceituno-Mata, L.; Morales, R.; Reyes-García, V.; Pardo-De-Santayana, M. Weeds and food diversity: Natural yield assessment and future alternatives for traditionally consumed wild vegetables. *J. Ethnobiol.* **2014**, *34*, 44–67. [CrossRef]
20. Paschoalinotto, B.H.; Polyzos, N.; Compocholi, M.; Rouphael, Y.; Alexopoulos, A.; Dias, M.I.; Barros, L.; Petropoulos, S.A. Domestication of wild edible species: The response of *Scolymus hispanicus* plants to different fertigation regimes. *Horticulturae* **2023**, *9*, 103. [CrossRef]
21. Herencia, J.F.; García-Galavís, P.A.; Dorado, J.A.R.; Maqueda, C. Comparison of nutritional quality of the crops grown in an organic and conventional fertilized soil. *Sci. Hortic.* **2011**, *129*, 882–888. [CrossRef]
22. Hallmann, E.; Sabała, P. Organic and conventional herbs quality reflected by their antioxidant compounds concentration. *Appl. Sci.* **2020**, *10*, 3468. [CrossRef]
23. Raei, Y.; Alami-milani, M. Organic cultivation of medicinal plants: A review. *J. Biodivers. Environ. Sci.* **2014**, *4*, 6–18.
24. Malik, A.A.; Suryapani, S.; Ahmad, J. Chemical vs organic cultivation of medicinal and aromatic plants: The choice is clear. *Int. J. Med. Aromat. Plants* **2011**, *1*, 5–13.
25. Lv, J.; Huang, H.; Yu, L.; Whent, M.; Niu, Y.; Shi, H.; Wang, T.T.Y.; Luthria, D.; Charles, D.; Yu, L.L. Phenolic composition and nutraceutical properties of organic and conventional cinnamon and peppermint. *Food Chem.* **2012**, *132*, 1442–1450. [CrossRef] [PubMed]
26. Ahuja, M.R.; Mohan Jain, S. (Eds.) *Genetic Diversity and Erosion in Plants: Indicators and Prevention*; Springer International Publishing: Cham, Switzerland, 2015; Volume 1, ISBN 9783319256351.
27. Nazari, M.; Zarinkamar, F.; Mohammad Soltani, B.; Niknam, V. Manganese-induced changes in glandular trichomes density and essential oils production of *Mentha aquatica* L. at different growth stages. *J. Trace Elem. Med. Biol.* **2018**, *50*, 57–66. [CrossRef]
28. Elhindi, K.; Al-Suhaibani, N.A.; Sharaf El-Din, A.F.; Yakout, S.M.; Al-Amri, S.M. Effect of foliar-applied iron and zinc on growth rate and essential oil in sweet basil (*Ocimum basilicum* L.) under saline conditions. *Prog. Nutr.* **2016**, *18*, 288–298.
29. Jahani, F.; Tohidi-Moghadam, H.R.; Larijani, H.R.; Ghooshchi, F.; Oveysi, M. Influence of zinc and salicylic acid foliar application on total chlorophyll, phenolic components, yield and essential oil composition of peppermint (*Mentha piperita* L.) under drought stress condition. *Arab. J. Geosci.* **2021**, *14*, 691. [CrossRef]
30. Shoormij, F.; Mirlohi, A.; Saeidi, G.; Shirvani, M. Combined foliar application of Zn and Fe increases grain micronutrient concentrations and alleviates water stress across diverse wheat species and ploidal levels. *Sci. Rep.* **2022**, *12*, 20328. [CrossRef]
31. Tzortzakis, N.; Neofytou, G.; Chrysargyris, A. Nitrogen fertilization coupled with zinc foliar applications modulate the production, quality, and stress response of *Sideritis cypria* plants grown hydroponically under excess copper concentrations. *Plants* **2025**, *14*, 691. [CrossRef]
32. Said-Al Ahl, H.A.H.; Hussein, M.S.; Abd El-Kader, A.A. Effect of nitrogen fertilizer and/or some foliar application on growth, herb yield, essential oil and chemical composition of dragonhead. *J. Med. Food Plants* **2010**, *2*, 12–28.
33. Chrysargyris, A.; Tzortzakis, N. Iron and zinc foliar spraying affected *Sideritis cypria* Post. growth, mineral content and antioxidant properties. *Plants* **2025**, *14*, 840. [CrossRef]
34. Hassanpouraghdam, M.B.; Mehrabani, L.V.; Tzortzakis, N. Foliar application of nano-zinc and iron affects physiological attributes of *Rosmarinus officinalis* and quietens NaCl salinity depression. *J. Soil Sci. Plant Nutr.* **2019**, *20*, 335–345. [CrossRef]
35. Nasiri, Y.; Zehtab-Salmasi, S.; Nasrullahzadeh, S.; Najafi, N.; Ghassemi-Golezani, K. Effects of foliar application of micronutrients (Fe and Zn) on flower yield and essential oil of chamomile (*Matricaria chamomilla* L.). *J. Med. Plants Res.* **2010**, *4*, 1733–1737. [CrossRef]
36. Ruiz-Torres, N.; Flores-Naveda, A.; Barriga-Castro, E.D.; Camposeco-Montejo, N.; Ramírez-Barrón, S.; Borrego-Escalante, F.; Niño-Medina, G.; Hernández-Juárez, A.; Garza-Alonso, C.; Rodríguez-Salinas, P.; et al. Zinc oxide nanoparticles and zinc sulfate impact physiological parameters and boosts lipid peroxidation in soil grown coriander plants (*Coriandrum sativum*). *Molecules* **2021**, *26*, 1998. [CrossRef]
37. Senanayake, S.P.J.N. Rosemary extract as a natural source of bioactive compounds. *J. Food Bioact.* **2018**, *2*, 51–57. [CrossRef]
38. Priss, O.; Yevlash, V. Technology of fresh herbs storage using hydrogel and antioxidant composition. *Food Environ. Saf.* **2017**, *16*, 256–261.
39. Lytra, K.; Tomou, E.; Chrysargyris, A.; Drouza, C.C.; Skaltsa, H.; Tzortzakis, N. Traditionally used *Sideritis cypria* Post: Phytochemistry, nutritional content, bioactive compounds of cultivated populations. *Front. Pharmacol.* **2020**, *11*, 650. [CrossRef]
40. Yildirim, E.; Turan, M.; Ekinci, M.; Ercisli, S.; Ozturk, H.I.; Aydin, M.; Ilhan, E.; Vicas, S.I.; Iancu, C.V.; Gitea, D.; et al. Composition of anthocyanins, specific sugars, and organic acids in wild edible aromatic and medicinal vegetables. *Horticulturae* **2025**, *11*, 145. [CrossRef]
41. Hanoğlu, D.Y.; Hanoğlu, A.; Güvenir, M.; Süer, K.; Demirci, B.; Başer, K.H.C.; Yavuz, D.Ö. Chemical composition and antimicrobial activity of the essential oil of *Sideritis cypria* Post endemic in Northern Cyprus. *J. Essent. Oil Res.* **2017**, *29*, 228–232. [CrossRef]

42. Karioti, A.; Vrahimi-Hadjilouca, T.; Droushiotis, D.; Rancic, A.; Hadjipavlou-Litina, D.; Skaltsa, H. Analysis of the essential oil of *Origanum dubium* growing wild in Cyprus. Investigation of its antioxidant capacity and antimicrobial activity. *Planta Med.* **2006**, *72*, 1330–1334. [CrossRef]
43. Xylia, P.; Chrysargyris, A.; Miltiadous, P.; Tzortzakis, N. *Origanum dubium* (cypriot oregano) as a promising sanitizing agent against *Salmonella enterica* and *Listeria monocytogenes* on tomato and cucumber fruits. *Biology* **2022**, *11*, 1772. [CrossRef] [PubMed]
44. Karousou, R.; Deirmentzoglou, S. The herbal market of Cyprus: Traditional links and cultural exchanges. *J. Ethnopharmacol.* **2011**, *133*, 191–203. [CrossRef]
45. Chrysargyris, A.; Tzortzakis, N. Optimizing nitrogen, phosphorus, and potassium requirements to improve *Origanum dubium* Boiss. growth, nutrient and water use efficiency, essential oil yield and composition. *Ind. Crop. Prod.* **2025**, *224*, 120291. [CrossRef]
46. Chrysargyris, A.; Tzortzakis, N. Nitrogen, phosphorus, and potassium requirements to improve *Sideritis cypria* growth, nutrient and water use efficiency in hydroponic cultivation. *Heliyon* **2025**, *11*, e40755. [CrossRef] [PubMed]
47. Chrysargyris, A.; Petrovic, J.D.; Tomou, E.; Kyriakou, K.; Xylia, P.; Kotsoni, A.; Gkretsi, V.; Miltiadous, P.; Skaltsa, H.; Sokovi, M.D.; et al. Phytochemical profiles and biological activities of plant extracts from aromatic plants cultivated in Cyprus. *Biology* **2024**, *13*, 45. [CrossRef]
48. Bergstrand, K.-J. Organic fertilizers in greenhouse production systems–a review. *Sci. Hortic.* **2022**, *295*, 110855. [CrossRef]
49. Tzortzakis, N.; Chrysargyris, A. Olive-mill and grape-mill residue impact the growth, physiology and nutrient status of grapevines young cuttings. *Sustain. Chem. Pharm.* **2024**, *37*, 101362. [CrossRef]
50. Richardson, A.D.; Duigan, S.P.; Berlyn, G.P. An evaluation of noninvasive methods to estimate foliar chlorophyll content. *New Phytol.* **2002**, *153*, 185–194. [CrossRef]
51. Meyers, K.J.; Watkins, C.B.; Pritts, M.P.; Liu, R.H. Antioxidant and antiproliferative activities of strawberries. *J. Agric. Food Chem.* **2003**, *51*, 6887–6892. [CrossRef]
52. Wojdyło, A.; Oszmiański, J.; Czemerys, R. Antioxidant activity and phenolic compounds in 32 selected herbs. *Food Chem.* **2007**, *105*, 940–949. [CrossRef]
53. Číž, M.; Čížová, H.; Denev, P.; Kratchanova, M.; Slavov, A.; Lojek, A. Different methods for control and comparison of the antioxidant properties of vegetables. *Food Control* **2010**, *21*, 518–523. [CrossRef]
54. Hajisolomou, E.; Neofytou, G.; Petropoulos, S.A.; Tzortzakis, N. The application of conventional and organic fertilizers during wild edible species cultivation: A case study of purslane and common sowthistle. *Horticulturae* **2024**, *10*, 1222. [CrossRef]
55. Loreto, F.; Velikova, V. Isoprene produced by leaves protects the photosynthetic apparatus against ozone damage, quenches ozone products, and reduces lipid peroxidation of cellular membranes. *Plant Physiol.* **2001**, *127*, 1781–1787. [CrossRef] [PubMed]
56. De Azevedo Neto, A.D.; Prisco, J.T.; Enéas-Filho, J.; De Abreu, C.E.B.; Gomes-Filho, E. Effect of salt stress on antioxidative enzymes and lipid peroxidation in leaves and roots of salt-tolerant and salt-sensitive maize genotypes. *Environ. Exp. Bot.* **2006**, *56*, 87–94. [CrossRef]
57. Carrascosa, A.; Pascual, J.A.A.; Ros, M.; Petropoulos, S.A.; Alguacil, M.d.M. Agronomical practices and management for commercial cultivation of *Portulaca oleracea* as a crop: A review. *Plants* **2023**, *12*, 1246. [CrossRef] [PubMed]
58. Kazimierczak, R.; Hallmann, E.; Rembiałkowska, E. Effects of organic and conventional production systems on the content of bioactive substances in four species of medicinal plants. *Biol. Agric. Hortic.* **2015**, *31*, 118–127. [CrossRef]
59. Peng, L.-C.; Ng, L.-T. Impacts of nitrogen and phosphorus fertilization on biomass, polyphenol contents, and essential oil yield and composition of *Vitex negundo* Linn. *Agriculture* **2022**, *12*, 859. [CrossRef]
60. Ahmadi, T.; Shabani, L.; Sabzalian, M.R. Improvement in drought tolerance of lemon balm, *Melissa officinalis* L. under the pre-treatment of LED lighting. *Plant Physiol. Biochem.* **2019**, *139*, 548–557. [CrossRef]
61. Ye, J.Y.; Tian, W.H.; Jin, C.W. Nitrogen in plants: From nutrition to the modulation of abiotic stress adaptation. *Stress Biol.* **2022**, *2*, 4. [CrossRef]
62. Adamczyk-Szabela, D.; Wolf, W.M. The influence of copper and zinc on photosynthesis and phenolic levels in basil (*Ocimum basilicum* L.), borage (*Borago officinalis* L.), common nettle (*Urtica dioica* L.) and peppermint (*Mentha piperita* L.). *Int. J. Mol. Sci.* **2024**, *25*, 3612. [CrossRef]
63. Amujoyegbe, B.J.; Opabode, J.T.; Olayinka, A. Effect of organic and inorganic fertilizer on yield and chlorophyll content of maize (*Zea mays* L.) and sorghum (*Sorghum bicolour* (L.) Moench). *African J. Biotechnol.* **2007**, *6*, 1869–1873. [CrossRef]
64. Aslam, W.; Arfan, M.; Shahid, S.A.; Anwar, F.; Mahmood, Z.; Rashid, U. Effects of exogenously applied Zn on the growth, yield, chlorophyll contents and nutrient accumulation in wheat line L-5066. *Int. J. Pharm. Chem. Biol. Sci* **2014**, *5*, 11–15.
65. Gao, D.; Ran, C.; Zhang, Y.; Wang, X.; Lu, S.; Geng, Y.; Guo, L.; Shao, X. Effect of different concentrations of foliar iron fertilizer on chlorophyll fluorescence characteristics of iron-deficient rice seedlings under saline sodic conditions. *Plant Physiol. Biochem.* **2022**, *185*, 112–122. [CrossRef] [PubMed]
66. Awad, A.A.M.; Rady, M.M.; Semida, W.M.; Belal, E.E.; Omran, W.M.; Al-Yasi, H.M.; Ali, E.F. Foliar nourishment with different zinc-containing forms effectively sustains carrot performance in zinc-deficient soil. *Agronomy* **2021**, *11*, 1853. [CrossRef]

67. Xue, Y.; Yan, W.; Gao, Y.; Zhang, H.; Jiang, L.; Qian, X.; Cui, Z.; Zhang, C.; Liu, S.; Wang, H.; et al. Interaction effects of nitrogen rates and forms combined with and without zinc supply on plant growth and nutrient uptake in maize seedlings. *Front. Plant Sci.* **2021**, *12*, 722752. [CrossRef]
68. Roosta, H.R.; Mohsenian, Y. Effects of foliar spray of different Fe sources on pepper (*Capsicum annum* L.) plants in aquaponic system. *Sci. Hortic.* **2012**, *146*, 182–191. [CrossRef]
69. Ahmed, R.; Uddin, M.K.; Quddus, M.A.; Samad, M.Y.A.; Hossain, M.A.M.; Haque, A.N.A. Impact of foliar application of zinc and zinc oxide nanoparticles on growth, yield, nutrient uptake and quality of tomato. *Horticulturae* **2023**, *9*, 162. [CrossRef]
70. Alinejad Elahshah, A.; Moradi, H.; Sadeghi, H. Boron and zinc foliar application enhanced the morphophysiological responses and mineral absorption in the hydroponically grown 'Aromas' strawberry. *J. Plant Nutr.* **2023**, *46*, 3487–3498. [CrossRef]
71. Raliya, R.; Tarafdar, J.C.; Biswas, P. Enhancing the mobilization of native phosphorus in the mung bean rhizosphere using ZnO nanoparticles synthesized by soil fungi. *J. Agric. Food Chem.* **2016**, *64*, 3111–3118. [CrossRef]
72. Neofytou, G.; Chrysargyris, A.; Xylia, P.; Botsaris, G.; Tzortzakis, N. Foliar iron and zinc modulate the qualitative and nutritional status of *Sideritis cypria* with diverse rates of phosphorus in hydroponic cultivation. *Agronomy* **2025**, *15*, 1178. [CrossRef]
73. Bagheri, H.; Ladan Moghadam, A.; Danaee, E.; Abdossi, V. Morphophysiological and phytochemical changes of *Mentha piperita* using calcium, potassium, iron and manganese nano-fertilizers. *Eur. J. Hortic. Sci.* **2021**, *86*, 419–430. [CrossRef]
74. Rouphael, Y.; Raimondi, G.; Paduano, A.; Sacchi, R.; Barbieri, G.; De Pascale, S. Influence of organic and conventional farming on seed yield, fatty acid composition and tocopherols of Perilla. *Aust. J. Crop Sci.* **2015**, *9*, 303–308.
75. Farnoosh, S.; Masoudian, N.; Safipour Afshar, A.; Nematpour, F.S.; Roudi, B. Foliar-applied iron and zinc nanoparticles improved plant growth, phenolic compounds, essential oil yield, and rosmarinic acid production of lemon balm (*Melissa officinalis* L.). *Environ. Sci. Pollut. Res.* **2024**, *31*, 36882–36893. [CrossRef]
76. Ochoa-Velasco, C.E.; Valadez-Blanco, R.; Salas-Coronado, R.; Sustaita-Rivera, F.; Hernández-Carlos, B.; García-Ortega, S.; Santos-Sánchez, N.F. Effect of nitrogen fertilization and *Bacillus licheniformis* biofertilizer addition on the antioxidants compounds and antioxidant activity of greenhouse cultivated tomato fruits (*Solanum lycopersicum* L. var. Sheva). *Sci. Hortic.* **2016**, *201*, 338–345. [CrossRef]
77. Pagare, S.S.; Bhatia, M.; Tripathi, N.; Pagare, S.S.; Bansal, Y.K. Secondary metabolites of plants and their role: Overview. *Curr. Trends Biotechnol. Pharm.* **2015**, *9*, 293–304.
78. Liu, Y.; Lan, X.; Hou, H.; Ji, J.; Liu, X.; Lv, Z. Multifaceted ability of organic fertilizers to improve crop productivity and abiotic stress tolerance: Review and perspectives. *Agronomy* **2024**, *14*, 1141. [CrossRef]
79. Mogazy, A.M.; Hanafy, R.S. Foliar spray of biosynthesized zinc oxide nanoparticles alleviate salinity stress effect on *Vicia faba* plants. *J. Soil Sci. Plant Nutr.* **2022**, *22*, 2647–2662. [CrossRef]
80. Klofac, D.; Antosovsky, J.; Skarpa, P. Effect of zinc foliar fertilization alone and combined with trehalose on maize (*Zea mays* L.) growth under the drought. *Plants* **2023**, *12*, 2539. [CrossRef]
81. Liu, Z.; Su, J.; Luo, X.; Meng, J.; Zhang, H.; Li, P.; Sun, Y.; Song, J.; Peng, X.; Yu, C. Nitrogen limits zinc-mediated stimulation of tillering in rice by modifying phytohormone balance under low-temperature stress. *Food Energy Secur.* **2022**, *11*, e359. [CrossRef]
82. Sida-Arreola, J.P.; Sánchez-Chávez, E.; Ávila-Quezada, G.D.; Zamudio-Flores, P.B.; Acosta Muñíz, C.H. Iron biofortification and its impact on antioxidant system, yield and biomass in common bean. *Plant Soil Environ.* **2015**, *61*, 573–576. [CrossRef]
83. Xie, G.; Zhang, X.; Duan, X.; Song, L.; Fan, M.; Wang, C.; Zhao, L. Effects of tillage and substitution of chemical fertilizers with organic fertilizers on leaf physiological characteristics and yield of maize. *Int. J. Plant Prod.* **2025**, *19*, 449–467. [CrossRef]
84. Nieves-Cordones, M.; López-Delacalle, M.; Ródenas, R.; Martínez, V.; Rubio, F.; Rivero, R.M. Critical responses to nutrient deprivation: A comprehensive review on the role of ROS and RNS. *Environ. Exp. Bot.* **2019**, *161*, 74–85. [CrossRef]
85. Lay-Pruitt, K.S.; Wang, W.; Prom-u-thai, C.; Pandey, A.; Zheng, L.; Rouached, H. A tale of two players: The role of phosphate in iron and zinc homeostatic interactions. *Planta* **2022**, *256*, 23. [CrossRef] [PubMed]
86. Anwar, S.; Khalilzadeh, R.; Khan, S.; Zaib-un-Nisa; Bashir, R.; Pirzad, A.; Malik, A. Mitigation of drought stress and yield improvement in wheat by zinc foliar spray relates to enhanced water use efficiency and zinc contents. *Int. J. Plant Prod.* **2021**, *15*, 377–389. [CrossRef]
87. Fan, X.; Zhou, X.; Chen, H.; Tang, M.; Xie, X. Cross-talks between macro- and micronutrient uptake and signaling in plants. *Front. Plant Sci.* **2021**, *12*, 663477. [CrossRef] [PubMed]
88. Toda, M.; Walder, F.; van der Heijden, M.G.A. Organic management and soil health promote nutrient use efficiency. *J. Sustain. Agric. Environ.* **2023**, *2*, 215–224. [CrossRef]

horticulturae

MDPI

Article

Peltate Glandular Trichomes in Relation to Their Parameters, Essential Oil Amount, Chemotype, Plant Sex and Habitat Characteristics in *Thymus pulegioides*

Kristina Ložienė

Laboratory of Economic Botany, State Scientific Research Institute Nature Research Centre, Akademijos St. 2, 08412 Vilnius, Lithuania; kristina.loziene@gamtc.lt

Abstract

The parameters and plant habitat characteristics of glandular trichomes could allow for faster and cheaper identification and selection of more essential oil-rich wild aromatic plants for further cultivation. This study aimed to establish relationships between the parameters of peltate glandular trichomes and essential oil content in commercially potential *Thymus pulegioides* in relation to plant sex, chemotype, and habitat characteristics. In total, 124 *T. pulegioides* plants belonging to different chemotypes and sexes and collected from 23 natural habitats were analysed. Essential oils were extracted by hydrodistillation, and a light microscope was used to investigate parameters of peltate glandular trichomes in upper and lower leaf epidermises. For investigation of the dynamics of the parameters of peltate glandular trichomes, *T. pulegioides* were growing in open ground under the same environmental conditions. Results demonstrated that the essential oil percentage in phenolic chemotype plants was higher than in plants of a non-phenolic chemotype. Females and hermaphrodites did not significantly differ according to essential oil percentage. Cover abundance of *T. pulegioides* negatively affects the density and diameter of peltate glandular trichomes and the essential oil percentage in *T. pulegioides*. The parameters of peltate trichomes in the upper leaf epidermis could be anatomical markers, helping to select *T. pulegioides* with higher essential oil contents from natural habitats as promising candidates as new crops.

Keywords: anatomical and environmental markers; chemical and sex polymorphism; *Thymus pulegioides*

1. Introduction

Although the collection of wild aromatic plants (which are most often also medicinal plants) is still an important method of obtaining raw material of these plants, it can cause the loss of their genetic diversity and habitat destruction [1]. Therefore, to meet the growing demand, to preserve genetic resources, and to guarantee the sustainable use of wild aromatic plants, the domestication and the cultivation of their wild forms are important. The content of essential oil, which is greatly influenced by the genetic characteristics of the plant, is one of the initial criteria for selecting aromatic plants from their natural habitats for their future cultivation [2].

Species of the *Thymus* (Lamiaceae) genus are pharmaceutically valuable essential oil-bearing medicinal and aromatic plants, many of which are commercially important as sources of secondary metabolites and phytochemicals [3,4]. The intraspecific chemical polymorphism—when plants of the same species accumulate essential oils with different chemical compositions—as well as the sexual dimorphism (gynodioecy)—when females (i.e., male-sterile individuals) coexist with hermaphroditic individuals in natural populations—is typical for species of this genus [5,6]. Plants of *Thymus* species possess two types of glandular trichomes—peltate and capitate. Literature data suggest that the peltate glandular trichomes (consisting of one basal epidermal cell; a wide, short stalk; and several secretory head cells) are more associated with the synthesis of essential oils in various members of the Lamiaceae genus because they produce most of monoterpenes and sesquiterpenes, the main compounds of essential oils [7,8]. When describing the raw material of *Thymi herba* in the European Pharmacopoeia, only peltate glandular trichomes are also described as secretory trichomes made up of secretory cells, the cuticle of which is generally raised by secretion to form a globule to avoid bladder-like covering [9]. Additionally, microscopic observations of peltate glandular trichome distribution and abundance in distinct plant organs of *Thymus albicans* showed their highest density in leaves, which is significantly distinct from their density in calyces, corollas, and bracts [10]. The morphological and chemical differences between females and hermaphrodites and different chemotypes in species of the *Thymus* genus have been studied [6,11]. However, the differences in the anatomical characteristics related to essential oil accumulation have not been investigated either between females and hermaphrodites or between different chemotypes.

Essential oil-bearing large thyme (*Thymus pulegioides* L.), widely occurring in Central Europe, Scandinavia, the Baltic region, and Southeast England, is a medicinal and aromatic plant characterised by different pharmaceutical properties, with a high intraspecific chemical polymorphism [12–15] and strongly expressed sexual dimorphism with high female frequency in natural habitats [16]. Due to the chemical polymorphism inherent to the species, it is possible to select wild-growing *T. pulegioides* clones belonging to different chemotypes with characteristic biologically active chemical compounds [16]. One of the initial selection criteria is the essential oil content, which, like other species of the *Thymus* genus, is highest during the flowering stage and varies significantly between different plants [9,17].

Determining the preliminary essential oil content in a plant not by classic hydrodistillation but by calculating its predicted content from the parameters of the peltate glandular trichomes would be a faster and cheaper method (it could be performed in nature with a portable microscope). Establishing the relationships between essential oil content and the parameters of peltate glandular trichomes and habitat characteristics could be helpful in order to facilitate the selection of *T. pulegioides* that accumulate more essential oil. Therefore, this study aimed to establish the relationships between the diameter and density of peltate glandular trichomes and essential oil content in relation to plant sex, chemotype, and some habitat characteristics.

2. Materials and Methods

2.1. Essential Oil Content, Chemotype, and Sex of Thymus pulegioides Plants

In total, 124 blooming *T. pulegioides* plants were randomly collected from 23 natural habitats (5–7 plants each, excluding one habitat with 4 investigated plants) (Table 1). Figure 1 demonstrates *T. pulegioides* growing wild in an investigated natural habitat. Eighty investigated individual plants were female, and 44 were hermaphrodites. A plant's sex was established by the morphology of the flowers: a female individual has a pistil in the flowers only, while a hermaphrodite has both pistils and stamens. Also, 79 individuals of *T. pulegioides* belonged to a group of phenolic chemotypes, and 45 individuals belonged to a group of non-phenolic chemotypes. Each chemotype has a specific odour [18,19], which enables the attribution of *T. pulegioides* individuals to the phenolic or non-phenolic chemotype group. Thus, the assignment of each *T. pulegioides* plant to one of several chemotype groups was performed organoleptically based on the odour of its aboveground parts in the following way: (1) for the group of phenolic chemotypes—according to the presence of the odour of monoterpene phenols thymol and carvacrol (both these isomers smell alike); (2) for the group of non-phenolic chemotypes—according to the presence of the odour of a sweet rose with a specific citrus tang or an "ester" odour with a lavender scent.

Figure 1. *Thymus pulegioides* in natural habitats (photos by A. Almonaitytė).

Essential oils were isolated by hydrodistillation in a Clevenger-type apparatus (Simax, Praha, Czech Republic) in 2 h from all 124 *T. pulegioides* plants [9]. The essential oils were isolated from dried above-ground parts of each plant separately in 2–3 replicates; the contents of essential oils were recalculated as percentages of dry matter (% *v*/*w*).

2.2. Analysis of Peltate Glandular Trichomes

The density and diameter of peltate glandular trichomes in the lower and upper leaf epidermis were investigated separately in all 124 *T. pulegioides* plants. The impress method and an ECLIPSE Ci-S Nikon light microscope (Nikon Europe B.V.) were used to investigate anatomical structures. Both sides of 5–7 fresh leaves, collected from the middle of different flowering stems of each plant, were spread with a thin layer of colourless nail polish, and the formed films were viewed with a light microscope. To assess the density of peltate glandular trichomes in plants, 30–50 view fields (6–7 view fields per leaf) on each leaf side in each plant were examined under a microscope. To assess the diameter of peltate glandular trichomes in the plants, 50 peltate glandular trichomes from each plant's lower and upper epidermis (7–10 peltate glandular trichomes in each analysed leaf) were investigated. The view fields used for analysis of the density and diameter of peltate glandular trichomes were chosen to avoid the margins, apex, and big veins of leaves.

2.3. Analysis of Habitats

A total of 23 different habitats of *T. pulegioides* were described in Lithuania (Table 1). The distances between habitats were not less than 20 km. The study was conducted in 16 m^2 habitat fields according to the methodology of Braun-Blanquet [20]. Plant communities were distinguished according to the vegetation classification systems of J. Balevičienė et al. [21]. Relative lightening of the habitats was defined visually.

Table 1. Characteristics of *Thymus pulegioides* habitats (*F-B*—vegetation class *Festuco-Brometea erecti* 1943; *T-G*—vegetation class *Trifolio-Geranietea sanguinei* 1961; *M-A*—vegetation class *Molinio-Arrhenatheretea elatioris* 1937; *—anthropogenic plant community; **—cover abundance of all *Thymus pulegioides* growing in habitat according to Braun-Blanquet scale [20]; ***—number of chemically and anatomically investigated *Thymus pulegioides* individuals in habitat.

No.	Locality	Coordinates WGS-84	Class of Vegetation	Relief	Inclination, °	Exposition	Lightning, %	Total Herb Cover, %	*Thymus pulegioides* **	*Thymus pulegioides* ***
1	Vilnius dist., Uosininkai	54.714975, 25.587978	*M-A*	wavy	–	–	95	90	1	4
2	Vilnius dist., Dunojai	54.620109, 25.708344	*T-G*	wavy	–	–	80	80	2	6
3	Vilnius dist., Vindžiūnai	54.563248, 25.703089	*F-B*	wavy	–	–	80	80	1	5
4	Vilnius dist., Grikieniai	54.777020, 25.114203	*T-G*	slope	30	S	100	98	2	7
5	Širvintos dist., Kernavė	54.879947, 24.846071	*T-G*	plane	–	–	100	85	+	7
6	Vilnius dist., Trečiokiškės	54.833401, 24.996916	*M-A*	wavy	–	–	98	90	2	5
7	Vilnius, Visoriai	54.753797, 25.261059	*M-A*	wavy	–	–	98	90	3	5
8	Elektrėnai mun., Paaliosė	54.795416, 24.885738	*F-B*	slope	10	S	90	80	1	6
9	Trakai dist., Meiriškės	54.725080, 24.878361	*F-B*	slope	20	S	100	95	1	7
10	Trakai dist., Maušiškės	54.605374, 24.869349	*M-A*	wavy	–	–	100	95	+	5
11	Trakai dist., Onuškis	54.522103, 24.616383	*M-A*	slope	20	SE	100	85	+	5
12	Trakai dist., Rūdiškės	54.512941, 24.783850	*T-G*	plane	–	–	100	90	1	5
13	Vilnius dist., Nemenčinė	54.908972, 25.322706	*M-A*	slope	30	S	100	95	+	5
14	Vilnius dist., Paberžė	55.016576, 25.401064	*M-A*	wavy	–	–	100	90	1	5
15	Molėtai dist., Dubingiai	55.091495, 25.432930	*M-A*	slope	15	SW	100	90	1	5
16	Vilnius, Mažieji Gulbinai	54.777692, 25.290565	*T-G*	plane	–	–	100	90	1	5
17	Vilnius dist., Minkeliai	54.947612, 25.489782	*M-A*	slope	10	SE	100	85	+	5
18	Vilnius dist., Bratoniškės	54.853805, 25.377456	*	plane	–	–	60	75	1	5
19	Šilalė dist., Rubinavas	55.285243, 22.811440	*F-B*	slope	10	SE	100	85	2	6
20	Šilalė dist., Medvėgalis	55.374361, 22231996	*T-G*	slope	30	SW	100	95	+	6
21	Šilalė dist., Kaltinėnai	55.340161, 22.260123	*T-G*	slope	10	S	100	70	+	5
22	Telšiai dist., Rainiai	55.574721, 22.174702	*M-A*	plane	–	–	100	80	1	5
23	Tauragė dist., Skaudvilė	55.406632, 22.598877	*T-G*	slope	20	S	100	85	1	5

The soil samples were collected from each habitat separately and dried at room temperature. Each sample of topsoil was prepared in the following way: 5 subsamples (subject to the area of habitat; each subsample ~100 g) were taken from a depth of 10–15 cm (from the plant rhizosphere) according to the envelope principle with a distance of 1 m from the central point of habitat and mixed (homogenized). The content of mobile phosphorus was estimated photoelectrocolorimetrically; the content of mobile potassium by flame photometry; the contents of mobile manganese, mobile iron, mobile calcium, mobile sulphur, and mobile magnesium by atomic absorption spectroscopy; and the soil pH electrometrically using a 1 M KCl solution. Soil analysis was performed at the Institute of Agriculture of the Lithuanian Research Centre for Agriculture and Forestry.

2.4. Dynamics of Density and Diameter of Peltate Glandular Trichomes

Eight plants of *T. pulegioides* belonging to different chemotypes and grown in the open ground of the field collection of the State Scientific Research Institute Nature Research Centre (Vilnius, Lithuania) were used for the investigation of the dynamics of the density and diameter of peltate glandular trichomes in leaves. These plants were grown under the same environmental conditions and not watered or fertilised. The diameter and density of peltate glandular trichomes were measured from the start of vegetation to the start of flowering every two weeks in the same way as described above.

2.5. Statistical Analysis

A *t*-test was used for the analysis of differences in pairs of various parameters of plants and plant groups (phenolic and non-phenolic chemotypes and females and hermaphrodites), Pearson correlation was used for analysis of correlation relationships between various parameters of plants, Spearman correlation was used for analysis of correlation relationships between parameters of plants groups, and the Kruskal–Wallis test was used for analysis of differences between habitats and habitat groups (belonging to three phytocoenological vegetation classes) according to the parameters of *T. pulegioides* and their habitats. The letter r denotes the correlation coefficients, and the letter N denotes the sample sizes (the number of individual *T. pulegioides* plants or the number of habitats, depending on the context of the content). Statistical data processing was carried out with STATISTICA® 7 and MS Excel 2016 software.

3. Results

3.1. Characteristics of Thymus pulegioides Habitats

Half of the investigated *T. pulegioides* habitats were located on south, southeast, or southwest slopes with 10–30° inclination (Table 1). Most habitats had good light, varying between 90% and 100%, and only three had light below 90%. Total herb cover in habitats varied from 70% to 98% and significantly positively correlated with habitat illumination ($r = 0.42$, $p < 0.05$, $N = 23$). According to the Braun-Blanquet scale, the value of cover-abundance of *T. pulegioides* was usually 1, and only one habitat reached a value of 3. Almost half of the investigated *T. pulegioides* habitats belonged to the grassland communities of phytocoenological vegetation class *Molinio-Arrhenatheretea*, a little less to the *Trifolio-Geranietea* vegetation class, and the least to the *Festuco-Brometea* vegetation class; one habitat in Vilnius district was anthropogenic (Table 1). Variation of soil pH and amounts of some micro- and macroelements in soils of the investigated habitats are presented in Table 2.

Table 2. Soil composition and pH of *Thymus pulegioides* habitats. A sample of topsoil of each habitat was prepared from 5 subsamples, which were taken from the plant rhizosphere according to the envelope principle at a distance of 1 m from the central point of the habitat and mixed. Therefore, the values presented in the table represent averages.

No.	Locality	pH_{KCl}	Phosphorus, mg/kg	Potassium, mg/kg	Manganese, mg/kg	Iron, mg/kg	Calcium, mg/kg	Magnesium, mg/kg	Sulphur, mg/kg
1	Vilnius dist., Uosininkai	7.2	76	72	116	544	8696	768	3.0
2	Vilnius dist., Dunojai	6.0	92	61	54.4	369	856	93	3.5
3	Vilnius dist., Vindžiūnai	6.0	115	78	72.7	477	1116	179	4.1
4	Vilnius dist., Grikieniai	6.7	57	122	176	1092	4776	1076	4.3
5	Širvintos dist., Kernavė	4.8	174	102	85.8	743	644	86	2.2
6	Vilnius dist., Trečiokiškės	7.1	116	78	122	547	6212	1096	7.9
7	Vilnius, Visoriai	7.2	119	86	139	777	12,768	1704	5.1
8	Elektrėnai mun., Paaliosė	6.8	132	52	101	785	1153	241	1.6
9	Trakai dist., Meiriškės	7.4	115	82	84.9	436	20,880	1252	2.2
10	Trakai dist., Maušiškės	7.3	310	63	96.1	501	9396	1412	5.0
11	Trakai dist., Onuškis	7.5	387	92	73.6	794	16,784	1624	3.1
12	Trakai dist., Rūdiškės	6.3	278	96	142	2560	4382	333	2.5
13	Vilnius dist., Nemenčinė	6.1	82	51	50.4	440	974	148	3.4
14	Vilnius dist., Paberžė	6.7	134	118	102	767	3588	1072	1.8
15	Molėtai dist., Dubingiai	7.0	115	92	122	703	3026	746	3.2
16	Vilnius, Mažieji Gulbinai	5.0	212	105	88.2	998	782	108	2.0
17	Vilnius dist., Minkeliai	7.6	141	58	103	766	8864	1792	3.3
18	Vilnius dist., Bratoniškės	6.5	202	79	105	700	1445	222	5.5
19	Šilalė dist., Rubinavas	7.0	180	53	142	808	4336	255	6.7
20	Šilalė dist., Medvėgalis	7.5	179	61	95.9	611	12,672	412	6.8
21	Šilalė dist., Kaltinėnai	5.1	91	78	59.5	611	661	91	5.2
22	Telšiai dist., Rainiai	7.4	110	101	87.2	937	19,900	1084	4.8
23	Tauragė dist., Skaudvilė	6.1	28	85	51.8	270	1657	148	5.4

3.2. Essential Oil Amounts of Thymus pulegioides in Different Habitats, Chemotypes, and Sexes

The essential oil percentage in the studied *T. pulegioides* plants was 0.82 ± 0.23% (N = 124) and varied between individuals within the limits of 0.12–1.68%. The average essential oil percentages of *T. pulegioides* investigated in different habitats are presented in Figure 2: the highest average essential oil percentages (1% and more) were determined in *T. pulegioides* of habitat nos. 11, 12, 18, and 19 and the lowest in those of habitat no. 21. However, only habitat no. 12 significantly ($p < 0.05$) differed from habitat no. 2, as well as habitat no. 21 significantly differing from habitat nos. 12 and 18, according to the essential oil percentage in *T. pulegioides*.

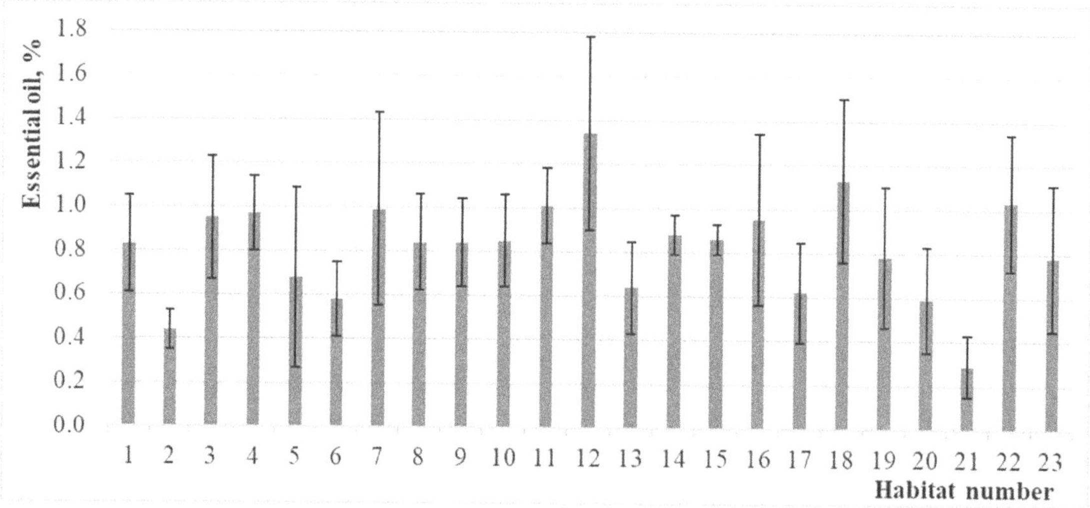

Figure 2. Variation of essential oil percentage in *Thymus pulegioides* between different habitats. Kruskal–Wallis test demonstrated that only habitat no. 12 significantly ($p < 0.05$) differed from habitat no. 2, as well as habitat no. 21 differing from habitat nos. 12 and 18, according to the essential oil percentage in *Thymus pulegioides*. The number of *Thymus pulegioides* individuals investigated in each habitat is presented in Table 1.

Statistically significant ($p < 0.05$) correlations were found between the average essential oil percentage of *T. pulegioides* in the habitat and the mobile potassium and mobile iron content in habitat soil (r = 0.58% and r = 0.57%, respectively; N = 23), as well as between the average essential oil percentage of *T. pulegioides* in the habitat and the cover abundance of *T. pulegioides* in the habitat (r = −0.48, N = 23). Although the average essential oil percentage of *T. pulegioides* in habitats belonging to grassland communities of the *Festuco-Brometea* vegetation class was 0.1% higher than in habitats belonging to grassland communities of the *Trifolio-Geranietea* vegetation class (0.85 ± 0.07% and 0.76 ± 0.33%, respectively), this difference was not statistically significant.

The essential oil percentage in *T. pulegioides* plants of the phenolic chemotypes was higher and significantly ($p < 0.05$) differed from the essential oil percentage in *T. pulegioides* plants of non-phenolic chemotypes (Table 3). Meanwhile, *T. pulegioides* females and hermaphrodites did not significantly differ according to the essential oil percentage, although in females, the average essential oil percentage was higher than in hermaphrodites (Table 3).

Table 3. Essential oil percentage, as well as the density and diameter, of peltate glandular trichomes in the upper and lower leaf epidermis in different chemotypes and sexes of *Thymus pulegioides*. The *t*-test demonstrated that only plants of phenolic chemotypes significantly ($p < 0.05$) differed from plants of no-phenolic chemotypes according to essential oil percentage (marked with different letters in the table). Significant differences in parameters between other groups were not detected.

Parameter		Chemotypes		Sexes	
		Phenolic (N = 79)	Non-Phenolic (N = 45)	Females (N = 80)	Hermaphrodites (N = 44)
Essential oil, %	Means ± SD	0.90 ± 0.32 a	0.67 ± 0.31 b	0.85 ± 0.34	0.76 ± 0.3
	Min	0.18	0.12	0.12	0.21
	Max	1.68	1.54	1.68	1.54
Density of peltate glandular trichomes in upper epidermis (in mm^2)	Means ± SD	7.0 ± 1.8	7.2 ± 1.6	7.2 ± 1.6	7.4 ± 1.9
	Min	3.3	4.3	3.3	4.1
	Max	10.2	11.7	10.9	11.9
Density of peltate glandular trichomes in lower epidermis (in mm^2)	Means ± SD	8.0 ± 2.1	8.1 ± 1.7	8.0 ± 2.1	7.5 ± 2.0
	Min	4.3	4.4	3.7	3.4
	Max	13.2	10.7	13.2	11.4
Diameter of peltate glandular trichomes in upper epidermis, µm	Means ± SD	64.7 ± 3.6	63.3 ± 4.5	64.1 ± 4.1	64.2 ± 3.9
	Min	52.2	51.1	51.1	53.3
	Max	73.7	74.2	73.7	74.2
Diameter of peltate glandular trichomes in lower epidermis, µm	Means ± SD	61.0 ± 5.5	59.6 ± 4.1	59.9 ± 3.4	61.6 ± 7.0
	Min	47.5	48.9	47.5	48.9
	Max	101.0	69.9	67.3	101.0

3.3. *Density of Peltate Glandular Trichomes in Thymus pulegioides Growing in Different Habitats and Belonging to Different Chemotypes and Sexes*

The peltate glandular trichome density was 7.0 ± 1.7 and 8.0 ± 2.0 trichomes in mm^2 (N = 124) and varied from 3.3 to 11.7 and from 4.3 to 13.2 trichomes in mm^2 in the upper and lower leaf epidermis of *T. pulegioides*, respectively. Although the upper and lower leaf epidermises significantly ($p < 0.05$) differed according to the peltate glandular trichome density, a significant correlation was not established between the peltate glandular trichome density in the upper and lower epidermis. Essential oil percentage positively correlated with peltate glandular trichome density in the upper and lower leaf epidermis. However, it was significant only in the upper epidermis (r = 0.34, $p < 0.05$, N = 124) (Figure 3A).

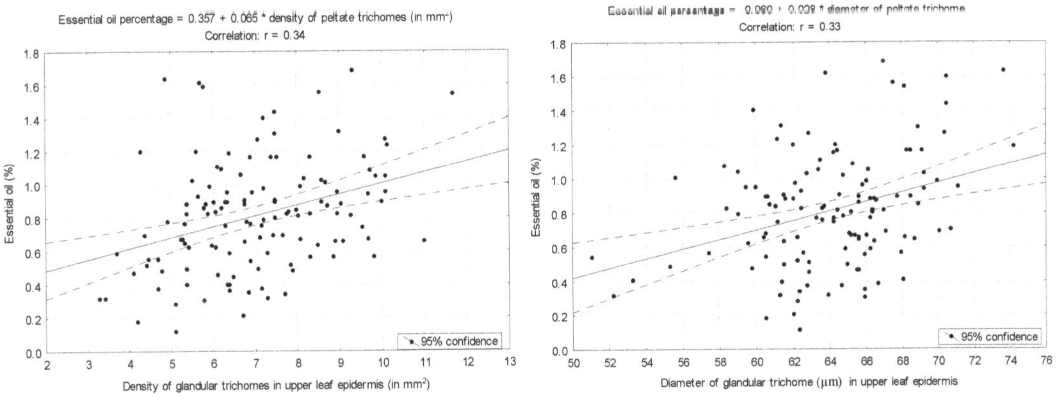

(A) **(B)**

Figure 3. Significant ($p < 0.05$) Spearman rank-order correlations between essential oil percentage and the density (**A**) and diameter (**B**) of peltate glandular trichomes in the upper leaf epidermis in *Thymus pulegioides* plants (N = 124).

Figure 4 represents the average density of peltate glandular trichomes of investigated *T. pulegioides* growing in different habitats. Although average values of peltate glandular trichome density in the upper and lower leaf epidermis of *T. pulegioides* growing in some habitats differed by as much as two times, only habitat nos. 18 and 6 differed significantly ($p < 0.05$) in peltate glandular trichome density in the upper epidermis.

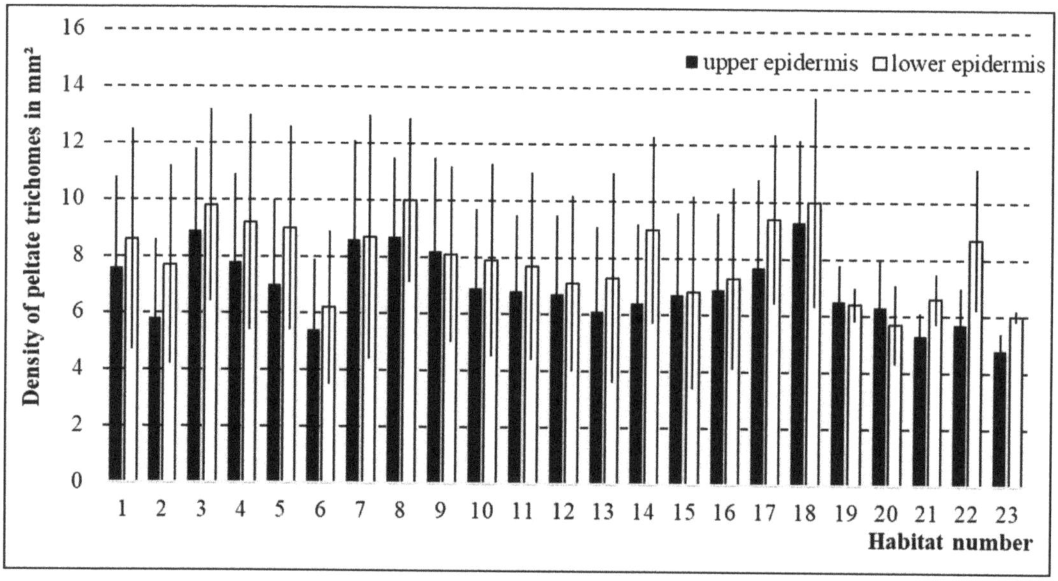

Figure 4. Variation of density of peltate glandular trichomes in the upper and lower leaf epidermis in *Thymus pulegioides* between different habitats. Kruskal–Wallis test demonstrated that only habitat nos. 18 and 6 differed significantly ($p < 0.05$) in terms of peltate glandular trichome density in the upper epidermis. The number of *Thymus pulegioides* individuals investigated in each habitat is presented in Table 1.

Significant correlations of peltate glandular trichome density with amounts of soil elements have not been established. The average density of peltate glandular trichomes in both *T. pulegioides* leaf epidermises in the habitat negatively correlated with habitat illumination and cover abundance of *T. pulegioides* in the habitat. Between the illumination of the habitat and the average density of peltate glandular trichomes in the lower epidermis, this relationship was significant ($r = -0.42$, $p < 0.05$, $N = 23$). *T. pulegioides* plants growing in habitats belonging to the *Molinio-Arrhenatheretea elatioris*, *Festuco-Brometea*, and *Trifolio-Geranietea* vegetation classes did not differ significantly in the density of peltate glandular trichomes in either the upper or lower epidermis.

In non-phenolic chemotypes, the peltate glandular trichome density in the upper and lower epidermis was higher than in phenolic chemotypes (Table 3). However, these differences were not significant. In phenolic and non-phenolic *T. pulegioides* individuals, significant ($p < 0.05$) positive correlations were found between essential oil percentage and density of peltate glandular trichomes in the upper and lower epidermis; these relationships were somewhat stronger in plants of phenolic chemotypes.

Females and hermaphrodites of *T. pulegioides* did not significantly differ in the peltate glandular trichome density in either the upper or lower epidermises. In females and hermaphrodites, the density of peltate glandular trichomes was higher in the lower epidermis than in the upper epidermis (Table 3). In both female individuals and hermaphrodites, statistically significant ($p < 0.05$) correlations were established between essential oil percentage and peltate glandular trichome density only in the upper leaf epidermis; this relationship was stronger in hermaphrodites than in female individuals (r = 0.23, N = 44 and r = 0.52, N = 80, respectively).

3.4. Diameter of Peltate Glandular Trichomes in Thymus pulegioides Growing in Different Habitats and Belonging to Different Chemotypes and Sexes

The diameter of peltate glandular trichomes was 64.2 ± 4.0 μm and 60.5 ± 5.0 μm (N = 124) and varied from 51.1 μm to 74.2 μm and from 47.5 μm to 101.0 μm in the upper and lower leaf epidermis of *T. pulegioides*, respectively (Figure 5). We established statistically significant correlations between the diameter of peltate glandular trichomes in the upper epidermis and the diameter of peltate glandular trichomes in the lower epidermis (r = 0.19, $p < 0.05$, N = 124). Also, significant ($p < 0.05$) positive relationships between the essential oil percentage in *T. pulegioides* and the diameter of peltate glandular trichomes both in the upper and lower epidermis were detected (r = 0.33 and r = 0.24, respectively, N = 124) (Figure 3B). Statistically significant relationships between the density and diameter of peltate glandular trichomes were not detected.

(**A**)

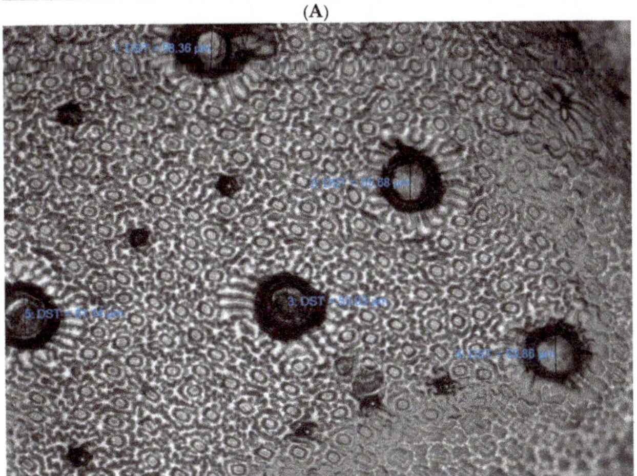

(**B**)

Figure 5. Peltate glandular trichomes and their diameters in the upper (**A**) and lower (**B**) epidermis of *Thymus pulegioides* leaf. In these images, it can be seen that the diameter of peltate trichomes is higher in the upper epidermis, and the density is higher in the lower epidermis (photos by R. Jurevičiūtė).

Diameters of peltate glandular trichomes in *T. pulegioides* growing in different habitats are represented in Figure 6. According to the diameter of peltate glandular trichomes in the lower epidermis, habitat nos. 2 and 3 differed significantly from habitat nos. 14 and 16, and habitat no. 2 also differed significantly from habitat nos. 17 and 23. According to the diameter of peltate glandular trichomes in the upper epidermis, habitat no. 3 differed significantly from habitat nos. 9, 12, and 16, and habitat no. 2 significantly differed from habitat no. 9.

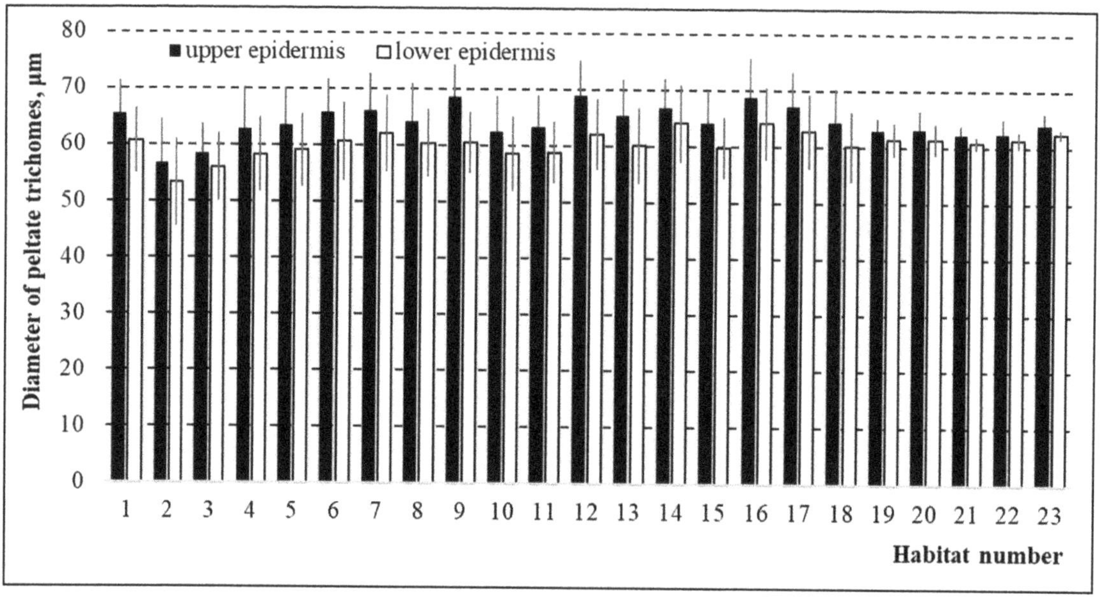

Figure 6. Variation of the diameter of peltate glandular trichomes in the upper and lower leaf epidermis in *Thymus pulegioides* between different habitats. Kruskal–Wallis test demonstrated that (1) according to the diameter of peltate glandular trichomes in the lower epidermis, habitat nos. 2 and 3 differed significantly ($p < 0.05$) from habitat nos. 14 and 16, and habitat no. 2 also significantly differed from habitat nos. 17 and 23; (2) according to the diameter of peltate glandular trichomes in the upper epidermis, habitat no. 3 differed significantly ($p < 0.05$) from habitat nos. 9, 12, and 16, and habitat no. 2 differed significantly from habitat no. 9. The number of *Thymus pulegioides* individuals investigated in each habitat is presented in Table 1.

A significant negative correlation was found between the average diameter of peltate glandular trichomes in the upper leaf epidermis of *T. pulegioides* and the mobile sulphur content of the soil in the habitat ($r = -0.43$, $p < 0.05$, N = 23). Like the density, the average diameter of peltate glandular trichomes in both leaf epidermises of *T. pulegioides* negatively correlated with the cover abundance of *T. pulegioides* in the habitat. *T. pulegioides* plants growing in habitats belonging to the *Molinio-Arrhenatheretea elatioris*, *Festuco-Brometea*, and *Trifolio-Geranietea* vegetation classes did not differ significantly according to the diameter of peltate glandular trichomes in the upper and lower epidermis.

As in the case of the peltate glandular trichome density, the peltate glandular trichome diameter in the upper epidermis did not differ significantly from the diameter of peltate glandular trichomes in the lower epidermis in *T. pulegioides* plants of phenolic chemotypes, but the parameters of these indicators, unlike in the case of peltate glandular trichome density, were higher in *T. pulegioides* plants of phenolic chemotypes than in no-phenolic chemotypes (Table 3).

In both sexes of *T. pulegioides* plants, the diameter of peltate glandular trichomes was higher in the upper epidermis than in the lower epidermis (Table 3), but females did not differ significantly from hermaphrodites in either the diameter of peltate glandular trichomes in the upper epidermis or the diameter of peltate glandular trichomes in the lower epidermis. Both females and hermaphrodites were found to have significant ($p < 0.05$) positive correlations between the essential oil percentage and the peltate glandular trichome diameter of both leaf epidermises.

3.5. Dynamics of Density and Diameter of Peltate Glandular Trichomes During Leaf Vegetation

After evaluating the dynamics of peltate glandular trichome density in *T. pulegioides* during the period from the beginning of leaf development to plant flowering, it was found that at the beginning of plant vegetation (in May), the density of peltate glandular trichomes in the upper leaf epidermis was higher than their density in the lower leaf epidermis (but did not differ significantly) or was it the same. Later, the density of peltate glandular trichomes in the upper leaf epidermis began to decrease and, before and during flowering, was lower than in the lower leaf epidermis; during flowering, the density of peltate glandular trichomes in the lower epidermis differed significantly ($p < 0.05$) from the density in the upper epidermis (Figure 7).

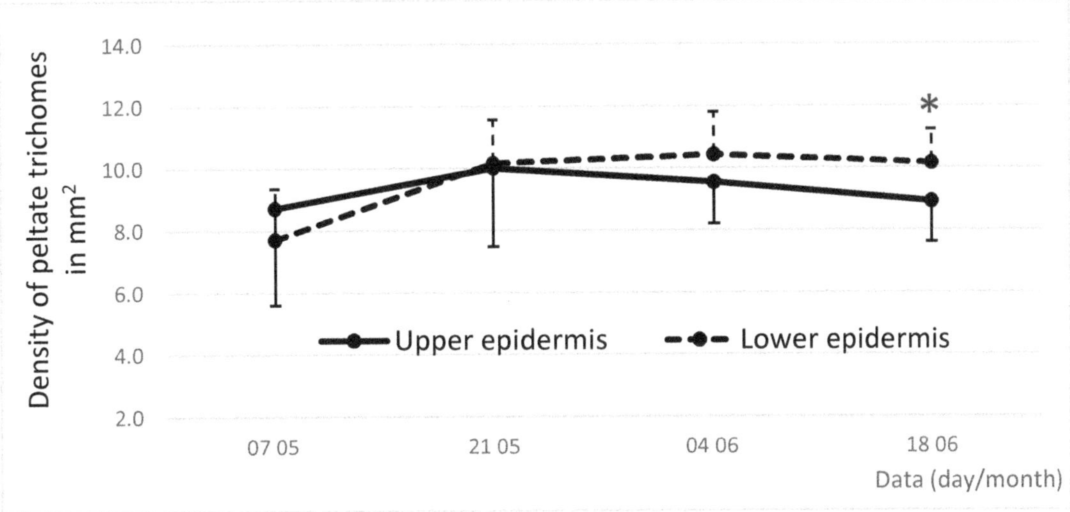

Figure 7. Variation in the density of peltate glandular trichomes in the upper and lower epidermis of *Thymus pulegioides* leaves from the start of plant vegetation to flowering. The Kruskal–Wallis test was used for analysis of differences in peltate glandular trichomes densities between the upper and lower epidermis (*—statistically significant ($p < 0.05$) differences between the density of peltate glandular trichomes in the upper and lower epidermis).

At the early beginning of vegetation (in the first half of May), the larger diameter of peltate glandular trichomes was characteristic of peltate glandular trichomes in the lower leaf epidermis. Later, starting in the second half of May, the situation changed, and until the flowering of the plants, the larger diameter was always recorded in peltate glandular trichomes of the upper epidermis; in the second half of May, the difference in the diameter of peltate glandular trichomes between the two epidermises was the highest and statistically significant ($p < 0.05$) (Figure 8).

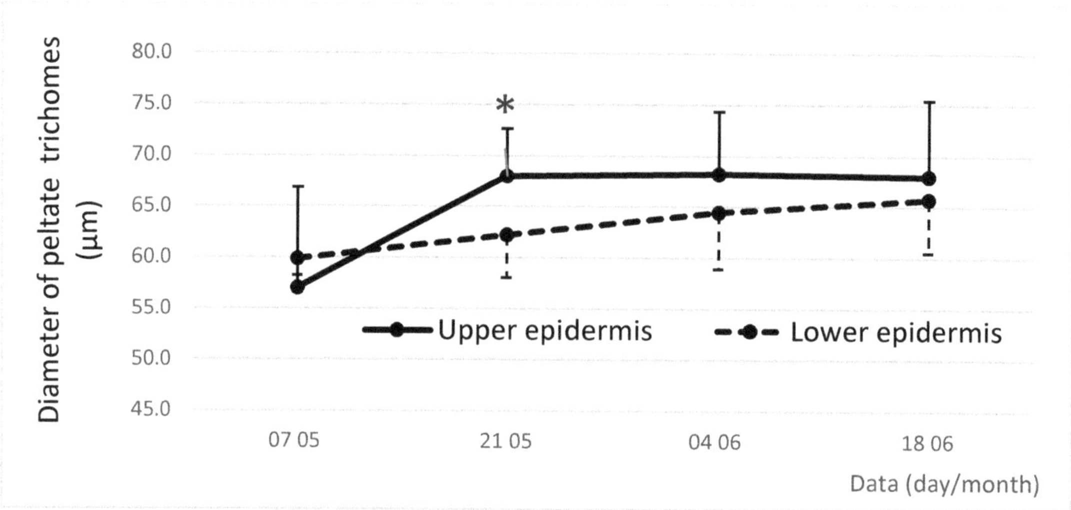

Figure 8. Variation in the diameter of peltate glandular trichomes in the upper and lower epidermis of *Thymus pulegioides* leaves from the start of plant vegetation to flowering. The Kruskal–Wallis test was used for analysis of differences in peltate glandular trichome diameter between the upper and lower epidermis (*—statistically significant ($p < 0.05$) differences between the diameter of peltate glandular trichomes in the upper and lower epidermis).

The raw material preparation of all plants of the *Thymus* genus is underway only during flowering, since this is when the plants accumulate the most essential oil and produce the largest aboveground biomass [9]. Therefore, dynamic changes of density and diameter in peltate glandular trichomes of *Thymus pulegioides* were made only prior to the flowering period.

4. Discussion

Among the studied *T. pulegioides* individuals, the essential oil percentage varied widely, from 0.12% to 1.68% (Figure 2). *T. pulegioides* is widespread almost throughout Europe, and intraspecific variation in essential oil content has been observed in many countries; for example, the essential oil content in *T. pulegioides* plants growing in Romania varied within the range of 0.7–1.1% [22], in Italy—0.9–1.1% [23], in Croatia—0.60–1.31% [24], in Portugal—it reached up to 1.8% [25], and in Kosovo—up to 1.58% [26]. Intraspecific differences in essential oil content may be influenced not only by the genetic characteristics of the plants but also by various abiotic environmental factors [27]. The studied *T. pulegioides* habitats were located in meadows that belonged to the *Molinio-Arrhenatheretea elatioris*, *Trifolio-Geranietea sanguinei*, and *Festuco-Brometea erecti* phytocoenological vegetation classes (Table 1). Also, habitats were characterised by good lighting and different soil pH and chemical composition values (Tables 1 and 2). However, results showed that although the essential oil percentage of *T. pulegioides* in habitats belonging to the *Festuco-Brometea* vegetation class was 0.1% higher than in habitats belonging to the *Trifolio-Geranietea* vegetation class, this difference was not statistically significant.

Meanwhile, the essential oil percentage of *T. pulegioides* in the habitats significantly ($p < 0.05$) correlated with the contents of mobile potassium and mobile iron in the habitat soil ($r = 0.58$ and $r = 0.57$, respectively, $N = 23$). It is not for nothing that *T. pulegioides* plants in habitat no. 12, where soil contained the highest (2–9 times more than in other habitats) mobile iron and high mobile potassium contents, were distinguished by the highest average essential oil percentage (Table 2 and Figure 2). According to literature data, the chemical composition of the soil can influence the synthesis of essential oil, but this effect may differ for different plant species. For example, potassium can affect the essential oil content differently, even in plants belonging to the same Lamiaceae family: when *Thymus mastichina* and *Origanum vulgare* plants were fertilized with potassium fertilizers, the essential oil content increased [28–30]. However, no reliable correlations were found between the amount of potassium in the soil and the amount of essential oil in *Origanum compactum* [31]; additionally, fertilization with potassium did not stimulate the synthesis of essential oil in *Ocimum basilicum* [32]. The previous studies also did not establish a reliable correlation between the essential oil content in *T. pulegioides* plants and the potassium content in habitat soil [33]. Chemical elements are required to synthesise and regulate enzymes involved in synthesising many essential oils' chemical compounds [34–36]. The same chemical element can be part of several enzymes responsible for the course of genetically programmed biosynthetic pathways of one or another compound. For example, manganese is important as a cofactor for synthesising terpene precursors and monoterpenes [37,38]. The same chemical element can affect the essential oil contents of different plant species differently through the genetically determined composition of volatile compounds synthesised in them. Also, interpreting the results of studies conducted in natural habitats is more complicated than interpreting the results of experiments conducted under controlled conditions, as some unknown habitat factors may have an additional influence. The results demonstrated that, although unreliably, the average percentage of *T. pulegioides* essential oil in habitats was negatively correlated only with the mobile sulphur amount in the soil. Negative correlations were also found between the mobile sulphur content in the soil and the studied peltate glandular trichome parameters (density and diameter). This suggests that mobile sulphur in the soil, negatively affecting the parameters of peltate glandular trichomes, may indirectly influence the essential oil percentage in *T. pulegioides* plants.

A statistically significant ($p < 0.05$) negative correlation between the average essential oil percentage in *T. pulegioides* in the habitat and the cover abundance of *T. pulegioides* in habitat ($r = -0.48$, $N = 23$) indicates that plants of this species accumulate lower amounts of essential oil in habitats with higher cover abundance of their own. This could be related to the intraspecific competition of plants in the habitat. Since this phenomenon negatively affects plant development, it may also suppress the development of peltate glandular trichomes. It is not for nothing that a negative relationship (although statistically insignificant) was also found between the cover abundance of *T. pulegioides* in the habitat and parameters of peltate glandular trichomes (density and diameter). Meanwhile, when *T. pulegioides* raw material is collected in natural habitats, it is recommended not to do so in habitats with a high cover abundance of *T. pulegioides*: due to possibly greater intraspecific plant competition, *T. pulegioides* in such habitats may accumulate lower amounts of essential oil. Also, for the same reason, it is recommended that optimal distances between plants be determined in crops of this species.

Although the significant correlation between the density of peltate glandular trichomes in the upper and lower epidermis was not established, the number of peltate glandular trichomes per unit area in the two leaf epidermises differed: the average density of peltate glandular trichomes in the lower leaf epidermis was significantly ($p < 0.05$) higher than in the upper epidermis (8.0 ± 2.0 and 7.0 ± 1.7 trichomes in mm^2, respectively; $N = 124$). Meanwhile, the diameter of peltate glandular trichomes was significantly higher in the upper epidermis. Similar trends were found in other Lamiaceae species: a higher number of glandular trichomes in the lower leaf epidermis was found in *Thymus serpyllum*, *Ocimum campechianum*, *O. basilicum*, *Mentha × piperita*, and *Salvia nemorosa*, and a larger diameter of glandular trichomes was found in the upper epidermis in *T. serpyllum*, *T. vulgaris*,

M. × *piperita*, *Salvia officinalis*, and *Origanum majorana* [23,39–43]. It is believed that essential oil accumulating in the glandular trichomes helps the plant defend itself against pests. Tiny insect pests (for example, aphids and whiteflies) that attack plants, protecting themselves from the sun, rain, and other environmental factors, also usually settle on the undersides of leaves [44]. Therefore, a higher density of glandular trichomes in the lower leaf epidermis may be a better way for the plant to protect itself directly from pests. However, the plants use volatile essential oils to repel pests (and herbivores) at a distance, too. In this case, the glandular trichomes located in the upper leaf epidermis are more useful. Perhaps this is why significant positive correlations were found between the essential oil percentage and the density and the diameter of peltate glandular trichomes in the upper leaf epidermis (Figure 3). However, it should be kept in mind that, although the peltate glandular trichomes in *Thymus albicans* plants were most abundant in leaves, a few of them were also in bracts, calyces, and corollas [10]. Also, a little essential oil is also accumulated in capitate trichomes (which consist of only one or a few head cells) [8]. Therefore, established correlations in *T. pulegioides* should be slightly adjusted.

However, as shown in an investigation of the dynamics of peltate glandular trichome density and diameter during *T. pulegioides* leaf development, the density of peltate glandular trichomes is not always higher in the lower epidermis, and the diameter of peltate glandular trichomes is not always greater in the upper leaf epidermis: until the second half of May, the density and diameter of peltate glandular trichomes were higher in the upper and in lower leaf epidermis, respectively (Figures 7 and 8). The number of glandular trichomes is closely related to plant ontogeny, and variations in their number on the upper and lower sides of leaves can be associated with changes in leaf orientation on the stem [39,45]. The density of glandular trichomes in *M.* × *piperita* at the beginning of leaf development, when the leaves grow vertically relative to the stem (orthotropic orientation) and the lower epidermis is not protected from direct sunlight, was higher in the upper epidermis of the leaf. Later, as the orthotropic orientation of the leaves changed to a plagiotropic orientation and more sunlight fell on the upper epidermis, the density of glandular trichomes in the upper epidermis steadily decreased, and in the lower epidermis, it reached a maximum when the leaves fully matured and exceeded the density of glandular trichomes in the upper epidermis [39]. Therefore, changes in leaf orientation during development, accompanied by the influence of light, could possibly inhibit the development of peltate glandular trichomes in the upper epidermis and promote them in the lower epidermis in *T. pulegioides*. This statement is also strengthened by the negative relationships found in our study between the density of peltate glandular trichomes and *T. pulegioides* habitat illumination, which, in the lower epidermis, was significant ($R = -0.42$, $p < 0.05$). A lower number of glandular trichomes at higher light intensity was found in *Perilla ocymoides* (Lamiaceae) plants, too [46].

As glandular trichomes mature, their secretory cells begin synthesising volatile compounds and secrete them into the subcuticular space, where they are stored. Therefore, the amount of essential oils synthesised by glandular trichomes can influence the increase in the diameter of glandular trichome heads [47,48]. This is confirmed by the significant ($p < 0.05$) positive correlation between the amount of essential oil and the diameter of peltate glandular trichomes in the upper and lower leaf epidermis of *T. pulegioides* ($R = 0.33$ and $R = 0.24$, respectively). Light positively affects the amount of essential oil synthesised in plants of the *Thymus* genus [49]. Therefore, at the beginning of vegetation, while *T. pulegioides* leaves were in an orthotropic position, the peltate glandular trichomes of the lower leaf epidermis received more direct sunlight and could synthesize essential oil more intensively; this could explain why the diameter of the peltate glandular trichomes was higher in the lower epidermis of the leaves in early May. Later, as the position of the leaves changes relative to the stem and the upper epidermis receives more light, the diameter of peltate glandular trichomes in the upper leaf epidermis increases (Figure 8).

5. Conclusions

This study showed that the density and/or diameter of peltate glandular trichomes in the upper leaf epidermis could be anatomical markers, helping to select in natural habitats with *T. pulegioides* individuals with higher amounts of essential oil more quickly and cheaply. Based on the established, reliable positive correlations, it is likely that individuals with a higher density and diameter of peltate glandular trichomes in the upper leaf epidermis accumulate more essential oil. However, this proposed method for rapid screening of essential oil-rich wild *T. pulegioides* plants based on peltate glandular trichome characteristics, due to the weak obtained correlation, can be considered more a trend than a significant conclusion, and it cannot be applied to other essential oil-bearing plants. Also, it would be valuable to extend observations to additional *Thymus* species, as well as different structural types of glandular trichomes and other parts of plants (for example, corollas and calyces) in which peltate glandular trichomes are also found, although in smaller quantities. A negative impact of high *T. pulegioides* cover abundance in natural habitats on the amounts of essential oils accumulating in *T. pulegioides* plants was detected; therefore, it is recommended that optimal distances between plants be determined in future crops of this species.

Funding: This research received no external funding.

Data Availability Statement: Data is contained within the article.

Acknowledgments: The author would like to thank A. E. Enonchong, R. Jurevičiūtė, L. Rožytė and A. Almonaitytė for assistance with the research material collection and some work in a laboratory, and the photos.

Conflicts of Interest: The author declares no conflict of interest. The funders had no role in the design of the study; in the collection, analyses or interpretation of data; in the writing of the manuscript; or in the decision to publish the results.

References

1. Zamani, S.; Fathi, M.; Ebadi, M.-T.; Máthé, Á. Global trade of medicinal and aromatic plants. A review. *J. Agric. Food Res.* **2025**, *21*, 101910. [CrossRef]
2. Franz, C.; Novak, J. Sources of Essential Oils. In *Handbook of Essential Oils: Science, Technology, and Applications*; Baser, K.H.C., Buchbauer, G., Eds.; CRC Press: Boca Raton, FL, USA, 2010; pp. 39–82.
3. Lawrence, B.M.; Tucker, A. The genus *Thymus* as a source of commercial products. In *Thyme: The Genus Thymus*; Stahl-Biskup, E., Sáez, F., Eds.; CRC Press: Boca Raton, FL, USA, 2010; pp. 252–262.
4. Nieto, G. A Review on applications and uses of *Thymus* in the food industry. *Plants* **2020**, *9*, 961. [CrossRef] [PubMed]
5. György, Z.; Incze, N.; Pluhár, Z. Differentiating *Thymus vulgaris* chemotypes with ISSR molecular markers. *Biochem. Syst. Ecol.* **2020**, *92*, 104118. [CrossRef]
6. Stakelienė, V.; Ložienė, K. Gynodioecy in *Thymus pulegioides* L., *T. serpyllum* L. and their hybrid *T.* × *oblongifolius* Opiz (Lamiaceae): Flower size dimorphism, female frequency and effect of environmental factors. *Plant Biosyst.* **2014**, *148*, 49–57. [CrossRef]
7. Gang, D.R.; Wang, J.; Dudareva, N.; Nam, K.H.; Simon, J.R.; Lewinsohn, E.; Pichersky, E. An investigation of the storage and biosynthesis of phenylpropenes in sweet basil. *Plant Physiol.* **2001**, *125*, 539–555. [CrossRef] [PubMed]
8. Bhatt, A.; Naidoo, Y.; Nicholas, A. The foliar trichomes of *Plectranthus laxiflorus* Benth [Lamiaceae]: An important medicinal plant. *N. Z. J. Bot.* **2010**, *48*, 55–61. [CrossRef]
9. The Councile of Europe. *European Pharmacopoeia*, 6th ed.; European Directorate for the Quality of Medicines & HealthCare (EDQM): Strasbourg, France, 2020; Volume 1.
10. Marquesa, N.T.; Filipe, A.; Pinto, P.; Barroso, J.; Trindade, H.; Power, D.M.; Figueiredo, A.C. Trichome density in relation to volatiles emission and 1,8-cineole synthase gene expression in *Thymus albicans* vegetative and reproductive organs. *Chem. Biodivers.* **2020**, *17*, e1900669. [CrossRef] [PubMed]

11. Thompson, J.D. Population structure and the spatial dynamics of genetic polymorphism in thyme. In *Thyme: The Genus Thymus*; Stahl-Biskup, E., Sáez, F., Eds.; CRC Press: Boca Raton, FL, USA, 2010; pp. 44–74.
12. Boz, I.; Gille, E.; Necula, R.; Dunca, S.; Zamfirache, M.M. Chemical composition and antibacterial activity of essential oils from five populations of *Thymus pulegioides* L. *Cellulose Chem. Technol.* **2015**, *49*, 169–174.
13. Ložienė, K.; Venskutonis, P.R.; Vaičiūnienė, J. Chemical diversity of essential oil of *Thymus pulegioides* L. and *Thymus serpyllum* L. growing in Lithuania. *Biologija* **2002**, *1*, 62–64.
14. Groendahl, E.; Ehlers, B.K.; Keefover-Ring, K. A New cis-Sabinene Hydrate Chemotype Detected in Large Thyme (*Thymus pulegioides* L.) Growing Wild in Denmark. *J. Essent. Oil Res.* **2008**, *20*, 40–41. [CrossRef]
15. Wester, P.; Möseler, B.M.; Knöss, W. Intra-population terpene polymorphism of *Thymus pulegioides* L.: Evidence for seven chemotypes in a German limestone grassland. *Biochem. Syst. Ecol.* **2020**, *93*, 104173. [CrossRef]
16. Ložienė, K. Selection of fecund and chemically valuable clones of thyme (*Thymus*) species growing wild in Lithuania. *Ind. Crops Prod.* **2009**, *29*, 502–508. [CrossRef]
17. Radušienė, J.; Janulis, V. Improvement of diversity, trade and conservation of medicinal and aromatic plants. *Medicina* **2004**, *40*, 705–709. [PubMed]
18. Mockutė, D.; Bernotienė, G. The α-terpinyl acetate chemotype of essential oil of *Thymus pulegioides* L. *Biochem. Syst. Ecol.* **2001**, *29*, 69–76. [CrossRef] [PubMed]
19. Mockutė, D.; Bernotienė, G. Chemical composition of the essential oils and the odor of *Thymus pulegioides* L. growing wild in Vilnius. *J. Essent. Oil Res.* **2005**, *17*, 415–441. [CrossRef]
20. Braun-Blanquet, J. *Pflanzensoziologie, Grundzuge der Vegetationskunde*, 3rd ed.; Springer: Berlin/Heidelberg, Germany, 1964.
21. Balevičienė, J.; Kizienė, B.; Lazdauskaitė, Ž.; Patalauskaitė, D.; Rašomavičius, V.; Sinkevičienė, Z.; Tučienė, A.; Venckus, Z. *Vegetation of Lithuania: 1. Meadows*; Šviesa: Vilnius, Lithuania, 1998.
22. Pavel, M.; Ristic, M.; Stevic, T. Essential oils of *Thymus pulegioides* and *Thymus glabrescens* from Romania: Chemical composition and antimicrobial activity. *J. Serb. Chem. Soc.* **2010**, *75*, 27–34. [CrossRef]
23. Martínez-Natarén, D.A.; Villalobos-Perera, P.A.; Munguía-Rosas, M.A. Morphology and density of glandular trichomes of *Ocimum campechianum* and *Ruellia nudiflora* in contrasting light environments: A scanning electron microscopy study. *Flora* **2018**, *248*, 28–33. [CrossRef]
24. Radonic, A.; Mastelic, J. Essential oil and glycosidically bound volatiles of *Thymus pulegioides* L. growing wild in Croatia. *Croat. Chem. Acta* **2008**, *81*, 599–606.
25. Pinto, E.; Pina–Vaz, C.; Salgueiro, L.; Concalves, M.J.; Costa-de-Oliveira, S.; Cavaleiro, C.; Palmaeira, A.; Rodrigues, A.; Martinez-de-Oliveira, J. Antifungal activity of the essential oil of *Thymus pulegioides* on *Candida*, *Aspergillus* and dermophyte species. *J. Med. Microbiol.* **2006**, *55*, 1367–1373. [CrossRef] [PubMed]
26. Ibrahimi, M.H.; Papajani, V.; Zelikovic, S.C.; Matevski, V. Essential oil analysis of Two *Thymus* spp. growing wild in Kosowo. *J. Essent. Oil Bear. Plants* **2014**, *17*, 832–837. [CrossRef]
27. Farhat, M.B.; Jordan, M.J.; Chaouch-Hamada, R.; Landoulsi, A.; Sotomayor, J.A. Phenophase effects of sage (*Salvia officinalis*) yield and composition of essential oil. *J. Appl. Res. Med. Aromat. Plants* **2016**, *3*, 87–93. [CrossRef]
28. Miguel, M.G.; Guerrero, C.; Rodrigues, C.; Broto, J.C.; Duarte, F.; Valencio, F.; Tavares, R. Main components of essential oils from wild Portuguese *Thymus mastichina* L. (L) spp. *mastichina* in different development stages or under culture conditions. *J. Essent. Oil. Res.* **2004**, *16*, 111–114. [CrossRef]
29. Said-Al Ahl, H.A.H.; Ayad, H.S.; Hendawy, S.F. Effect of potassium humate and nitrogen fertilizer on herb and essential oil of oregano under different irrigation intervals. *Ozean J. Appl. Sci.* **2009**, *2*, 319–323.
30. Said-Al Ahl, H.A.H.; Wahby, M.S. Effect of water stress and potassium humate on the productivity of oregano plant using saline and fresh water irrigation. *Ozean J. Appl. Sci.* **2010**, *3*, 125–141.
31. Bakhy, K.; Benlhabib, O.; Bighelli, A.; Casanova, J.; Tomi, F.; Faiz, C.A. Yield and chemical variability of essential oil isolated from aerial parts of wild *Origanum compactum* Benth. from Moroccan Western Rif. Am. *J. Essent. Oil Nat. Prod.* **2014**, *4*, 9–17.
32. Mansoorkhani-Roghaye, A.; Shahriari, Z.; Mohaselli, V.; Osfoori, M.; Shahriari, A.G. Effect of graded levels of NPK on herb oil, yield and oil composition of basil (*Ocimum basilicum* L.). *Glob. J. Res. Med. Plants Indig. Med.* **2012**, *1*, 258–264.
33. Vaičiulytė, V.; Ložienė, K.; Taraškevičius, R. Impact of Edaphic and Climatic Factors on *Thymus pulegioides* Essential Oil Composition and Potential Prevalence of Chemotypes. *Plants* **2022**, *11*, 2536. [CrossRef] [PubMed]
34. Koeduka, T.; Fridman, E.; Gang, D.R.; Vassao, D.G.; Jackson, B.L.; Kosh, C.M.; Orlova, I.; Spassova, S.M.; Lewis, N.G.; Noel, J.P.; et al. Eugenol and isoeugenol, characteristic aromatic constituents of spices, are biosynthesized via reduction of a coniferyl alcohol ester. *Proc. Natl. Acad. Sci. USA* **2006**, *103*, 10128–10133. [CrossRef] [PubMed]
35. Malenci, D.R.; Kevesan, Z.S.; Popovic, M.T. Mineral composition of selected *Salvia* species growing wild in the Vojvodina province. *Zb. Matice Srp. Prir. Nauk.* **2003**, *105*, 25–33. [CrossRef]
36. Gul, H.; Said, A.; Saeed, B.; Mohammad, F.; Ahmad, I. Effects of foliar application of nitrogen, potassium and zinc on wheat growth. *J. Agric. Biol. Sci.* **2011**, *6*, 56–58.

37. Tabata, M. Genetics of monoterpene biosynthesis in perilla plants. *Plant Biotechnol. J.* **2000**, *17*, 237–280. [CrossRef]
38. Singh, P.; Misra, A.; Srivastava, N.K. Influence of Mn deficiency on growth, chlorophyll content, physiology and essential monoterpene oils in genotypes of spearmint (*Mentha spicata*). *Photosynthetica* **2001**, *39*, 473–476. [CrossRef]
39. Maffei, M.; Chialva, F.; Sacco, T. Glandular trichomes and essential oils in developing peppermint leaves. *New Phytol.* **1989**, *111*, 707–716. [CrossRef] [PubMed]
40. Kamašina, V.; Ložienė, K. The evaluation of phenotypic diversity of *Thymus* × *oblongifolius* according to some anatomical characters and comparison with parent species. *Acta Bot. Hung.* **2009**, *51*, 85–97. [CrossRef]
41. Dhawan, S.S.; Shukla, P.; Gupta, P.; Lal, R.K. A cold-tolerant evergreen interspecific hybrid of *Ocimum kilimandscharicum* and *Ocimum basilicum*: Analyzing trichomes and molecular variations. *Protoplasma* **2016**, *253*, 845–855. [CrossRef] [PubMed]
42. Talebi, S.M.; Mahdiyeh, M.; Nohooji, M.G.; Akhani, M. Analysis of trichome morphology and density in *Salvia nemorosa* L. (Lamiaceae) of Iran. *Botanica* **2018**, *24*, 49–58. [CrossRef]
43. Kowalski, R.; Kowalska, G.; Jankowska, M.; Nawrocka, A.; Kaława, K.; Pankiewicz, U.; Włodarczyk-Stasiak, M. Secretory structures and essential oil composition of selected industrial species of Lamiaceae. *Acta Sci. Pol. Hortorum Cultus* **2019**, *18*, 53–69. [CrossRef]
44. Singh, R.; Singh, G. Aphids. In *Polyphagous Pests of Crops*; Omkar, Ed.; Springer: Singapore, 2021; pp. 105–182.
45. Lee, Y.L.; Ding, P. Production of essential oil in plants: Ontogeny, secretory structures and seasonal variations. *Pertanika J. Sch. Res. Rev.* **2016**, *2*, 1–10.
46. Kimura, M.; Ishii, M.; Yoshimi, M.; Ichimura, M.; Tomitaka, Y. Essential oils and glandular trichomes of perilla seedlings as affected by light intensity. *Acta Hortic.* **2000**, *515*, 219–226. [CrossRef]
47. Boz, I.; Lobiuc, A.; Tănase, C. Chemical composition of essential oils and secretory hairs of *Thymus dacicus* Borbás related to harvesting time. *Cellul. Chem. Technol.* **2017**, *51*, 813–819.
48. Rodrigues, L.; Monteiro, P.; Póvoa, O.; Teixeira, G.; Moldão, M.; Figueiredo, A.C.; Monteiro, A. Morphology of secretory structures and essential oil composition in *Mentha cervina* L. from Portugal. *Flavour Fragr. J.* **2008**, *23*, 340–347. [CrossRef]
49. Tohidi, B.; Rahimmalek, M.; Arzani, A.; Sabzalian, M.R. Thymol, carvacrol, and antioxidant accumulation in *Thymus* species in response to different light spectra emitted by light-emitting diodes. *Food Chem.* **2020**, *307*, 125521. [CrossRef] [PubMed]

horticulturae

MDPI

Article

Difference Analysis of Non-Volatile and Volatile Components in Kujingcha (*Ilex dabieshanensis*) Compared with Green Tea (*Camellia sinensis*)

Linlong Ma [1,†], Yanan Peng [2,†], Dan Cao [1,†], Ping Fan [3], Lingyi Wang [4], Guobiao Feng [5], Aimin Lei [6], Baisong Hu [6], Yijin Liu [7], Yanli Liu [1] and Xiaofang Jin [1,*]

1 Key Laboratory of Tea Resources Comprehensive Utilization of Ministry of Agriculture and Rural Affairs, Fruit and Tea Research Institute, Hubei Academy of Agricultural Sciences, Wuhan 430064, China; malinlong@hbaas.com (L.M.); skyiswide@163.com (D.C.); zkslyl@163.com (Y.L.)
2 Tuanfeng Daqishan Kujingcha Co., Ltd., Huanggang 438810, China; p732463214@163.com
3 Huanggang Agricultural Technology Promotion Center, Huanggang 438000, China; fp455161736@163.com
4 Huanggang Academy of Agricultural Sciences, Huanggang 438000, China; w1440855821@126.com
5 Tuanfeng Agricultural and Rural Bureau, Huanggang 438800, China; fgb13307255668@163.com
6 Tuanfeng Agricultural Technology Promotion Center, Huanggang 438800, China; lam1007472525@126.com (A.L.); myth1367101@126.com (B.H.)
7 Tuanfeng Kujingcha Research and Development Center, Huanggang 438810, China; ly276848@163.com
* Correspondence: jxf1130@126.com
† These authors contributed equally to this work.

Abstract

Kujingcha (KJC) is a widely consumed substitute tea due to its unique flavor quality and health benefits. However, the biochemical basis for the formation of KJC's unique flavor quality and health benefits remain unclear. In this study, using *Camellia sinensis* green tea and its processed fresh leaves as a control, we systematically analyzed the non-volatile and volatile components in KJC and its processed fresh leaves. The results indicate that high levels of flavonoids and water-soluble sugars, and low levels of amino acids and water-soluble proteins, are important biochemical foundations for the formation of taste quality in KJC. High aldehyde, alkene and heterocyclics contents contribute significantly to the aroma of KJC, among which heterocyclics are the key components for the formation of KJC's rich pan-fried bean-like aroma. Flavonoids such as neohesperidin, hyperoside, rutin, astilbin and morin are important components for the formation of KJC's health benefits.

Keywords: Kujingcha; green tea; non-volatile component; volatile component

1. Introduction

Kudingcha is an important traditional substitute tea in China and is widely consumed due to its unique flavor quality and many health benefits [1]. A wide variety of plant species are used for processing Kudingcha, and nearly 20 plant species from different families in different regions of China have been named Kudingcha because of their similar appearance, flavor and traditional usage [2]. Plants of the genus *Ilex*, such as *Ilex kudingcha*, *I. latifolia*, *I. cornuta*, *I. pentagona*, *I. chinensis* and *I. rotunda*, are regarded as the main plant sources of Kudingcha and are called "large-leaved Kudingcha" [3]. Among them, *I. kudingcha* and *I. latifolia* are the main plant sources for processing Kudingcha products in the current market, while few reports exist on Kudingcha made from other *Ilex* genus plants [2,4].

In the Dabie Mountain area of Huanggang city, Hubei Province, local people have been using the fresh leaves of *I. dabieshanensis* to process Kudingcha according to the traditional

https://doi.org/10.3390/horticulturae11070804

green tea (*Camellia sinensis*) processing technique for a long time, which is called Kudingcha (KJC) locally and also known as "Tuanfeng Kucha" [5,6]. In terms of its flavor characteristics, KJC has a strong bitter taste followed by a sweet aftertaste, a clear and mellow taste and a rich pan-fried bean-like aroma [5]. Moreover, it has been proven in practice that KJC has health benefits such as clearing away internal heat, detoxifying the body, refreshing the mind, helping digestion and resolving food stagnation, reducing fat and facilitating weight loss, lowering blood sugar and blood pressure, reducing inflammation, and killing bacteria [7]. Due to its unique flavor characteristics and health benefits, it is deeply favored by consumers, and the products are sold to regions such as Guangdong, Hong Kong and Macau. Today, KJC has become a featured product in local areas, especially "Tuanfeng Kujingcha", which is a protected product with the geographical indication of the People's Republic of China.

Although the processing technique of KJC is similar to that of *C. sinensis* green tea, vast differences exist in their flavor characteristics and health benefits. The reason lies in the large differences in their biochemical and volatile components. To date, a large number of research reports have focused on the formation of the flavor quality and health benefits of *C. sinensis* green tea. For example, tea polyphenols, catechins, caffeine, theanine, etc., are the main substances that impart the important taste quality and health benefits of *C. sinensis* green tea [8,9]. Linalool, nonanal, dimethyl sulfide, geraniol, caproicacidhexneylester, etc., are the key components that constitute the aroma of *C. sinensis* green tea [10,11].

However, few systematic studies focus on the main biochemical components of KJC, and the biochemical basis for the formation of KJC's unique flavor characteristics and health benefits remains unclear, which greatly affects the promotion and application of KJC and the development of new products. Metabolomics technologies serve as important tools for studying plant metabolites, playing a significant role in elucidating the formation mechanisms of flavor quality in plant-derived foods, as well as in the discovery and utilization of functional components [12,13]. Therefore, in this study, using traditional green tea and its processed fresh leaves as controls, we systematically analyzed the main non-volatile and volatile components in KJC and its processed fresh leaves through high-performance liquid chromatography (HPLC), ultra-high-performance liquid chromatography coupled with triple quadrupole mass spectrometry (UPLC-QQQ-MS) and head space solid-phase microextraction with gas chromatography and mass spectrometry (HS-SPME-GC-MS). This approach reveals the material basis underlying the formation of KJC's flavor quality and health benefits. The results are conductive to comprehensively understanding the flavor components and health-active components of KJC and provide an important basis for the further application and new product development of KJC in the future.

2. Materials and Methods

2.1. Experimental Materials

In May 2023, the fresh leaves (one bud and two leaves) of *I. dabieshanensis* (KJCFL) and *C. sinensis* (GTFL) were picked at the tea garden of Tuanfeng Daqishan Kujingcha Co., Ltd. (Huanggang, China) Next, KJC and green tea (GT) were processed according to the technological processes of spreading, fixation, rolling, drying and re-drying. Specifically, the spreading process involved placing fresh leaves at a density of 0.75 kg/m^2 under conditions of a room temperature of 18 °C~20 °C and approximately 75% relative humidity until the moisture content in the leaves decreased to around 60~65%. For fixation, the spread leaves were processed in a drum fixation machine (6CST-80, Hangzhou Chunjiang Tea Machinery Co., Ltd., Hangzhou, China) at 280 °C for 2 min. Rolling was carried out using a rolling machine (6CR-65, Zhejiang Chunjiang Tea Machinery Co., Ltd., Hangzhou, China) for 30 min. Drying was carried out using a hot-air dryer (6CHM-901, Zhejiang Fuyang Tea

Machinery Co., Ltd., Hangzhou, China) at 110~120 °C for 10~15 min, followed by 80 °C for 15~20 min, and the moisture content of all tea samples was controlled to below 6%. Finally, 500 g of each sample was randomly selected, and this was repeated three times; the samples were frozen quickly with liquid nitrogen, then freeze-dried by freeze dryers, and placed in a refrigerator at −80 °C for storage until further analysis.

2.2. Chemicals and Reagents

The chemicals and reagents used in the study are listed in Table S1.

2.3. Experiment Methods

2.3.1. Sensory Evaluation and Quantitative Descriptive Analysis (QDA)

The infusion color, aroma and taste characteristics of the tea samples were subjected to sensory evaluation according to GB/T 23776-2018 [14]. Meanwhile, six experienced evaluators (three females and three males) conducted QDA of the tea samples. In brief, precisely 3.0 g of each tea sample was weighed and brewed with 150 mL of boiling water for 5 min. After assessing the aroma and taste characteristics of all tea samples, six primary aroma types (floral, sweet, nutty, roasted and faint scent) and six primary taste types (sweetness, sourness, astringency, bitterness umami) were selected for QDA. A 10-point scale was used to rate the intensity of these six aroma attributes in the tea infusion, where 0 = absent, 3 = weak, 5 = moderate, 7 = strong, and 10 = very strong.

2.3.2. Determination of Main Conventional Biochemical Components

The total free amino acid content was determined by the ninhydrin colorimetric method, with L-glutamic acid as the reference substance (GB/T 8314-2013) [15]. The content of tea polyphenols was determined by the Folin phenol colorimetric method, with gallic acid as the reference substance (GB/T 8313-2018) [16]. The determination of water extract content was carried out with reference to GB/T 8305-2013 [17]. The contents of water-soluble protein, water-soluble sugar and total flavones were determined according to the method of Li (2018) [18].

2.3.3. Determination of Flavonoid Content

Flavonoid content was determined by UPLC-QQQ-MS according to the following method, which was described by Xue (2025) [19]: Accurately weigh 0.10 g of the sample and place it in a 2 mL centrifuge tube, add 0.80 mL of 80% methanol–water solution, shake it thoroughly for 1 min, and perform ultrasonic treatment at 4 °C for 30 min. Let it stand still at 4 °C for 60 min, and centrifuge it for 10 min. The supernatant is then removed for solid-phase extraction. Place the supernatant in a concentrator and concentrate it to dryness, then add 0.20 mL of 80% methanol–water solution to redissolve it, shake it thoroughly for 5 min, centrifuge it at 4 °C for 10 min, and take the supernatant for testing using the testing instrument.

UPLC-QQQ-MS analysis was performed using a Waters Acquity UPLC system (Waters, Milford, MA, USA) with a UPLC HSS T3 (1.8 μm, 2.1 mm × 100 mm, Waters, Milford, MA, USA) coupled to a QQQ-MS (AB SCIEX 5500, Waters, Milford, MA, USA). Chromatographic conditions: The column temperature was 40 °C, the flow rate was 0.3 mL·min^{-1} and the injection volume was 5 μL. Mobile phase A was water with 0.1% formic acid, and mobile phase B was acetonitrile with 0.1% formic acid. The elution gradients were as follows: 0 min, 90% A and 10% B; 0–1 min, 90% A and 10% B; 1–14 min, 10% A and 90% B; 14–15 min, 10% A and 90% B; 15–15.1 min, 90% A and 10% B; 15.1–18 min, 90% A and 10% B. Mass spectrum conditions: ESI ion source, Curtain Gas at 35 arb, Collision GAS at 9 arb, IonSpray voltage at 4000 V, ion source temperature at 400 °C, IonSource Gas1 at 55 arb, IonSource Gas2 at 55 arb, Spray voltage at 3.1 kV (negative).

The content of each flavonoid component was quantified using the internal standard method with quercetin D5 as the reference standard. The calculation was performed according to the formula $f = (Ms/As)/(Ar/Mr)$, where As and Ar represent the peak areas of the internal standard and reference standard, respectively, and Ms and Mr indicate the amounts of internal standard and reference standard added.

2.3.4. Determination of Catechins and Caffeine Content

Catechin and caffeine content was determined by HPLC according to the following method, which was described in our previous study [13]: Accurately weigh 0.05 g of the sample and put it into a 10 mL centrifuge tube. Add 5 mL of 70% methanol–water solution, place it in a water bath at 70 °C for 20 min, and centrifuge it at 3500 rpm for 5 min, and take the supernatant and place it in a 10 mL volumetric flask. Repeat the above steps once to extract water. Finally, make up the volume to 10 mL with a 70% methanol-water solution, filter it through a 0.22 μm filter membrane and conduct detection using the testing instrument.

The conditions and apparatus used in HPLC were a Waters 2695 HPLC system (Waters, Milford, MA, USA) equipped with a 2998 PDA detector, and a symmetry C18 column (18.5 μm, 4.6 mm × 250 mm, Waters, Milford, MA, USA). Chromatographic conditions: The column temperature was 40 °C, the flow rate was 1.0 mL·min^{-1}, and the injection volume was 10 μL. Mobile phase A was 2% acetic acid, and mobile phase B was acetonitrile. The elution gradients were as follows: 0 min, 93.5% A and 6.5% B; 0–4 min, 93.5% A and 6.5% B; 4–16 min, 85% A and 15% B; 16–28 min, 75% A and 25% B; 28–35 min, 93.5% A and 6.5% B; 35–40 min, 93.5% A and 6.5% B.

2.3.5. Determination of Amino Acid Content

Amino acid content was determined according to the method described in our previous study [13], but slight modifications were made to the extraction method. The procedure is as follows: Accurately weigh 0.10 g of the sample and put it into a 10 mL centrifuge tube. Add 5 mL of water, place it in a water bath at 100 °C for 20 min, centrifuge it at 3500 rpm for 5 min, and take the supernatant and place it in a 10 mL volumetric flask. Repeat the above steps once to extract water. Finally, make up the volume to 10 mL with water, filter it through a 0.22 μm filter membrane and conduct detection using the testing instrument.

2.3.6. Determination of Volatile Content

Volatile component content was determined by HS-SPME-GC-MS according to the following method, which was described by in our previous study [13]. Accurately weigh 3.0 g of the tea sample and put it into an extraction flask. Add 90.0 mL of boiling water with 2.00 μL of the internal standard (90 mg·L^{-1} ethyl decanoate). Meanwhile, use a 50/30 μm DVB/CAR/PDM extraction probe (the extraction probe was aged for 30 min at 250 °C prior to the experiment) to adsorb the volatiles. Keep it in a water bath at 60 °C for 60 min, then take it out and immediately conduct detection using the testing instrument. All extractions were performed in triplicate to ensure analytical reproducibility.

The conditions and apparatus used in GC-MS were a 7890A gas chromatograph and a 5975C mass spectrometer (Agilent, Santa Clara, CA, USA), an HB-5MS (30 m × 0.32 mm × 0.25 μm), and an elastic quartz capillary column (Agilent, Santa Clara, CA, USA). Chromatographic conditions: The temperature of the injection port was 240 °C, the carrier gas was helium, and the flow rate was 1.0 mL·min^{-1}. The column temperature was kept at 50 °C for 5 min, increased to 180 °C at a rate of 3 °C·min^{-1} and maintained for 2 min, and increased to 250 °C at a rate of 10 °C·min^{-1} and maintained for 3 min. Mass spectrum conditions: The EI ionization energy was 70 eV. The mass scanning range was 50–600 amu. The temperature of the ion source was 230 °C. The temperature of the quadrupole was 150 °C. The temperature of the mass spectrum transmission line was 280 °C.

Tandem retrieval and manual analysis were performed on the obtained mass spectra using the NIST11.L spectral library, and a mass spectrum matching degree greater than 90% was taken as the standard for substance identification. The concentration of each volatile component was calculated according to the formula $C_i = C_{is} \times A_i/A_{is}$, where C_i represents the concentration of a certain component (μg·L^{-1}), C_{is} represents the concentration of the internal standard (μg·L^{-1}), A_i represents the peak area of a certain component, and A_{is} represents the peak area of the internal standard.

2.3.7. Data Analysis

The analyses of all samples (KJC, KJCFL, GT and GTFL) were conducted with three biological replicates, and no pooling was performed during sample preparation. GraphPad Prism 5 was used to draw histograms. Principal component analysis (PCA) was performed separately on the contents of 35 flavonoids and 58 volatile components across the four sample groups using SIMCA 14.1 software, and autoscaling was applied to all variables prior to modeling. Difference analysis was performed using SPSS 19.0 software, with one-way analysis of variance followed by significance testing at $p < 0.05$. Heat maps and radar charts were drawn through the online website https://www.chiplot.online (accessed on 8 April 2025). The data were subjected to Z-score normalization prior to heatmap analysis to enable the comparative visualization of metabolite expression patterns.

3. Results

3.1. *Sensory Evaluation of KJC and GT*

As shown in Figure 1A, KJC and GT exhibited distinct differences in sensory characteristics. KJC displayed a yellowish green and bright infusion color, with a strong bitter taste followed by a sweet aftertaste, and a rich pan-fried bean-like aroma. In contrast, GT presented a soft green and bright infusion color, featuring a fresh and mellow taste with a refreshing chestnut-like aroma. To further characterize the taste and aroma profiles of KJC and GT in detail, QDA was performed to evaluate five taste and aroma attributes for each sample. As shown in Figure 1B, KJC showed significantly higher bitterness intensity than GT, whereas GT exhibited more pronounced umami and astringency. Additionally, no significant differences were observed between the two samples in terms of sweetness and sourness intensities. Regarding aroma attributes (Figure 1C), KJC demonstrated a significantly stronger roasty attribute compared to GT, while GT had more distinct nutty and faint-scent characteristics. No significant differences were found between the two samples in terms of the floral and sweet aroma profiles.

3.2. *Analysis of Non-Volatile Component Content in Tea Samples*

3.2.1. Main Conventional Biochemical Components Analysis

The type and content of biochemical components largely explain the taste quality and health benefits of tea [9,20]. Considering the obvious differences in the taste quality and health benefits of KJC, the main conventional biochemical components, which include water extracts, tea polyphenols, caffeine, free amino acids, water-soluble protein, and water-soluble sugar, of four tea samples (KJC, KJCFL, GT and GTFL) were analyzed. As shown in Figure 2, the water extract content was highest in KJC (50.53 ± 0.36%), and was significantly higher than that in GT. In contrast, KJCFL showed the lowest water extract content (43.07 ± 0.64%). The tea polyphenol contents in KJC and KJCFL were 11.08 ± 0.67% and 10.00 ± 0.78%, respectively, both significantly lower than those in GT and GTFL. Caffeine, a key contributor to the bitter taste of green tea [9], was detected at

4.34 ± 0.16% in GT and 3.87 ± 0.26% in GTFL, but was undetectable in both the KJC and KJCFL samples. The total free amino acid contents in KJC and KJCFL were 2.36 ± 0.03% and 1.43 ± 0.05%, and the water-soluble protein contents were 0.84 ± 0.07% and 0.74 ± 0.08%, significantly lower than those in GT and GTFL. The water-soluble sugar contents in KJC and KJCFL were 9.26 ± 0.48% and 11.67 ± 0.15%, both significantly higher than those in GT and GTFL. Meanwhile, the change rules of the main conventional biochemical components were extremely similar after KJCFL and GTFL were processed into KJC and GT. Specifically, the water extracts, tea polyphenols, amino acids, and water-soluble proteins all demonstrate varying degrees of increase, while water-soluble sugars indicate the opposite change rules.

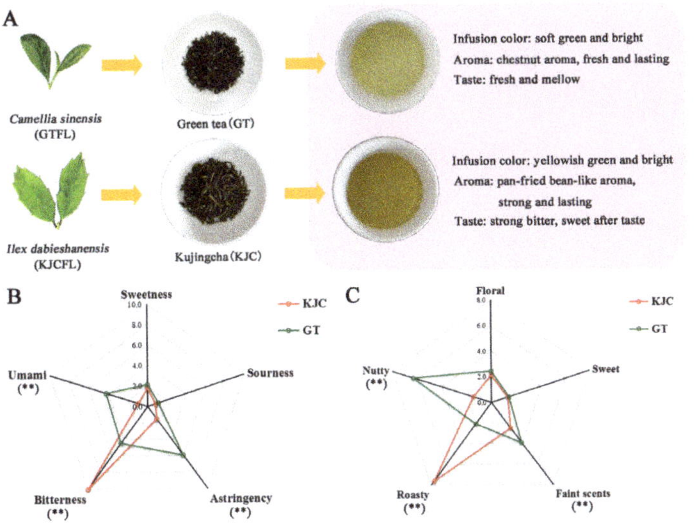

Figure 1. Sensory evaluation of KJC and GT. (**A**) Sensory description of infusion color, aroma and taste. (**B**) QDA radar chart of taste. (**C**) QDA radar chart of aroma. "**" indicates highly significant differences ($p < 0.01$).

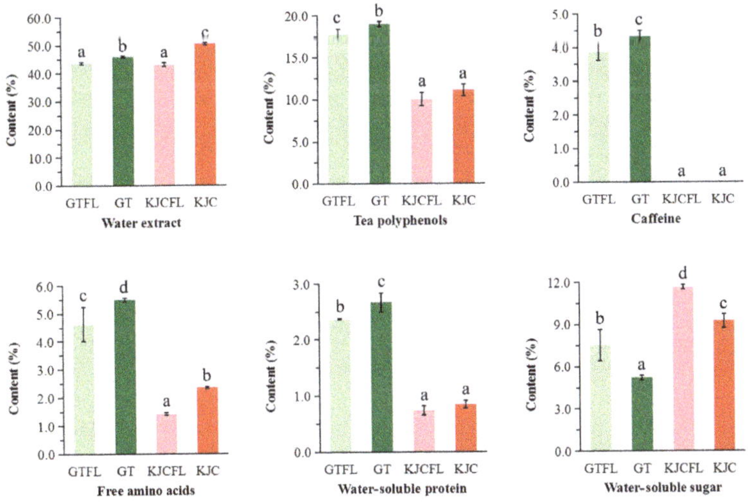

Figure 2. Contents of main conventional biochemical components in tea samples. Different lowercase letters indicate significant differences ($p < 0.05$).

3.2.2. Catechin Analysis

Catechins are the main components of tea polyphenols in green tea, accounting for about 70% of total tea polyphenols [8], and are important substances in the formation of the bitter and astringent taste and health benefits of green tea [20,21]. To further clarify the differences between KJC and GT in terms of their taste and health benefits, seven catechins were determined by HPLC in the four tea samples. As shown in Figure 3, the total catechin contents in KJC and KJCFL were only 3.92 ± 0.16 mg·g^{-1} and 1.46 ± 0.21 mg·g^{-1}, and the ratios to tea polyphenols were only 0.04 and 0.01, much lower than those in GT and GTFL. Only C and GCG were detected in KJC and KJCFL, and both showed varying levels of increase after processing. EGCG had the highest content of catechins in green tea and was significantly related to the tea's taste and health benefits [8,20]. The EGCG contents in GT and GTFL were 60.86 ± 2.07 mg·g^{-1} and 67.50 ± 2.48 mg·g^{-1}, while it was not detected in either KJC or KJCFL. Therefore, the low content of catechins is the main reason why the content of tea polyphenols in KJC is lower than that in GT, which further shows that catechins are not the main components in the formation of the taste and health benefits of KJC.

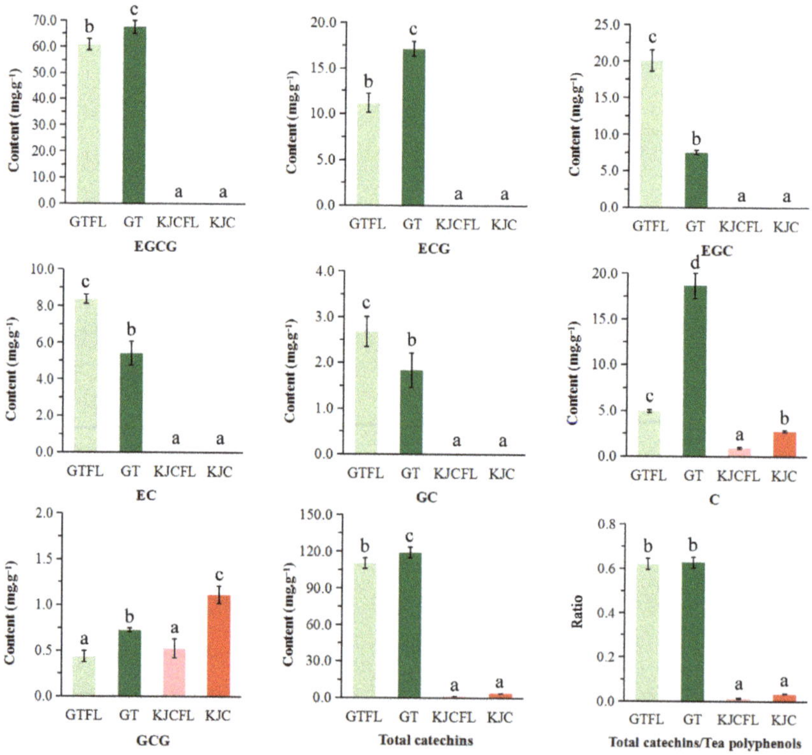

Figure 3. Contents of catechins in tea samples. Different lowercase letters indicate significant differences ($p < 0.05$).

3.2.3. Flavonoid Analysis

Flavonoids are an important component of polyphenols in tea and are closely related to the taste quality and health benefits of tea [8,22]. As shown in Figure 4A, the total flavonoids in KJC and KJCFL were 2.33 ± 0.02% and 1.81 ± 0.10%, significantly higher than those in GT (0.67 ± 0.01%) and GTFL (0.70 ± 0.05%). Meanwhile, the ratios of total flavonoids to tea polyphenols in KJC and KJCFL were 0.21 and 0.18, while a ratio of only 0.04 was found in GT and GTFL (Figure 4B). To further clarify the main reasons for the formation of high flavonoid content in KJC and KJCFL, LC-MS was used to determine the contents of 35 flavonoids in the four samples. PCA revealed distinct clustering patterns among the four tea samples based on their flavonoid profiles (Figure 4C). The first two principal components accounted for 81.0% of the total variance (PC1 = 50.5%, PC2 = 30.5%), effectively capturing the majority of variation in the 35 flavonoid compounds analyzed. Notably, all biological replicates of each tea sample clustered tightly together, while clear separations were observed between different tea groups, indicating substantial differences in their flavonoid composition. It is known from Figure 4D that the contents of five flavonoids, including neohesperidin, hyperside, rutin, astilbin and morinin, were relatively high in KJC, and their contents were 0.18 ± 0.00 mg·g^{-1}, 0.12 ± 0.01 mg·g^{-1}, 0.10 ± 0.00 mg·g^{-1}, 0.05 ± 0.00 mg·g^{-1} and 0.05 ± 0.00 mg·g^{-1}, higher than those in GT and GTFL. The contents of three flavonoids, neohesperidin, rutin and hyperside, were relatively high in KJCFL, with contents of 0.23 ± 0.02 mg·g^{-1}, 0.17 ± 0.01 mg·g^{-1} and 0.06 ± 0.00 mg·g^{-1}, and all higher than those of GT and GTFL.

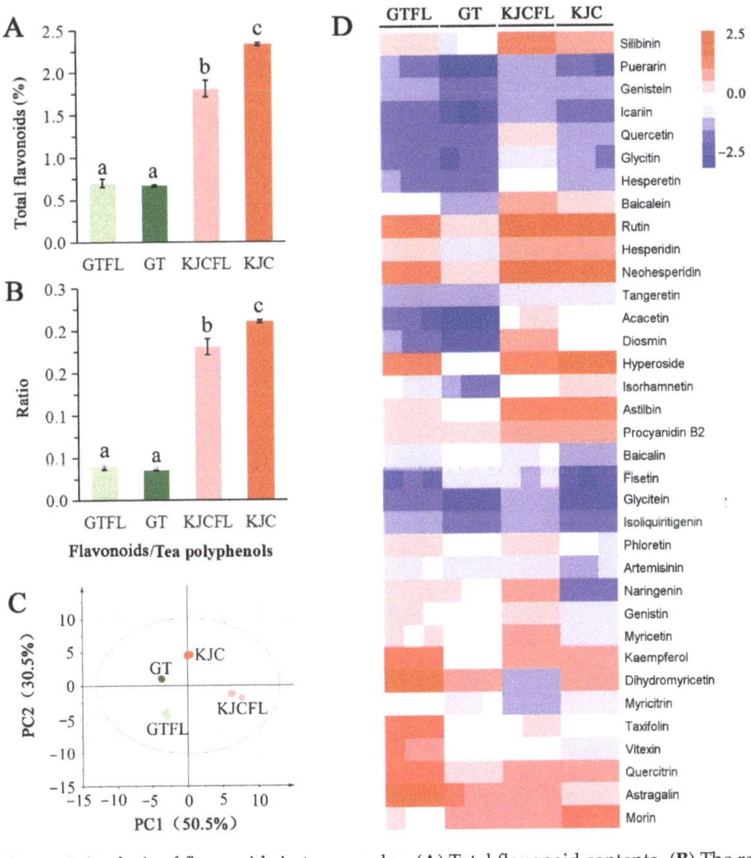

Figure 4. Analysis of flavonoids in tea samples. (**A**) Total flavonoid contents. (**B**) The ratios of total flavonoids to tea polyphenols. (**C**) PCA derived from flavonoid contents. (**D**) A heat map of the relative flavonoid contents. Different lowercase letters indicate significant differences ($p < 0.05$).

3.2.4. Amino Acid Analysis

Amino acids are an important component of the umami taste of green tea, and have a strong positive correlation with the quality of tea, their content can also effectively reduce the bitter taste [8,23]. To further clarify the reasons for the formation of low amino acid contents in KJC and KJCFL, the content of amino acids in the four tea samples was determined by HPLC. Theanine is a characteristic amino acid in tea plants and is significantly related to tea quality and health benefits [8,23]. It is observed from Figure 5A that the theanine contents in GT and GTFL were 22.47 $\pm$ 0.35 mg·g^{-1} and 20.96 $\pm$ 0.69 mg·g^{-1}, while it was not detected in KJC and KJCFL, which is also the main factor responsible for the low amino acid content in KJC. In addition, the main amino acids in *C. sinensis* green tea, such as aspartic acid, serine, glutamic acid, glutamine and arginine, were significantly lower in KJC than in GT. The other amino acids with relatively low content in *C. sinensis* green tea were also mostly low in KJC, but the contents of cysteine and valine in KJC were 1.25 $\pm$ 0.10 mg·g^{-1} and 1.44 $\pm$ 0.03 mg·g^{-1}, significantly higher than those in GT, and cysteine was not detected in either GT or GTFL (Figure 5B). Meanwhile, it is observed from Figure 5A,B that after KJCFL and GTFL were processed into KJC and GT, most amino acid contents were increased to varying degrees, which may have been caused by the hydrolysis of proteins and polypeptides in fresh leaves during processing.

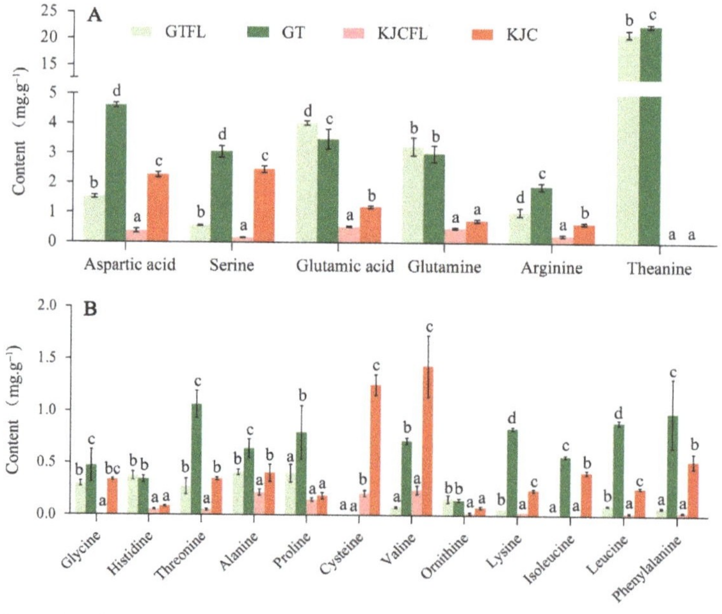

Figure 5. Content of amino acids in tea samples. (**A**) Relatively high-content amino acids. (**B**) Relatively low-content amino acids. Different lowercase letters indicate significant differences ($p < 0.05$).

3.3. Analysis of Volatile Component Content in Tea Samples

A total of 58 volatile components were detected, including 7 alcohols, 12 aldehydes, 7 esters, 18 alkenes, 7 aromatic hydrocarbons, 4 heterocycles and 3 other components. PCA revealed distinct clustering patterns among the four tea samples based on their volatile component profiles (Figure 6A). The first two principal components accounted for 81.0% of the total variance (PC1 = 50.1%, PC2 = 27.4%), effectively capturing the majority of variation in the 58 volatile components analyzed. Notably, all biological replicates of each tea sample clustered tightly together, while clear separations were observed between different tea groups, indicating substantial differences in their volatile components. A Venn diagram analysis showed that 32, 33, 22 and 23 volatile components were detected in KJC, GT, KJCFL and GTFL, respectively. Five common components (linalool, nonanal, (+)-dipentene, δ-cadinene and trans-calamenene) were found in all samples. Notably, 17 common components were found in KJC and GT, and 11 common components were found in KJCFL and GTFL (Figure 6B). In addition, after KJCFL and GTFL were processed into KJC and GT, the number of volatile components increased significantly.

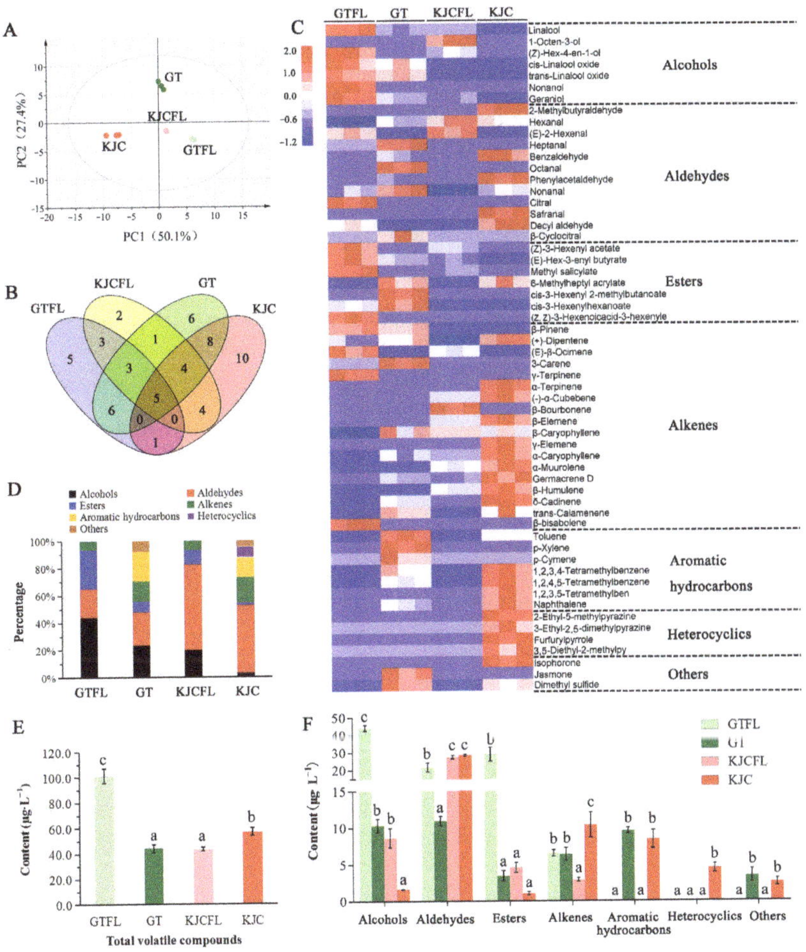

Figure 6. The analysis of volatile components in tea samples. (**A**) PCA derived from volatile component contents. (**B**) A Venn diagram of the volatile components in tea samples. (**C**) A heat map of the relative volatile component contents. (**D**) The percentages of the volatile component types. (**E**) The total volatile component contents. (**F**) The contents of each volatile component type. Different lowercase letters indicate significant differences ($p < 0.05$).

To further understand the changes in the volatile components of the four tea samples, a heat map analysis was performed based on the content of volatile components. As shown in Figure 6C, the volatile component contents of the four tea samples showed large differences, and obvious differences were detected in the types of volatile components. The proportions of aldehydes, alkenes, and heterocyclics components in KJC were significantly higher than those in GT, while the opposite pattern was found in alcohols and esters, which may explain the difference in aroma between KJC and GT (Figure 6D). Meanwhile, after KJCFL and GTFL were processed into KJC and GT, the proportions of alkenes, aromatic hydrocarbons and other volatile components increased significantly.

The total volatile component contents of the four tea samples were GTFL (101.43 ± 5.72 μg·L^{-1}), KJC (57.03 ± 3.01 μg·L^{-1}), GT (44.21 ± 3.03 μg·L^{-1}) and KJCFL (43.60 ± 1.30 μg·L^{-1}). The total volatile component content in KJC prepared from KJCFL increased significantly, while it showed an opposite pattern in GT prepared from GTFL (Figure 6E). As shown in Figure 6F, KJC had the highest contents of aldehydes, alkenes, and heterocyclics components, which were 28.51 ± 0.54 μg·L^{-1}, 10.37 ± 1.70 μg·L^{-1} and 4.48 ± 0.60 μg·L^{-1}. Among the aldehydes, phenylacetaldehyde, 2-Methylbutyraldehyde and benzaldehyde contents were relatively high in KJC, and phenylacetaldehyde content (17.06 ± 1.43 μg·L^{-1}) was the highest. Among the alkenes, δ-Cadinene and (+)-Dipentene contents were relatively high in KJC. Heterocyclics were only detected in KJC, among which the contents of 2-Ethyl-5-methylpyrazine and 3-Ethyl-2,5-dimethylpyrazine were relatively high. Among the tea samples, KJC had the lowest content of alcohols and esters, of which only linalool was detected for alcohols, and only 6-Methylheptyl acrylate was detected for esters. Moreover, aromatic hydrocarbons and other volatile components were relatively high in KJC and GT, while they were not detected in KJCFL and GTFL. Among them, the aromatic hydrocarbon contents in KJC and GT were 8.42 ± 1.23 μg·L^{-1} and 9.58 ± 0.40 μg·L^{-1}, both of which were dominated by toluene; the other volatile component contents in KJC and GT were 2.55 ± 0.49 μg·L^{-1} and 3.46 ± 0.88 μg·L^{-1}, both of which were dominated by dimethyl sulfide.

4. Discussion

4.1. Flavonoids Are Significantly Associated with the Overall Flavor and Health Benefits of KJC

Polyphenols are the main components responsible for the bitter and astringent taste of tea, and are also important functional components of tea [8,24]. Both KJC and KJCFL contain relatively high levels of tea polyphenols, which are closely related to the flavor quality and health benefits of KJC. Tea polyphenols primarily consist of catechins, flavonoids, anthocyanins, leucoanthocyanidins, phenolic acids, and depsides [8,20]. Among these, catechins are the most abundant, accounting for approximately 70% of total polyphenol content [8]. Sensorily, catechins contribute predominantly to the bitter and astringent taste profile of green tea, with their concentration and composition directly determining the flavor quality [21]. More importantly, catechins serve as the key bioactive components responsible for green tea's various health benefits, including its antioxidant, antimicrobial, antiviral, radioprotective, antiaging, and anticancer effects [8,24]. However, the composition of tea polyphenols exhibited obvious differences in KJC and GT, whereby the total catechin content in KJC and KJCFL was extremely low, while the total flavonoid content was significantly higher than that in GT and GTFL, accounting for about 20% of the total tea polyphenols. KJC contained relatively higher levels of flavonoids, including neohesperidin, hyperoside, rutin, astilbin and morin, compared to GT, all of which have been extensively demonstrated to possess various physiological and pharmacological functions. For instance, neohesperidin exhibits potential blood glucose-regulating effects [25], hyperoside shows significant neuroprotective and antioxidant activity [26], rutin exhibits significant antioxidant, anti-inflammatory and cardioprotective effects [27], astilbin possesses antioxidant, hypoglycemic and anticancer properties [28], and morin demonstrates anti-inflammatory and antioxidant effects [29]. Moreover, most flavonoids show obvious bitter and astringent characteristics. For example, neohesperidin is an important bitter substance in citrus fruit [30], and a high content of flavonoid glycosides like rutin is an important reason for the bitter taste of green tea in summer and autumn [31]. Therefore, it is speculated that high flavonoid content is an important basis for the overall taste and health benefits of KJC.

4.2. Umami and Sweet Components Are Important Supplements to Regulate the Overall Taste of KJC

The umami and sweet tastes account for a small proportion of the taste of green tea, but they can supplement and regulate the overall taste [32]. Amino acids and water-soluble protein are the main components of the umami taste of green tea and can effectively relieve the bitter and astringent taste of tea [23]. The albino tea cultivar, due to its higher amino acid content, produces green tea with a notably superior fresh and mellow taste compared to conventional tea varieties, along with significantly less bitterness and astringency [33]. The amino acid and water-soluble protein contents in KJC were relatively low, 57.0% and 68.5% lower than those in GT. *C. sinensis* green tea is rich in various amino acids, but not all amino acids have an umami taste, and some amino acids even have a bitter or astringent taste [23]. Theanine, glutamic acid and aspartic acid, which are the main umami amino acids in *C. sinensis* green tea, were all found at low levels in KJC. Notably, theanine, the characteristic amino acid of green tea, was not even detected. Therefore, the low contents of amino acids and water-soluble protein are an important reason for the formation of the intense bitterness of KJC. Water-soluble sugars such as fructose, glucose and sucrose are important sweet-tasting components in tea [34]. They serve as the primary substances responsible for the lingering sweet aftertaste of tea infusions and participate in aroma formation through Maillard reactions [35]. KJC contains significantly higher levels of water-soluble sugars, showing a 77.7% increase compared to GT. Therefore, the high content of water-soluble sugars forms the material basis for KJC's distinctive sweet aftertaste. Furthermore, complex polysaccharides composed of sugars combined with proteins, mineral elements and other units exhibit multiple health benefits, including weight management, lipid reduction, blood pressure regulation, and free radical scavenging, and these components demonstrate remarkable efficacy in preventing and treating metabolic diseases such as obesity and diabetes [35,36]. The elevated water-soluble sugar content in KJC may be associated with these health-promoting effects.

4.3. Heterocyclics Contribute to the Formation of the Bean-Like Aroma of KJC

The aroma of tea, resulting from the comprehensive effect of tea volatile components in different concentration combinations on the human olfactory nerve, is an important indicator for evaluating the quality of tea [8,37]. Heterocyclics are nitrogen-containing compounds, such as pyrazines and pyrroles, generated by the Maillard reaction between reducing sugars in fresh leaves and amino acids and proteins at relatively high temperatures during the tea processing process, and they are the key components in the formation of the roasted aroma and bean-like aroma of tea [38,39]. Pyrazines are important aroma components in oolong tea and mainly contribute to the roasted aroma, nutty aroma and popcorn aroma [40]. A total of 4 heterocyclics were detected in KJC, namely 2-Ethyl-5-methylpyrazine, 3-Ethyl-2,5-dimethylpyrazine, furfurylpyrrole and 3,5-Diethyl-2-methylpy. Among them, 2-Ethyl-5-methylpyrazine and 3,5-Diethyl-2-methylpyrazine are the key substances in the formation of the roasted aroma of Huangdacha [41], 3,5-Diethyl-2-methylpy is the key substance in the formation of the roasted aroma of crispy rice [42], and furfurylpyrrole is the key substance in the formation of the roasted aroma of rougui Wuyi rock tea [43]. Therefore, heterocyclics are important components for the formation of the bean-like aroma of KJC. Meanwhile, KJC exhibited significantly higher proportions of aldehydes and alkenes compared to GT, which synergistically complemented and enriched the dominant aroma profile of the tea, resulting in more complex olfactory characteristics. Among the aldehydes, phenylacetaldehyde was the most abundant compound in KJC, accounting for approximately 60% of the total aldehyde content. Notably, phenylacetaldehyde serves as the key aromatic component responsible for the chestnut-like fragrance characteristic of green tea [43]. Additionally, olefinic compounds such as (+)-dipentene and δ-cadinene, known for their low odor thresholds, made positive contributions to the overall tea aroma quality [44].

5. Conclusions

In this study, we found that the high tea polyphenol content in KJC was closely associated with its flavor quality and health benefits. The tea polyphenols in KJC were mainly composed of flavonoids, with relatively high levels of neohesperidin, hyperoside, rutin, astilbin and morin potentially serving as important bioactive compounds contributing to its health effects. The intense bitterness followed by a sweet aftertaste and the clean, mellow taste of KJC are primarily attributed to its unique composition featuring high levels of flavonoids and water-soluble sugars combined with low contents of amino acids and water-soluble proteins. This study identified high concentrations of aldehydes, olefins and heterocyclic compounds as key aroma components in KJC. Among these, the heterocyclic compounds 2-ethyl-5-methylpyrazine, 3-ethyl-2,5-dimethylpyrazine, furfurylpyrrole and 3,5-diethyl-2-methylpyrazine were particularly important for the formation of KJC's bean-like aroma flavor. These results will provide a theoretical basis for studies on the flavor quality and health benefits of KJC and scientific support for its further development and utilization.

Supplementary Materials: The following supporting information can be downloaded at: https://www.mdpi.com/article/10.3390/horticulturae11070804/s1, Table S1: chemicals and reagents source information.

Author Contributions: Conceptualization, L.W.; Data Curation, L.M. and D.C.; Formal Analysis, L.M. and L.W.; Funding acquisition, X.J.; Investigation, P.F., L.W. and B.H.; Methodology, L.M. and Y.P.; Project Administration, X.J. and Y.P.; Resources, P.F. and Y.L. (Yijin Liu); Software, D.C. and Y.L. (Yanli Liu); Supervision, G.F. and A.L.; Validation, A.L. and B.H.; Visualization, P.F. and G.F.; Writing—Original Draft, L.M. and Y.P.; Writing—Review and Editing, D.C. and X.J. All authors have read and agreed to the published version of the manuscript.

Funding: The authors acknowledge the financial support of the Agricultural Science and Technology Innovation Project of Hubei Province (2025-620-000-001-020), and Hubei Seed Industry High-Quality Development Project of China.

Institutional Review Board Statement: In compliance with the "Ethics Review Measures for Life Sciences and Medical Research Involving Humans" issued by the National Health Commission of China in 2023 (Document No. 4), specifically Article 32, our study qualifies for exemption from ethical review as it does not present any potential risk to the participants.

Informed Consent Statement: Informed consent was obtained from all subjects involved in the study.

Data Availability Statement: The original contributions presented in the study are included in the article. Further inquiries can be directed to the corresponding authors.

Conflicts of Interest: Author Yanan Peng was employed by the company Tuanfeng Daqishan Kujingcha Co., Ltd. The remaining authors declare that the research was conducted in the absence of any commercial or financial relationships that could be construed as a potential conflict of interest.

References

1. Wüpper, S.; Lüersen, K.; Rimbach, G. Chemical composition, bioactivity and safety aspects of Kuding tea-from beverage to herbal extract. *Nutrients* **2020**, *12*, 2796. [CrossRef] [PubMed]
2. Li, L.; Xu, L.; Ma, G.; Dong, Y.; Peng, Y.; Xiao, P. The large-leaved kudingcha (*Ilex latifolia* Thunb and *Ilex kudingcha* C.J. Tseng): A traditional Chinese tea with plentiful secondary metabolites and potential biological activities. *J. Nat. Med.* **2013**, *67*, 425–437. [CrossRef] [PubMed]
3. Li, L.; Peng, Y.; Ma, G.; He, C.; Feng, Y.; Lei, Q.; Xiao, P. Quantitative analysis of five kudinosides in the large-leaved Kudingcha and related species from the genus *Ilex* by UPLC-ELSD. *Phytochem. Anal.* **2012**, *23*, 677–683. [CrossRef]
4. Gan, R.; Zhang, D.; Wang, M.; Corke, H. Health benefits of bioactive compounds from the genus *Ilex*, a source of traditional caffeinated beverages. *Nutrients* **2018**, *10*, 1682. [CrossRef]
5. Lei, G.; Wu, Z.; Xu, Y.; Bao, Y. Effects of different processing methods on the contents of polyphenols in Kujingcha tea. *Lishizhen Med. Mater. Med. Res.* **2003**, *14*, 345–346.
6. Zhang, F.; Lu, X.; Chu, D.; Yin, R.; Wang, C.; Zhou, Y.; Li, Y.; Li, N.; Cai, X.; Chen, H. A New *Ilex dabieshanensis* Cultivar "*Ilex dabieshanensis* No.1". *Acta Hortic. Sin.* **2018**, *45*, 2827–2828.
7. Jiang, A.; Wang, X.; Xiang, J.; Huang, W. Research on instant Kujing tea. *Sci. Technol. Food Ind.* **2012**, *33*, 280–282.
8. Wan, X. *Tea Biochemistry*, 3rd ed.; China Agriculture Press: Beijing, China, 2003. (In Chinese)
9. Zhang, L.; Cao, Q.; Granato, D.; Xu, Y.; Ho, C. Association between chemistry and taste of tea: A review. *Trends Food Sci. Technol.* **2020**, *101*, 139–149. [CrossRef]
10. Ma, L.; Cao, D.; Liu, Y.; Gong, Z.; Liu, P.; Jin, X. A comparative analysis of the volatile components of green tea produced from various tea cultivars in China. *Turk. J. Agric. For.* **2019**, *43*, 451–463. [CrossRef]
11. Zhai, X.; Zhang, L.; Granvogl, M.; Ho, C.; Wan, X. Flavor of tea (*Camellia sinensis*): A review on odorants and analytical techniques. *Compr. Rev. Food Sci. Food Saf.* **2022**, *21*, 3867–3909. [CrossRef]
12. Zhao, Y.; Li, F.; Zhou, H.; Wang, Y.; Bian, J.; Sun, Y.; Du, X. Utilizing nontargeted metabolomics integrated with quality component quantitation via differential analysis to reveal the taste differences between and metabolite characteristics of hawk black tea and hawk white tea. *Food Res. Int.* **2024**, *197*, 115216. [CrossRef] [PubMed]
13. Ma, L.; Yang, Y.; Cao, D.; Liu, Y.; Gong, Z.; Ercisli, S.; Luo, Z.; Jin, X. Analysis of the biochemical and volatile components of Qianlincha and Qiandingcha prepared from *Eurya alata Kobuski* and *Camellia cuspidate*. *Agronomy* **2021**, *11*, 657. [CrossRef]
14. *GB/T 23776-2018*; Methodology for Sensory Evaluation of Tea. Standardization Administration of China: Beijing, China, 2018.
15. *GB/T 8314-2013*; Tea—Determination of Free Amino Acids Content. Standardization Administration of China: Beijing, China, 2013.
16. *GB/T 8313-2018*; Determination of Total Polyphenols and Catechins Content in Tea. Standardization Administration of China: Beijing, China, 2018.
17. *GB/T 8305-2013*; Tea—Determination of Water Extracts Content. Standardization Administration of China: Beijing, China, 2013.
18. Li, Y. *Comprehensive Experiment of Tea Science*; China Light Industry Press: Beijing, China, 2018. (In Chinese)
19. Xue, J.; Liu, P.; Xu, Y.; Gui, A.; Wang, X.; Ye, F.; Feng, L.; Wang, S.; Teng, J.; Gao, S.; et al. Effect of cultivar and process on the astringency of matcha based on flavonoids-targeted metabolomic analysis. *Food Res. Int.* **2025**, *204*, 115954. [CrossRef] [PubMed]
20. Lin, F.; Wei, X.; Liu, H.; Li, H.; Xia, Y.; Wu, D.; Zhang, P.; Gandhi, G.; Li, H.; Gan, R. State-of-the-art review of dark tea: From chemistry to health benefits. *Trends Food Sci. Technol.* **2021**, *109*, 126–138. [CrossRef]
21. Narukawa, M.; Kimata, H.; Noga, C.; Watanabe, T. Taste characterisation of green tea catechins. *Int. J. Food Sci. Technol.* **2010**, *45*, 1579–1585. [CrossRef]
22. Atmani, D.; Chaher, N.; Atmani, D.; Berboucha, M.; Debbache, N.; Boudaoud, H. Flavonoids in human health: From structure to biological activity. *Curr. Nutr. Food Sci.* **2009**, *5*, 225–237. [CrossRef]
23. Yu, Z.; Yang, Z. Understanding different regulatory mechanisms of proteinaceous and non-proteinaceous amino acid formation in tea (*Camellia sinensis*) provides new insights into the safe and effective alteration of tea flavor and function. *Crit. Rev. Food Sci. Nutr.* **2020**, *60*, 844–858. [CrossRef]
24. Khan, N.; Mukhtar, H. Tea polyphenols in promotion of human health. *Nutrients* **2018**, *11*, 39. [CrossRef]
25. Zhang, J.; Sun, C.; Yan, Y.; Chen, Q.; Luo, F.; Zhu, X.; Li, X.; Chen, K. Purification of naringin and neohesperidin from Huyou (*Citrus changshanensis*) fruit and their effects on glucose consumption in human HepG2 cells. *Food Chem.* **2012**, *135*, 1471–1478. [CrossRef]
26. Cao, J.; Tang, C.; Gao, M.; Rui, Y.; Zhang, J.; Wang, L.; Wang, Y.; Xu, B.; Yan, B. Hyperoside alleviates epilepsy-induced neuronal damage by enhancing antioxidant levels and reducing autophagy. *J. Ethnopharmacol.* **2020**, *257*, 112884. [CrossRef]
27. Semwal, R.; Joshi, S.; Semwal, R.; Semwal, D. Health benefits and limitations of rutin-A natural flavonoid with high nutraceutical value. *Phytochem. Lett.* **2021**, *46*, 119–128.
28. Xia, H.; Lv, C.; Lu, Y.; Zeng, C.; Qin, S.; Shi, M. Natural deep eutectic ready to use extract of astilbin: Super high in vitro bioaccessibility, α-amylase and α-glucosidase enzyme inhibition kinetics. *Food Res. Int.* **2023**, *173*, 113368. [CrossRef] [PubMed]

29. Liu, X.; Han, L.; Wang, M.; Zhang, Y. Inhibitory Mechanism of Morin on Pancreatic Lipase. *Food Sci.* **2025**, *46*, 40–48.
30. Rouseff, R.; Martin, S.; Youtsey, C. Quantitative survey of narirutin, naringin, hesperidin, and neohesperidin in citrus. *J. Agric. Food Chem.* **1987**, *35*, 1027–1030. [CrossRef]
31. Qi, D.; Dai, W.; Tan, L.; Peng, Q.; Zhang, Y.; Lin, Z. Study on the effects of the fixation methods on the chemical components and taste quality of summer green tea. *J. Tea Sci.* **2016**, *36*, 18–26.
32. Shi, Z. *Chaye Shenping Yu Jianyan*; China Agricultural Press: Beijing, China, 2010. (In Chinese)
33. Li, Y.; Han, Z.; Wang, M.; Yan, Y.; Ma, R.; Wang, H.; Deng, W. Metabolomics and sensory evaluation reveal the influence of four albino tea cultivars on the quality of processed green tea. *Food Res. Int.* **2025**, *209*, 116180. [CrossRef]
34. Li, W.; Zhang, Z.; Chen, R.; Sun, L.; Lai, X.; Li, Q.; Hao, M.; Zhang, S.; Li, Q.; Sun, S.; et al. Metabolomics-based analysis of the effects of differences in soluble sugars on the sweetness quality of six major tea types in China. *Food Funct.* **2025**, *16*, 3707–3720.
35. Lu, T.; Dong, X.; Lan, W.; Zhou, B.; Teng, J.; Xia, N.; Khan, M.; Huang, L. From microbial perspective: Manufacturing process, chemical composition and health benefis of Liupao tea-A comprehensive review. *Food Res. Int.* **2025**, *206*, 116088. [CrossRef]
36. Xiao, J.; Jiang, H. A review on the structure-function relationship aspect of polysaccharides from tea materials. *Crit. Rev. Food Sci. Nutr.* **2015**, *55*, 930–938. [CrossRef]
37. Yang, Z.; Baldermann, S.; Watanabe, N. Recent studies of the volatile compounds in tea. *Food Res. Int.* **2013**, *53*, 585–599. [CrossRef]
38. Van Boekel, M. Formation of flavour compounds in the Maillard reaction. *Biotechnol. Adv.* **2006**, *24*, 230–233. [CrossRef] [PubMed]
39. Guo, X.; Song, C.; Ho, C.; Wan, X. Contribution of L-theanine to the formation of 2,5-dimethylpyrazine, a key roasted peanutty flavor in Oolong tea during manufacturing processes. *Food Chem.* **2018**, *263*, 18–28. [PubMed]
40. Ho, C.; Zheng, X.; Li, S. Tea aroma formation. *Food Sci. Hum. Wellness* **2015**, *4*, 9–27. [CrossRef]
41. Pei, Z.; He, G.; Liu, Y.; Hu, Y.; Li, M.; Wan, X.; Zhai, X. Characterization of the key aroma-active compounds responsible for the rice cruse-like aroma of large-leafed yellow tea (*Camellia sinensis*). *Food Sci.* **2023**, *44*, 289–297.
42. Yang, P.; Yu, M.; Song, H.; Xu, Y.; Lin, Y.; Granvogl, M. Characterization of key aroma-active compounds in rough and moderate fire rougui wuyi rock tea (*Camellia sinensis*) by sensory-directed flavor analysis and elucidation of the influences of roasting on aroma. *J. Agric. Food Chem.* **2022**, *70*, 267–278.
43. Zhu, Y.; Lv, H.; Shao, C.; Kang, S.; Zhang, Y.; Guo, L.; Dai, W.; Tan, J.; Peng, Q.; Lin, Z. Identification of key odorants responsible for chestnut-like aroma quality of green teas. *Food Res. Int.* **2018**, *108*, 74–82. [CrossRef]
44. Schuh, C.; Schieberle, P. Characterization of the key aroma compounds in the beverage prepared from Darjeeling black tea: Quantitative differences between tea leaves and infusion. *J. Agric. Food Chem.* **2006**, *54*, 916–924.

horticulturae

MDPI

Article

The Effect of Cropping System and Irrigation Regime on the Plant Growth and Biochemical Profile of *Cichorium spinosum*

Beatriz H. Paschoalinotto [1,†], Nikolaos Polyzos [2,†], Vasiliki Liava [2], Filipa Mandim [1], Tânia C. S. P. Pires [1], Mikel Añibarro-Ortega [1], Isabel C. F. R. Ferreira [1], Maria Inês Dias [1], Lillian Barros [1] and Spyridon A. Petropoulos [2,*]

1 Centro de Investigação de Montanha (CIMO), LA SusTEC, Instituto Politécnico de Bragança, Campus de Santa Apolónia, 5300-253 Bragança, Portugal; paschoalinotto@ipb.pt (B.H.P.); filipamandim@ipb.pt (F.M.); tania.pires@ipb.pt (T.C.S.P.P.); mikel@ipb.pt (M.A.-O.); iferreira@ipb.pt (I.C.F.R.F.); maria.ines@ipb.pt (M.I.D.); lillian@ipb.pt (L.B.)

2 Laboratory of Vegetable Production, Department of Agriculture, Crop Production and Rural Environment, University of Thessaly, Fytokou Street, 384 46 N. Ionia, Greece; npolyzos@uth.gr (N.P.); vasiliki.liava@gmail.com (V.L.)

* Correspondence: spetropoulos@uth.gr

† These authors contributed equally to this work.

Abstract: This study evaluated the effects of three irrigation treatments (control (rain-fed plants), deficit irrigation (DI: 50% of maximum field capacity), full irrigation (FI: 100% of maximum field capacity)), and two crop-management treatments (with or without crop rotation with bean, CR, and NCR, respectively) on the plant growth and chemical composition of *C. spinosum*. The results indicated that deficit irrigation combined with crop rotation increased the weight of leaves per plant, followed by rain-fed plants for the same crop-management treatment. Additionally, these two factors significantly influenced the nutritional profile, free sugars, and organic acid content in a variable manner. Moreover, the control treatment and deficit irrigation increased the content of K, Na, and Mg, which are highly mobile nutrients, whereas the levels of moderately mobile nutrients such as Fe, Mn, Cu, and Zn decreased. Deficit irrigation without crop rotation significantly increased the content of total tocopherols, followed by deficit irrigation with crop rotation and full irrigation without crop rotation. The main fatty acids were α-linolenic acid (C18:3n3), followed by palmitic acid (C16:0) and linoleic acid (C18:2n6), while the control and deficit irrigation treatments combined with crop rotation increased PUFA and decreased SFA content. Furthermore, deficit irrigation and crop rotation induced the accumulation of phenolic compounds, flavonoids, and phenolic acids, especially the content of the major compounds (e.g., chicoric acid, quercetin-*O*-hexurunoside, and luteolin-*O*-hexurunoside). The leaf extracts exhibited varied antioxidant activity (assessed by TBARS and OxHLIA assays), and antimicrobial activity. On the other hand, no antifungal, antiproliferative (except for AGS cell line), hepatotoxic, or anti-inflammatory effects were recorded. In conclusion, the combination of deficit irrigation and crop rotation with bean positively affected the quality traits and the fresh weight of leaves, thus suggesting that such eco-friendly practices could have beneficial effects in the cultivation of *C. spinosum* plants within the context of climate-change mitigation strategies.

Keywords: wild edible species; deficit irrigation; antimicrobial properties; crop rotation; cytotoxic effects; bioactive properties; phenolic compounds; antioxidant activity

1. Introduction

The long history of wild edible plants in the Mediterranean culture can be traced back to ancient times, and a strong bond has been created between human societies and wild plants, which have been exploited for food, medicinal, and religious purposes from time to time [1,2]. Wild greens were highly appreciated during famine periods because they offered a valuable and nutritive food source [3,4]. During recent times and as the standards of living improved, consumers tended to neglect these species and shifted towards unhealthy and "modern" dietary habits; however, nowadays, there is a renewed interest for the exploitation of these plants, which are considered functional foods due to their rich phytochemical content and their high nutritional value [5].

Cichorium spinosum L. is a wild edible species valued for its tender edible leaves and is often consumed in many local Mediterranean recipes [6]. The species thrives in many Mediterranean countries, especially in coastal and semi-arid regions, and it is highly acknowledged as a culinary trademark of the Cretan diet [7,8]. Regarding the chemical composition and nutritional value of *C. spinosum* leaves, reports in the literature highlight its increased content in vitamins (namely C, E, and K1), phenolic acids (chicoric and 5-*O*-Caffeoylquinic acid), total glutathione, proteins, fatty acids, caretonoids (β-carotene and lutein) and minerals [9,10]. The leaves of the plant are usually hand-picked from the wild and consumed raw in leafy salads, boiled with meat or even consumed in pickled form [11]. However, the seasonal availability, limited access to wild habitats, and the increasing market needs for high-added-value and functional food products have created a niche for the commercial exploitation of the plant due to its rich phytochemical properties and its high nutritional value [12]. For this purpose, several recent research studies have focused on optimizing the agronomic practices that will facilitate its commercial exploitation without compromising the health benefits of the edible product [13].

The implications of the ongoing climate change have raised serious obstacles to the cultivation of conventional crops with a severe impact on crop production [14,15]. Water is a key natural resource for food production, being essential for high-yield crops due to its involvement in plant functions and physiological processes [16,17]. Net crop water-use requirements are expected to increase globally by 25% at the end of 2080 due to changing environmental conditions and increased irrigation cost [18]. Likewise, the recharging of groundwater in the Mediterranean countries is expected to decrease by approximately 30% in the near future, while crop water-use requirements are expected to increase due to climate change [19]. To overcome these challenges, it is of great importance to exploit and valorize alternative crops that can adapt to harsh conditions but also to improve water-use efficiency and develop new technological irrigation tools [20,21]. In this context, deficit irrigation (DI) could be a valuable tool to address irrigation water scarcity [22]. This strategy may have a beneficial effect on crop production by improving the marketable yield per unit of water used, reducing nutrient loss near to the root area and excessive vegetative vigor, as well as by decreasing the incidence of diseases related to high humidity [23].

Recent research has focused on highlighting the importance of suggesting alternative/complementary crops that can withstand adverse environmental conditions, especially in arid and semi-arid regions that are mostly affected by climate adversities and drought in particular, as well as to reveal the mechanisms involved in this adaptation [24,25]. In this context, the commercialization of wild edible species as high-added-value substitutes of conventional crops could be a key pillar of a sustainable strategy aiming to mitigate the severe impact of climate change on food security [26–29], as well as in the diversification of food systems [30–32]. Moreover, Polyzos et al. [33] found that deficit irrigation could be used as a sustainable tool to manage irrigation water in *Scolymus hispanicus* cultivation without compromising crop yield and the bioactive profile of the edible portion of plants.

However, it should be noted that this irrigation technique (e.g., deficit irrigation) should be tailored to each species, since a varied response has been recorded and various mechanisms of adaptation and tolerance have been reported in the literature [22,29,34].

On the other hand, crop rotation is a very old agronomic practice with profound advantages such as improved climate resilience of crops through the enhancement of water dynamics by increased soil water storage and crop water-use efficiency, the improvement of physical and chemical properties of soil, the increase in soil enzyme activity and nutrient use efficiency, the improvement of soil microbiome, the reduction in weed competitiveness, effective pest and disease management, and the increased biodiversity of agroecosystem [35–40].

So far, there is a scarcity in the literature reports regarding the integration of wild edible plants in modern cropping systems. Therefore, the aim of the present study was to evaluate the potential inclusion of *C. spinosum* plant in a rotation scheme with common bean (*Phaseolus vulgaris* L.) under deficit irrigation conditions and further evaluate the effects of cropping system and irrigation regime on crop performance, and the chemical profile and bioactive properties of the edible leaves of the species. Our results could contribute to the commercial exploitation of wild edible greens and also highlight the importance of implementing sustainable agronomic tools in modern agriculture as a means to mitigate climate change and ensure food security.

2. Materials and Methods

2.1. Experimental Conditions

At the experimental farm of the University of Thessaly in Velestino (Greece; 39°37′18.6″ N, 22°22′55.1″ E), young seedlings of *C. spinosum* L. (Asteraceae family) were transplanted to the field when they reached the stage of 3–4 true leaves. Seedlings were prepared by seeds obtained from Geniki Fytotechniki S.A. (Athens, Greece) after sowing them in seed trays containing peat (pH 5.5–6.5, base substrate with no fertilizers added; Klasmann-Deilmann GmbH, Geeste, Germany). The species is depicted in Figure 1. In the current study, there were 3 irrigation treatments, namely deficit irrigation (DI: 50% of field capacity), full irrigation (FI: 100% of field capacity) and the control treatment where the plants were rain-fed. Moreover, the effect of crop rotation was also tested by applying two treatments, namely crop rotation (CR) with plants of *Phaseolus vulgaris* L. and no crop rotation (NCR). The plants were placed at distances of 0.50 cm between the rows and 0.33 cm within the rows, in which each treatment included 3 rows with a plant density of 45 plants per treatment. The plants were fertigated with a base dressing of N-P-K (Atlas 20-20-20 + TE; Gavriel S.A., Volos, Greece) in amounts varying based on the plant growth and the environmental circumstances; whereas, during the growing period of the plants, irrigation was performed once or twice per week, depending on the environmental conditions via a drip irrigation system. Weed management was carried out manually, while pests and pathogens were controlled with chemical management according to the recommended cultivation practices. The soil was sandy clay loam (38% sand, 36% silt, and 26% clay), with pH = 7.4 (1:1 soil/H_2O) and organic matter content = 1.3%. The meteorological data throughout the experimental period (mean air temperature, mean highest and lowest temperature, relative humidity and rainfall) are presented in Table 1. The harvest of the plants took place in 14 May 2021 and the fresh weight of leaves/plant (g) and number of leaves/plant were calculated per treatment.

Figure 1. *Cichorium spinosum* plants cultivated in the field before harvest (**left** photo) and at the end of the growth cycle (**right** photo). (Source: personal record of Dr. Spyridon A. Petropoulos).

Table 1. Meteorological data during the cultivation period of *C. spinosum* L. plants.

Month	Mean Temperature (°C)	Mean Highest Temperature (°C)	Mean Lowest Temperature (°C)	Relative Humidity (%)	Rainfall (mm)
March	9.15	15.47	3.07	51.75	55.90
April	13.56	20.16	7.08	52.63	21.90
May	20.95	28.52	12.91	45.81	21.50
Mean	14.55	21.38	7.69	50.06	33.10

2.2. Nutritional Characterization

The nutritional characterization was determined following official methods of analysis (AOAC methods) in terms of total fat (n° 989.05), crude protein (n° 991.02), ash (n° 935.42), total dietary fiber (n° 991.43 and 992.16) and carbohydrates by difference. The results were expressed in g/100 g dw. Energy value in the dried powder samples of *C. spinosum* leaves was calculated following the European Regulation n° 1169 as described below: energy (kcal/100 g dw) = ((9 × total fat content) + (4 × (crude protein content + carbohydrate content)) + (2 × total dietary fiber content). The outcomes were expressed in kcal/100 g dw.

2.3. Organic Acids

For the extraction of organic acids, 1 g of *C. spinosum* samples were placed in a 25 mL solution of metaphosphoric acid (4.5% *v*/*v*) under magnetic stirring at room temperature for 20 min and subsequently filtered, following the protocol as detailed by Polyzos et al. [41]. The identification of organic acids and their respective quantification was determined by comparing retention times, spectra with commercial standards, and respective safety rates (five-level lines). The results were expressed in mg/100 g dw.

2.4. Tocopherols

The extraction procedure and chromatographic characterization of tocopherols were carried out in accordance with the methods outlined by Polyzos et al. [33]. The chromatographic analysis was performed using an HPLC system, comprising a quaternary pump, degasser, automatic sampler, and fluorescence detector. The separation process was accomplished on a Polyamide II normal-phase column (250 × 4.6 mm, 5 µm) at 35 °C, with the detector set to 290 nm excitation and 330 nm emission. The mobile phase comprised a mixture of hexane and ethyl acetate (70:30, *v*/*v*), with a flow rate of 1 mL/min. Identification and quantification were conducted using Clarity 2.4 software.

2.5. Sugars Composition

Sugars composition was determined using high-performance liquid chromatography (HPLC) with a refractive index (RI) detector, according to the methodology described by Polyzos et al. [41]. The analysis was carried out by HPLC-RI at 35 °C, using a Knauer Smartline system equipped with an RI detector (Knauer Smartline 2300, Knauer, Berlin, Germany) and a Eurospher NH2 100-5 column (4.6 × 250 mm, 5 µm, Knauer). The mobile phase consisted of a mixture of acetonitrile–deionized water (70:30, *v*/*v*) at a flow rate of 1 mL/min. The compounds were identified by chromatographic comparison with authentic standards, while quantification was carried out using the internal standard method. The sugar content was expressed in g/100 g dw.

2.6. Mineral Composition

The mineral composition of *C. spinosum* leaves was determined in accordance with the AOAC protocol [42]. In summary, 10 mL of nitric acid was utilized for the digestion of the dry powder leaves in a microwave system at 200 °C and 1600 watts for a duration of 30 min. Thereafter, the solution was diluted to a volume of 50 mL with distilled water. Concentrations of potassium (K), sodium (Na), calcium (Ca), magnesium (Mg), iron (Fe), manganese (Mn), zinc (Zn), and copper (Cu) were then quantified by atomic absorption spectrophotometry (Perkin Elmer 1100B, Waltham, MA, USA).

2.7. Fatty-Acid Composition

The fatty-acid profile was determined by gas chromatography coupled to a flame ionization detector (GC-FID, DANI instrument model GC 1000, Milan, Italy), in accordance with the methodology described by Polyzos et al. [33]. The identification of the fatty acids was based on the comparison of the relative retention times of the peaks of the standard mixture of 37 FAMEs with those of the samples. The data were processed using Clarity 4.0.1.7 software (DataApex, Podohradska, Czech Republic) and expressed as a relative percentage of each fatty acid.

2.8. Phenolic Compounds and Bioactive Properties

To evaluate the phenolic compounds and bioactive properties, the leaves of *C. spinosum* were extracted with ethanol–water (80:20 *v/v*). Briefly, 1 g of *C. spinosum* leaves was added to 30 mL of ethanol–water solution (80:20, *v*/*v*; 30 mL) at room temperature, under constant magnetic stirring (150 rpm) for 1 h. The solution was filtered, and the process was repeated with the same conditions. The alcoholic fraction was evaporated under reduced pressure, while the aqueous fraction was frozen and freeze-dried, and the extract was kept at room temperature until further analysis.

2.8.1. Phenolic Compounds Profile

The lyophilized hydroethanolic extracts of *C. spinosum* leaves were redissolved in an ethanol–water mixture (20:80 *v/v*) to yield a final concentration of 10 mg/mL for chromatographic analysis. The phenolic profile was evaluated through chromatographic analysis following the procedure described by Polyzos et al. [33]. To identify the phenolic compounds, a comparison of the retention times was conducted with the literature and/or, when possible, with the UV–21 Vis mass spectra of authentic commercial standards. The results were expressed in mg/g of extract.

2.8.2. Antioxidant Properties

Two in vitro antioxidant activities were carried out, namely the inhibition of lipid peroxidation (TBARS) and inhibition of oxidative hemolysis (OxHLIA). Trolox was used as positive control in both antioxidant activities evaluated. The antioxidant potential was evaluated via the TBARS assay, with hydroethanolic extracts of *C. spinosum* leaves following the method described by Polyzos et al. [33]. The results obtained were expressed as IC_{50} values (μg/mL, the concentration of the extract required to provide 50% of the antioxidant activity).

For the OxHLIA assay, a 2.8% (*v/v*) erythrocyte solution (200 μL) in phosphate-buffered saline (PBS, pH 7.4) was combined with 400 μL of one of the following: extract solution (20–770 μg/mL), Trolox (3.91–4000 μg/mL), PBS (negative control), or distilled water (baseline) [33]. Optical density was measured at 690 nm using an ELx800 microplate reader until hemolysis was complete. IC_{50} values (μg/mL) were calculated for a 60 min interval using GraphPad Prism® (version 8.1). These values indicate the extract concentration required to protect 50% of the blood-cell population from oxidative hemolysis within the specified time frame.

2.8.3. Antibacterial and Antifungal Properties

To evaluate the antibacterial and antifungal activities, the obtained extract of *C. spinosum* leaves from hydroethanolic extraction was prepared at a stock concentration of 10 mg/mL and serially diluted [33]. The evaluation focused on the antibacterial activity of the extracts against a range of Gram-positive and Gram-negative bacterial strains, against clinical isolated strains obtained from patients hospitalized in various departments at the Hospital Center of Trás-os-Montes and Alto Douro (Vila Real, Portugal), including *Escherichia coli* (VRU12881), *Klebsiella pneumoniae* (VRI17214), *Morganella morganii* (VRU14272), *Proteus mirabilis* (VRU17684), *Pseudomonas aeruginosa* (VRU14123), *Enterococcus faecalis* (VRU14041), *Listeria monocytogenes* (VRU17684), Methicillin-resistant *Staphylococcus aureus* (MRSA;VRI17654), as well as foodborne pathogens (*Enterobacter cloacae* ATCC 49741, *Escherichia coli* ATCC 25922, *Pseudomonas aeruginosa* ATCC 9027, *Salmonella enterica* ATCC 13076; *Yersinia enterocolitica* ATCC 8610, *Bacillus cereus* ATCC 11778, *Listeria monocytogenes* ATCC 19111, *Staphylococcus aureus* ATCC 11632), that were purchased at Frilabo, Porto, Portugal. The minimum inhibitory concentration (MIC) values were defined as the lowest concentration that indicate bacterial growth inhibition, while the minimum bactericidal concentration (MBC) values corresponded to the concentration required to kill the bacteria. Ampicillin and Streptomycin were used for all bacteria tested, and Methicillin was also used for Staphylococcus aureus as a positive control.

Antifungal assays were performed using two microfungi (*Aspergillus fumigatus* (ATCC 204305) and *Aspergillus brasiliensis* (ATCC 16404); both obtained from Frilabo, Porto, Portugal)), according to the method previously described in the literature [33]. The Minimum Fungicidal Concentration (MFC) was established through a process of subculturing the tested compounds and subsequent observation for a duration of 72 h at a temperature of 25 °C. The concentration at which no visible growth was observed was designated as the MFC, thereby indicating a 99.5% eradication of the initial inoculum. Ketoconazole was utilized as positive control.

2.8.4. NO-Production Inhibition, Antiproliferative and Hepatotoxic Properties

The nitric oxide (NO)-production inhibition of hydroethanolic extracts was evaluated by lipopolysaccharide (LPS)-stimulated murine macrophage cell line (RAW 264.7) by measuring NO production [41]. The quantification of NO levels was performed using the Griess reagent and a nitrite calibration curve. The results obtained were expressed as IC_{50} values, representing the concentration required to inhibit 50% of NO production.

The antiproliferative and hepatotoxic potentials were also evaluated using the sulforhodamine B (SRB) colorimetric assay as described by Polyzos et al. [41]. The cell lines utilized included four tumor cell lines (AGS—gastric adenocarcinoma, Caco-2—colorectal adenocarcinoma, MCF-7—breast adenocarcinoma, and NCI-H460—non-small cell lung cancer) and one non-tumor cell line (PLP2—porcine liver primary culture for hepatotoxicity). The results were expressed as the extract concentration required to inhibit 50% of cell proliferation (GI_{50}). Ellipticine was utilized as a positive control, and cells without extracts as a negative control. All procedures were conducted under aseptic conditions in a vertical laminar flow chamber.

2.9. Statistical Analysis

The experimental design was a split plot, with each plot consisting of water-deficit stress with fully randomized sub-plots comprising the crop-rotation treatments. Statistical analysis was conducted with the aid of JMP v 16.1 (SAS Institute Inc., Cary, NC, USA). Data were evaluated for normal distribution according to the Shapiro–Wilk test, whereas the means of values were compared by Duncan's Multiple Range test at $p = 0.05$.

3. Results and Discussion

3.1. Growth Parameters

The results regarding the effects of irrigation treatments and crop management on growth parameters of *C. spinosum* L. are presented in Figure 2. Deficit irrigation (DI) achieved the highest weight of leaves/plant (18.4 g) followed by full irrigation (FI) (17.2 g) and the control treatment (C; rain-fed plants) (16.1 g), while the highest number of leaves/plants was recorded for the control treatment (31.1 g), followed by FI (29.4 g) and DI (29.0 g) treatments without any significant differences among the treatments. On the other hand, crop rotation had a beneficial effect on the weight of leaves/plant (CR; 19.2 g), while no significant differences were recorded for the number of leaves/plant between crop rotation and the control treatment (no rotation implemented). The interaction of the tested factors showed that deficit irrigation and crop rotation resulted in the highest weight of leaves (20.5 g) being significantly different from all the treatments where no rotations were applied, whereas the lowest overall weight of leaves was recorded for the rain-fed plants combined with no rotation (12.6 g). On the other hand, the highest number of leaves was recorded for fully irrigated plants subjected to crop rotation (32.2), whereas the lowest values were recorded for DI × CR and FI × NCR treatments (27.3 and 27.1, respectively).

Our findings are in total accordance with the study of Guarise et al. [28], who evaluated the potential of growing *Sisymbrium officinale* (L.) Scop. as a novel leafy vegetable under drought conditions, and they reported that reduction in irrigation water by 50% did not compromise plant yield. Similar observations have also been reported by Polyzos et al. [33] who assessed the crop performance of golden thistle (*Scolymus hispanicus* L.) under drought-stress conditions, and they suggested that deficit irrigation did not significantly affect fresh yield compared to the rain-fed plants, where a significant reduction was recorded, while the number of leaves/plant varied among the studied irrigation treatments. According to the study of Mohamed et al. [43], drought-stress conditions at 80% of field capacity also did not have a detrimental effect on the growth of purslane plants (*Portulaca oleracea L.*) compared to the plants grown at 100% of field capacity, while Saheri et al. [44] mentioned a gradual reduction in plant growth and yield of purslane with increasing drought stress. Furthermore, Delfine et al. [45], who studied the effect of drought stress on chicory plants (*C. intybus* L.), observed a significant reduction in fresh biomass with increasing drought stress, especially at 50 days after transplantation; nonetheless, the authors suggest the cultivation of the species under proper rain-fed conditions as soon as the plants are subjected to a moderate water stress.

Crop rotation with *P. vulgaris* L. had a clear positive effect on the fresh biomass yield of *C. spinosum* plants, a finding which has also been recorded in other studies [40,46,47].

According to the meta-analysis of Zhao et al. [48], the integration of legumes in crop-rotations schemes increased the yield by up to 20%, for a large variety of crops including wheat (*Triticum aestivum* L.), vegetables, and oilseed crops. Jensen et al. [49] also observed that crop rotation of legumes and oat (*Avena sativa* L.) with winter barley (*Hordeum vulgare* L.) had positive effects on barley yield, although they suggested that this increase varied depending on the irrigation regime and the soil type (sand or loam). Similarly, Sánchez-Navarro et al. [50] evaluated the effect of rotating two different varieties of cowpea (*Vigna unguiculata* (L.) Walp.) with broccoli (*Brassica oleracea* var. *italica*) and they reported that legumes not only enhanced the yield of broccoli but they also increased the soil nutrient content and reduced the fertilizer inputs by 20%. Moreover, Cui et al. [51] studied the effect of a long-term crop-rotation scheme using flax (*Linum usitatissimum* L.), wheat (*Triticum aestivum* L.), and potato (*Solanum tuberosum* L.) in various combinations and they suggested that the effect of rotation on the annual crop yield and water-use efficiency varied depending on crop combinations. Crop rotation was suggested by Ghadirnezhad Shiade et al. [52] as a sustainable agronomic tool towards the mitigation of negative effects of drought stress on crops, since the resilience of crops against weather extremities may increase through the increase in ecosystem diversification and the improvement of soil properties which are associated with rotational schemes.

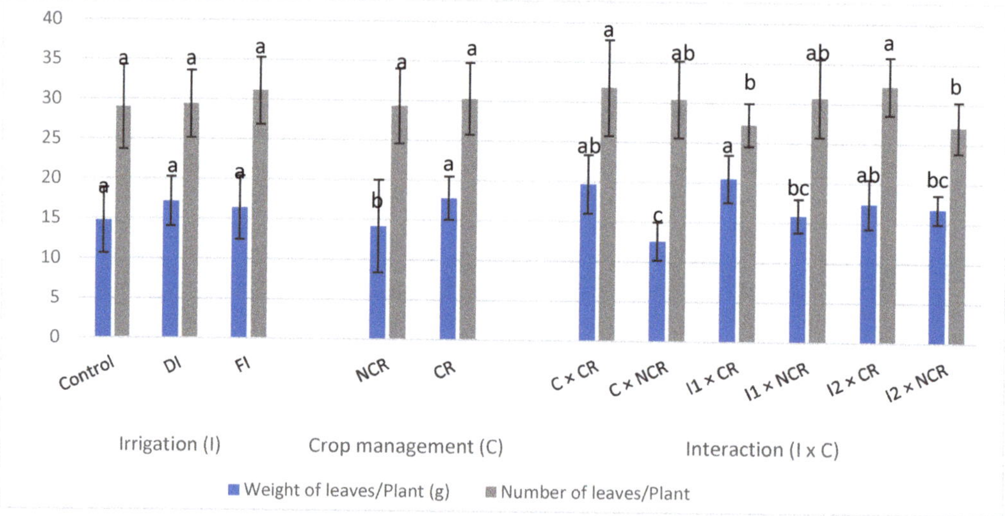

Figure 2. Growth parameters of *C. spinosum* L. plants in relation to irrigation and crop management (mean ± SD). Different Latin letters above bars of the same color and for the same factor or their interaction are significantly different according to Duncan's Multiple Range test (DMRT) at $p = 0.05$. DI: deficit irrigation; FI: full irrigation; NCR: no crop rotation; CR: crop rotation.

3.2. Chemical Composition

3.2.1. Nutritional Characterization

The nutritional profile of *C. spinosum* leaves is displayed in Table 2. Crop management did not significantly affect most parameters, except for crude protein, which was higher when no crop rotation was implemented, whereas irrigation significantly all the nutritional parameters apart from crude protein. In particular, rain-fed conditions increased total fat and fiber content (3.34 g/100 g dw and 49.37 g/100 g dw, respectively), while DI and FI treatments increased total carbohydrate and ash content (15.5 g/100 g dw and 19.1 g/100 g dw, respectively). Finally, FI treatment resulted in a decrease in the energetic value of leaves being significantly lower from control and DI treatments. Regarding the combinatorial effect of the tested factors, significant interactions were recorded (Table 2). In terms of total fat and total fiber content, the highest values (3.5 g/100 g dw and 50.4 g/100 g dw, respectively) were observed in rain-fed plants grown in a crop-rotation scheme, while deficit irrigation (DI) for the same crop management resulted in the highest carbohydrate content and energetic value (20.3 g/100 g dw and 276.3 kcal/100 g dw, respectively). Finally, plants grown under rain-fed conditions and full irrigation with no crop rotation recorded the highest values of crude protein and ash content (23.4 g/100 g dw and 19.76 g/100 g dw, respectively).

A previous study by Polyzos et al. [41], reported similar values for the nutritional profile of *C. spinosum* leaves. Moreover, according to Polyzos et al. [33], deficit irrigation also had a significant effect on the nutritional profile of *S. hispanicus*, thus suggesting that irrigation management can be a practical tool towards the improvement of the nutritional profile of wild edible greens. In the study of Yousefvand et al. [53], it was reported that drought-stress conditions significantly reduced ash content in shallot plants (*Allium hirtifolium*), while similar findings were recorded by Allamine et al. [54] for the ash content in grain sorghum (*Sorghum bicolor* L.). The same authors also observed that water stress did not affect protein content, a finding which is in accordance with our study [54]. Similarly to our study, drought stress led to increased carbohydrate content in black cumin plants (*Nigella sativa* L.) acting as osmoprotectants against the adverse effects of water stress [55]. Moreover, in previous studies from Petropoulos et al. [56,57], it was reported that abiotic stress conditions may significantly affect the nutritional profile of wild edible species such as *C. spinosum* and *Centaurea raphanina* subsp. *mixta*, either due to the concentration effect and the consequent increase in dry matter content or due to the induced biosynthesis of osmolytes as part of plant's stress-defense mechanism.

3.2.2. Organic Acids

The main organic acids detected in the studied samples were quinic acid, oxalic acid, and succinic acid in descending order, while shikimic acid and ascorbic acid were only detected in traces (Table 3). According to the literature, the abovementioned compounds have been previously detected in *C. spinosum* leaves [41,58,59], while these reports also detected malic and citric acid, which were not detected in our study. In contrast, succinic acid was recorded in *C. spinosum* leaves for the first time in our study. Full irrigation resulted in a decrease in the content of the major organic acids (e.g., quinic and oxalic acid) and consequently in total organic acid content, while no significant differences were recorded between the control and DI treatments. Similarly, Muhairi et al. [60] and Skrypnik et al. [61], who evaluated the effect of drought stress on lettuce (*Lactuva sativa* L.), salvia (*Salvia officinalis* L.), and oregano (*Origanum vulgare* L.), reported an increase in organic acid content in response to increasing stress levels, as part of the plants' defense mechanism. Zhao et al. [62] also reported an increase in titratable acidity and vitamin C content in tomato fruit (*Solanum lycopersicum* L.) when plants were subjected to moderate drought-stress conditions due to the concentration effect. Regarding the effect of crop rotation on the quality traits of *C. spinosum* leaves, our results suggest that this agronomic practice did not significantly affect the content of the major and total organic acids (Table 3). However, other cultivation practices such as fertilization and growth stage may have a significantly impact on organic acid content as already reported by Petropoulos et al. [63] and Polyzos et al. [41], while the genotype may also affect the content of individual and total organic acid in the leaves of the species [64].

Table 2. Nutritional profile (g/100 g dw) and energy content (kcal/100 g dw) of *Cichorium spinosum* leaves in relation to irrigation and crop management (mean ± SD).

Treatments		Total Fat	Crude Protein	Ash	Total Dietary Fiber	Carbohydrates	Energy (kcal/100 g dw)
Irrigation (I)	Control (C)	3.34 ± 0.16 a	21.25 ± 2.45 a	15.22 ± 0.69 b	49.37 ± 1.22 a	10.82 ± 0.78 b	257.10 ± 4.16 a
	Deficit Irrigation (DI)	3.04 ± 0.24 b	21.81 ± 0.15 a	15.22 ± 0.13 b	44.40 ± 5.51 b	15.5 ± 5.2 a	265.5 ± 11.7 a
	Full Irrigation (FI)	2.75 ± 0.06 c	20.37 ± 0.71 a	19.1 ± 0.8 a	48.5 ± 1.3 a b	9.32 ± 0.37 b	240.5 ± 1.2 b
Crop Management (CM)	Crop Rotation (CR)	3.17 ± 0.32 a	20.23 ± 1.26 b	16.49 ± 1.42 a	46.47 ± 5.34 a	13.6 ± 5.0 a	256.93 ± 15.47 a
	No Crop Rotation (NCR)	2.91 ± 0.22 a	22.06 ± 1.17 a	16.5 ± 2.5 a	48.37 ± 0.92 a	10.15 ± 0.89 a	251.81 ± 9.38 a
I × CM	C × CR	3.5 ± 0.1 a	19.1 ± 0.9 d	15.8 ± 0.2 c	50.4 ± 1.1 a	11.21 ± 0.03 b	253.4 ± 0.8 c
	C × NCR	3.19 ± 0.03 b	23.4 ± 1.1 a	14.6 ± 0.5 e	48.3 ± 0.2 c	10.44 ± 1.03 b	260.8 ± 1.3 b
	DI × CR	3.2 ± 0.1 b	21.7 ± 0.2 bc	15.3 ± 0.1 d	39.4 ± 0.1 e	20.3 ± 0.1 a	276.3 ± 0.5 a
	DI × NCR	2.8 ± 0.1 c	21.8 ± 0.2 b	15.1 ± 0.1 d	49.4 ± 0.1 b	10.8 ± 0.2 b	254.8 ± 0.3 c
	FI × CR	2.8 ± 0.1 cd	19.7 ± 0.2 d	18.4 ± 0.2 b	49.7 ± 0.4 b	9.4 ± 0.5 c	241.1 ± 1.5 d
	FI × NCR	2.7 ± 0.1 d	20.9 ± 0.6 c	19.76 ± 0.01 a	47.3 ± 0.3 d	9.2 ± 0.1 c	239.8 ± 0.2 d

Means in the same column for the same factor (irrigation and crop management) or their interaction followed by different Latin letters are significantly different according to Duncan's Multiple Range test (DMRT) at $p = 0.05$.

Table 3. Organic acid composition (mg/100 g dw) of *Cichorium spinosum* leaves in relation to irrigation and crop management (mean ± SD).

Treatments		Oxalic Acid	Quinic Acid	Shikimic Acid	Ascorbic Acid	Succinic Acid	Total Organic Acids
Irrigation (I)	Control (C)	4.90 ± 0.23 a	6.65 ± 0.19 a	tr	tr	1.55 ± 0.04 a	13.11 ± 0.27 a
	Deficit Irrigation (DI)	5.04 ± 0.08 a	6.8 ± 0.21 a	tr	tr	1.40 ± 0.19 a	13.25 ± 0.24 a
	Full Irrigation (FI)	4.2 ± 0.6 b	4.32 ± 1.75 b	tr	tr	1.20 ± 0.48 a	9.71 ± 2.84 b
Crop Management (CM)	Crop Rotation (CR)	4.83 ± 0.14 a	6.43 ± 0.43 a	tr	tr	1.60 ± 0.05 a	12.86 ± 0.49 a
	No Crop Rotation (NCR)	4.60 ± 0.72 a	5.41 ± 2.03 a	tr	tr	1.18 ± 0.34 b	11.19 ± 3.05 a
I × CM	C × CR	4.8 ± 0.2 bc	6.6 ± 0.2 a	tr	tr	1.57 ± 0.01 b	12.97 ± 0.42 b
	C × NCR	5.04 ± 0.3 a	6.7 ± 0.3 a	tr	tr	1.52 ± 0.05 b	13.26 ± 0.03 a
	DI × CR	4.97 ± 0.02 ab	6.8 ± 0.3 a	tr	tr	1.57 ± 0.04 b	13.3 ± 0.3 a
	FI × NCR	5.11 ± 0.04 a	6.8 ± 0.4 a	tr	tr	1.2 ± 0.1 c	13.11 ± 0.3 ab
	DI × CR	4.7 ± 0.1 c	5.9 ± 0.3 b	tr	tr	1.64 ± 0.05 a	12.24 ± 0.3 c
	FI × NCR	3.65 ± 0.04 d	2.69 ± 0.09 c	tr	tr	0.75 ± 0.01 d	7.09 ± 0.07 d

Organic acids (mg/mL)—oxalic acid ($y = 8 \times 10^6x + 331{,}789$, $R^2 = 0.9912$); quinic acid ($y = 692{,}575x + 11{,}551$; $R^2 = 0.9983$); shikimic acid ($y = 5 \times 10^7x + 567119$, $R^2 = 0.9903$); ascorbic acid ($y = 5 \times 10^7x + 449{,}262$, $R^2 = 0.9813$); succinic acid ($y = 493{,}834 - 33106$, $R^2 = 0.9986$). Means in the same column for the same factor (irrigation and crop management) or their interaction followed by different Latin letters are significantly different according to Duncan's Multiple Range test (DMRT) at $p = 0.05$.

Regarding the interaction effects between the studied factors on organic acid content, a varied response was recorded (Table 3). However, it should be noted that reduced irrigation in the form of deficit irrigation or the rain-fed conditions resulted in an increase in the content of major organic acids and consequently in total organic acid content, whereas full irrigation, especially when no crop rotation was implemented, recorded the lowest overall values for the content of all the detected organic acids.

3.2.3. Tocopherols

The only detected tocopherols in *C. spinosum* leaves were α- and β-tocopherol, while deficit irrigation resulted in the highest content of β-tocopherol and consequently of total tocopherols (Table 4). On the other hand, crop management affected only β-tocopherol, which was significantly higher when no crop rotation was implemented. Regarding the combinatorial effect of the tested factors, the highest concentration of β- and total tocopherols (0.158 and 0.214 mg/100 g dw, respectively) was observed in plants grown under deficit irrigation without crop rotation, while α-tocopherol was the highest in rain-fed plants grown in the crop-rotation scheme. The literature reports showed a varied tocopherol composition in *C. spinosum* leaves, including α- and δ-tocopherol [56,64], α-, β-, and γ-tocopherols [41], or all the forms of tocopherols [65], suggesting that several factors such as the genotype, the growing conditions, or the harvesting stage may affect the tocopherol profile. Muhairi et al. [60], who studied the effect of drought stress on the content of antioxidant compounds in lettuce (*L. sativa* L.), also reported that α-tocopherol content increased when the irrigation interval increased up to 96 h and it served as a scavenger of reactive oxygen species that may disrupt the membranes of plant leaves. Similarly, Polyzos et al. [33] observed that deficit irrigation significantly increased α-tocopherol content in the leaves of *S. hispanicus* plants and they also highlighted the important its protective role against abiotic stressors. According to several studies, tocopherols are essential for plant adaption to stress conditions, since they may effectively scavenge lipid peroxides, oxygen radicals, and singlet oxygen, thereby enhancing plant tolerance to stressors [66–68]. Furthermore, tocopherols are not only involved in the defense mechanisms of plants but they are also essential in several metabolic processes and functions that promote plant growth and development under stress or non-stress conditions [69].

Table 4. Tocopherol composition (mg/100 g dw) of *Cichorium spinosum* leaves in relation to irrigation and crop management (mean ± SD).

Treatments		α-Tocopherol	β-Tocopherol	Total Tocopherols
Irrigation (I)	Control (C)	0.047 ± 0.016 a	0.107 ± 0.003 b	0.153 ± 0.013 b
	Deficit Irrigation (DI)	0.04 ± 0.01 a	0.15 ± 0.01 a	0.19 ± 0.02 a
	Full Irrigation (FI)	0.031 ± 0.005 b	0.109 ± 0.041 b	0.140 ± 0.036 b
Crop Management (CM)	Crop Rotation (CR)	0.044 ± 0.013 a	0.10 ± 0.03 b	0.149 ± 0.032 a
	No Crop Rotation (NCR)	0.038 ± 0.13 a	0.138 ± 0.022 a	0.176 ± 0.031 a
I × CM	C × CR	0.0617 ± 0.0003 a	0.104 ± 0.003 e	0.165 ± 0.003 c
	C × NCR	0.0319 ± 0.0001 d	0.11 ± 0.003 d	0.142 ± 0.003 d
	DI × CR	0.0354 ± 0.0004 c	0.141 ± 0.005 c	0.177 ± 0.005 b
	FI × NCR	0.055 ± 0.001 b	0.158 ± 0.007 a	0.214 ± 0.008 a
	DI × CR	0.035 ± 0.001 c	0.072 ± 0.001 f	0.107 ± 0.003 e
	FI × NCR	0.027 ± 0.001 e	0.147 ± 0.004 b	0.173 ± 0.003 b

Means in the same column for the same factor (irrigation and crop management) or their interaction followed by different Latin letters are significantly different according to Duncan's Multiple Range test (DMRT) at $p = 0.05$.

3.2.4. Sugars Composition

Glucose and sucrose were the main sugars detected in *C. spinosum* leaves, followed by fructose and trehalose which were detected in lesser amounts (Table 5). A variable response to the irrigation system and crop management was observed. In particular, deficit irrigation and rain-fed conditions increased the content of glucose, sucrose, and total sugars content, while the highest content of fructose and trehalose was detected in FI treatment. Moreover, implementing crop rotation resulted in increased amounts of fructose and sucrose, also contributing to a higher total sugars content, whereas trehalose increased under full irrigation. A significant interaction between the studied factors was also recorded, with DI × CR treatment showing the highest content of sucrose and total sugars, while glucose content increased for C × NCR treatment. Moreover, fructose content was the highest for FI × CR treatment, while trehalose content did not differ between C × NCR and full irrigation (regardless of crop management) (Table 5). A similar free sugars composition in *C. spinosum* leaves was recently reported by Liava et al. [59], as well as in previous studies [58], while Polyzos et al. [41] detected the same compounds and higher amounts of fructose compared to our study. All these studies highlighted that agronomic practices, such as fertilization regime, harvesting stage, or the genotype may affect free sugars composition in the leaves of the species and eventually have an impact on the organoleptic properties and the quality of the edible product. Similarly to our study, Polyzos et al. [33], who assessed the effect of different irrigation regimes on *S. hispanicus* plants, also reported that full irrigation resulted in decreased values of glucose and sucrose and increased fructose content. Moreover, Kapur et al. [70] suggested that drought stress (50% of crop-pan coefficient value) increased fructose, glucose, and sucrose contents in strawberry fruit (*Fragaria × ananassa* Duch.), while Abdou et al. [55] detected an increase in total soluble sugars under deficit irrigation (50% of crop evapotranspiration) in black cumin (*N. sativa* L.) leaves, highlighting their role as osmoprotectants. Ibrahim et al. [71] also noted an increase in total soluble-solid content in broccoli (*B. oleracea* var. *italica*) flowers under water-deficit stress as part of the osmoprotective mechanism of plants, a finding which was also confirmed by Thomas et al. [72], who suggested an increase in the expression of sugar metabolism genes under water-stress conditions. Finally, Xu et al. [73] reported that deficit irrigation induces the redistribution of photosynthetic assimilates to the most metabolically active tissues to support plant growth. On the other hand, Götze et al. [74] reported that the monoculture of sugar beet (*Beta vulgaris* subsp. *vulgaris* (var. *saccharifera*)) decreased the sugar content, while crop rotation resulted in higher sugar content regardless of the rotation scheme (e.g., winter wheat (*T. aestivum* L.), alfalfa (*Medicago sativa* L.), grain maize (*Zea mays* L.) and potato (*S. tuberosum* L.) combined in different combinations).

Table 5. Free sugars composition (mg/100 g dw) of *Cichorium spinosum* leaves in relation to irrigation and crop management (mean ± SD).

Treatments		Fructose	Glucose	Sucrose	Trehalose	Total Sugars
Irrigation (I)	Control (C)	0.73 ± 0.06 ab	2.67 ± 0.08 a	2.4 ± 0.2 ab	0.38 ± 0.02 a	6.15 ± 0.17 a
	Deficit Irrigation (DI)	0.69 ± 0.09 b	2.68 ± 0.09 a	2.75 ± 0.57 a	0.33 ± 0.06 b	6.45 ± 0.52 a
	Full Irrigation (FI)	0.82 ± 0.12 a	2.23 ± 0.24 b	2.06 ± 0.24 b	0.41 ± 0.01 a	5.51 ± 0.13 b
Crop Management (CM)	Crop Rotation (CR)	0.83 ± 0.08 a	2.4 ± 0.3 a	2.70 ± 0.45 a	0.35 ± 0.06 b	6.29 ± 0.56 a
	No Crop Rotation (NCR)	0.67 ± 0.04 b	2.64 ± 0.16 a	2.09 ± 0.19 b	0.39 ± 0.0 a	5.8 ± 0.3 b
I × CM	C × CR	0.79 ± 0.02 b	2.6 ± 0.1 b	2.55 ± 0.07 b	0.36 ± 0.01 c	6.31 ± 0.01 b
	C × NCR	0.68 ± 0.02 d	2.73 ± 0.04 a	2.20 ± 0.02 c	0.40 ± 0.02 ab	6.0 ± 0.1 c
	DI × CR	0.78 ± 0.04 b	2.61 ± 0.01 b	3.3 ± 0.1 a	0.27 ± 0.01 d	6.9 ± 0.1 a
	FI × NCR	0.61 ± 0.01 e	2.75 ± 0.11 a	2.23 ± 0.09 c	0.39 ± 0.01 b	5.97 ± 0.01 c
	DI × CR	0.94 ± 0.04 a	2.01 ± 0.11 d	2.28 ± 0.06 c	0.41 ± 0.03 a	5.63 ± 0.01 d
	FI × NCR	0.71 ± 0.01 c	2.44 ± 0.08 c	1.84 ± 0.04 d	0.40 ± 0.01 ab	5.39 ± 0.04 e

Means in the same column for the same factor (irrigation and crop management) or their interaction followed by different Latin letters are significantly different according to Duncan's Multiple Range test (DMRT) at $p = 0.05$.

3.2.5. Mineral Composition

The mineral composition of *C. spinosum* leaves is displayed in Table 6. The irrigation regime significantly affected macro- and micro-mineral content in a varied manner. In particular, K, Na, Ca and Mg content significantly decreased under the full irrigation regime, while the same treatment resulted in the highest content of Fe and Mn. On the other hand, Cu and Zn content was the lowest under rain-fed conditions. Regarding the crop-management effect, only K, Na and Mg content was affected, where K increased when crop rotation was implemented, whereas Na and Mg showed the highest content when no rotation was applied. The combination of the two factors also resulted in significant differences between the treatments in a variable manner. For example, the highest content Ca was recorded for the C × CR treatment, while Na increased for the C × NCR treatment; K content was the highest for C and DI treatments under crop rotation, while Mg content increased in rain-fed conditions (for both crop-management treatments) and for DI × NCR treatment. Finally, full irrigation and no crop rotation resulted in the highest content of Fe, Mn and Zn, while Cu increased when deficit irrigation was combined with crop rotation. According to literature reports, *C. spinosum* leaves are rich in macro- and micro-nutrients, while their content can be affected by several factors including the genotype and the growing conditions [64,75]. Previously, Polyzos et al. [33] who studied the effect of different irrigation regimes on the mineral composition of S. *hispanicus* plants reported that deficit irrigation did not affect K, Fe, and Mn, while it increased the content of Ca, Mg, Cu, and Zn. Moreover, the same authors observed that rain-fed plants had the lowest content of Ca, Mg, Cu, and Zn and the highest Na value [33]. In contrast, deficit irrigation did not affect the content of Ca, Cu, Fe, Ca, Mg, Na, and N in bean (*P. vulgaris* L.) pods, although it resulted in increased P and Zn values [76], while drought stress reduced N, P, and K content in purslane (*P. oleracea* L.) leaves, tomato (*S. lycopersicum* L.) fruits, and broccoli (*B. oleracea* var. *italica*) leaves [43,71,77]. The negative effects of deficit irrigation on the content of specific nutrients could be attributed to a limited uptake of moderately mobile elements due to the severe effects of water stress on soil enzyme activity and soil microbiome, which interfere in the soil–plant interaction [78], as well as to the reduced uptake of water from roots due to increased values of soil-matrix potential [79,80]. Regarding the effect of crop-rotation ion nutrients' status, Υfantopoulos et al. [81], who studied the effect of crop rotation on a cabbage (*B. oleracea* var. *capitata*) crop, observed that nutrient content was not influenced by rotation, while Haruna and Nkongolo [82] reported that soil nutrients were not affected by the continuous monocropping of corn (*Z. mays* L.) and soybean (*Glycine max* (L.) Merr.), as well as by corn–soybean and soybean–corn rotation. According to Adesanya et al. [83], the positive effects of crop rotation on nutrient availability are nutrient-specific and they may also vary in the long term due to weather conditions.

3.2.6. Fatty-Acid Composition

In terms of the profile of fatty acids, twenty individual compounds were identified as presented in Table 7. The most abundant fatty acid was α-linolenic acid (C18:3n3) with values ranging from 37.1% to 45.4%, followed by palmitic acid (C16:0; 27.84% to 20.88%), and linoleic acid (C18:2n6; 16.7% to 21.1%). The leaves also had a high content of polyunsaturated fatty acids (PUFA; 54.6% to 65.6%), followed by saturated fatty acids (SFA; 29.5% to 39.6%) and monounsaturated fatty acids (MUFA; 4.8% to 5.9%). Irrigation regime had a variable effect on the major fatty acids, where rain-fed and deficit irrigation conditions increased the content of α-linolenic acid, while the highest content of linoleic acid was recorded for full irrigation. In contrast, irrigation regime had no effect on palmitic acid content. Additionally, crop rotation resulted in a reduction in palmitic acid and an increase in α-linolenic and linoleic acid content, which contributed to higher PUFA/SFA ratios. The combined effect of the tested factors also showed a variable response to the composition of fatty acids. In particular, crop rotation under rain-fed and deficit irrigation conditions resulted in the highest α-linolenic acid content, whereas palmitic acid increased when no crop rotation was implemented and plants were treated with deficit and full irrigation. Moreover, linoleic acid benefitted from the DI × NCR treatment. Finally, rain-fed and deficit irrigation treatments combined with crop rotation had the highest and lowest values of PUFA and SFA, respectively.

A similar fatty-acid profile was recorded by Petropoulos et al. [64], who suggested genotypic differences among various *C. spinosum* ecotypes, as well as a significant impact of agronomic practices and harvesting stage [64]. Moreover, Liava et al. [59] highlighted the high ratios of PUFA/SFA in *C. spinosum* leaves, which are associated with high nutritional value and health benefits. In contrast to our study, water deficit did not affect fatty-acid content in grain sorghum (*Sorghum bicolor* (L.) Moench) [54] and groundnut (*Arachis hypogaea* L.) [84], whereas Ghaffari et al. [85] reported a significant increase in palmitic and linoleic acid content in sunflower (*Helianthus annuus* L.) seeds under drought conditions. Similarly, deficit irrigation (50% of evapotranspiration) decreased the oil quality in chia seeds due to increased palmitic acid and decreased linoleic and α-linolenic acid content compared to full irrigation (100% of evapotranspiration) [86], while Abdou et al. [55] suggested an increase in SFA and a decrease in polyunsaturated fatty acids in *N. sativa* seed oil due to the activation of the lipase enzyme under water-stress conditions. According to Upchurch [87], the acclimatization of plants to stress conditions is associated with increases in unsaturated fatty-acid content which help cells to retain membrane fluidity and avoid oxidative damage. This is also evident in our study with the increased content of α-linolenic acid under rain-fed and deficit irrigation conditions, whereas the lack of effect of drought stress on the overall content of PUFA suggests the drought tolerance of the species [88]. Regarding the effect of the cropping system, Stepien et al. [89] suggested that integrating canola (*B. napus* L.) in a crop-rotation scheme resulted in increased content of linoleic and α-linolenic acid, while Mohammadi et al. [90] reported a variable effect of crop rotation on the canola fatty-acid profile depending on the rotation scheme (e.g., the succession and diversification of crops included in the scheme). On the other hand, Wacal et al. [91] reported an increase in α-linolenic and linoleic acids in sesame (*Sesamum indicum* L.) seeds grown under continuous monocropping, which indicates the variable impact of cropping system depending on the crop, the rotation scheme, and the growing conditions.

Table 6. Mineral composition (mg/100 g dw) of *Cichorium spinosum* leaves in relation to irrigation and crop management (mean ± SD).

Treatments		K (g/Kg)	Na (mg/Kg)	Ca (g/Kg)	Mg (g/Kg)	Fe (mg/Kg)	Mn (mg/Kg)	Cu (mg/Kg)	Zn (mg/Kg)
Irrigation (I)	Control (C)	36.6 ± 6.6 ab	6004 ± 1052 a	11.98 ± 0.49 a	3.70 ± 0.07 a	849 ± 47 b	105.7 ± 18.6 c	9.37 ± 0.54 b	45.8 ± 1.3 b
	Deficit Irrigation (DI)	37.8 ± 6.6 a	5259 ± 694 a	11.1 ± 0.3 b	3.5 ± 0.2 a	537 ± 32 b	135.7 ± 9.7 b	12.9 ± 0.5 a	58.4 ± 1.4 a
	Full Irrigation (FI)	29.5 ± 3.4 b	3481 ± 572 b	9.5 ± 0.7 c	3.1 ± 0.4 b	2915 ± 645 a	183.3 ± 18.4 a	12.3 ± 0.4 a	58.8 ± 9.9 a
Crop Management (CM)	Crop Rotation (CR)	39.6 ± 5.4 a	4212 ± 959 b	10.7 ± 1.5 a	3.3 ± 0.4 b	1260 ± 113 a	133.3 ± 14.9 a	12.3 ± 3.1 a	51.8 ± 5.9 a
	No Crop Rotation (NCR)	29.6 ± 2.5 b	5617 ± 1300 a	11.1 ± 0.7 a	3.6 ± 1.2 a	1608 ± 142 a	149.8 ± 17.6 a	10.8 ± 1.5 a	56.8 ± 9.8 a
I × CM	C × CR	43 ± 2 a	5049 ± 170 c	12.3 ± 0.4 a	3.7 ± 0.1 a	888 ± 44 c	88.7 ± 0.4 f	8.9 ± 0.4 e	46 ± 1 d
	C × NCR	30.6 ± 0.4 c	6960 ± 154 a	11.7 ± 0.6 b	3.6 ± 0.1 a	809 ± 4 d	122.7 ± 1.3 e	9.8 ± 0.4 d	45 ± 2 d
	DI × CR	44 ± 3 a	4627 ± 82 d	10.78 ± 0.06 c	3.35 ± 0.13 b	565 ± 25 e	144.5 ± 0.3 c	16.1 ± 0.5 a	59 ± 1 b
	FI × NCR	31.9 ± 1.1 bc	5891 ± 62 b	11.4 ± 0.1 b	3.7 ± 0.2 a	510 ± 10 f	126.9 ± 2 d	9.8 ± 0.2 d	57 ± 1 b
	DI × CR	32.6 ± 0.5 b	2962 ± 117 f	8.9 ± 0.2 e	2.7 ± 0.1 c	2327 ± 74 b	166.7 ± 5.7 b	11.9 ± 0.3 c	50 ± 1 c
	FI × NCR	26.3 ± 0.4 d	4000 ± 89 e	10.2 ± 0.3 d	3.45 ± 0.05 b	3504 ± 20 a	199.9 ± 2.3 a	12.7 ± 0.03 b	68 ± 1 a

Means in the same column for the same factor (irrigation and crop management) or their interaction followed by different Latin letters are significantly different according to Duncan's Multiple Range test (DMRT) at $p = 0.05$.

Table 7. Fatty-acid composition (relative percentage) of *Cichorium spinosum* leaves in relation to irrigation and crop management (mean ± SD).

	Irrigation (I)			Crop Management (CM)		I × CM					
	C	DI	FI	CR	NCR	C × CR	C × NCR	DI × CR	DI × NCR	FI × CR	FI × NCR
C6:0	tr	tr	tr	tr	tr	tr	tr	tr	tr	tr	tr
C11:0	0.139 ± 0.016 a	0.26 ± 0.17 a	0.193 ± 0.063 a	0.123 ± 0.012 a	0.276 ± 0.118 a	0.125 ± 0.003 e	0.154 ± 0.003 v	0.109 ± 0.003 f	0.423 ± 0.004 a	0.136 ± 0.004 d	0.251 ± 0.008 b
C12:0	0.11 ± 0.05 a	0.175 ± 0.101 a	0.122 ± 0.063 a	0.071 ± 0.009 b	0.201 ± 0.051	0.066 ± 0.002 e	0.156 ± 0.001 c	0.084 ± 0.001 d	0.268 ± 0.007 a	0.065 ± 0.002 e	0.181 ± 0.005 b
C13:0	0.79 ± 0.24 b	1.010 ± 0.015 a	0.662 ± 0.054 b	0.875 ± 0.197 a	0.769 ± 0.196 a	1.01 ± 0.02 a	0.576 ± 0.001 d	1.002 ± 0.011 a	1.02 ± 0.03 a	0.613 ± 0.01 c	0.712 ± 0.003 b
C14:0	0.994 ± 0.073 a	1.116 ± 0.328 a	1.146 ± 0.087 a	0.94 ± 0.11 b	1.232 ± 0.156 a	0.927 ± 0.004 d	1.061 ± 0.001 c	0.82 ± 0.03 e	1.42 ± 0.03 a	1.07 ± 0.03 c	1.22 ± 0.06 b
C14:1	0.33 ± 0.12 b	0.437 ± 0.039 a	0.261 ± 0.015 b	0.37 ± 0.07 a	0.31 ± 0.12 a	0.44 ± 0.01 b	0.218 ± 0.005 f	0.4 ± 0.2 c	0.47 ± 0.02 a	0.27 ± 0.01 d	0.248 ± 0.007 e
C15:0	0.168 ± 0.038 a	0.154 ± 0.074 a	0.209 ± 0.027 a	0.151 ± 0.065 b	0.203 ± 0.016 a	0.134 ± 0.001 e	0.204 ± 0.005 c	0.086 ± 0.001 f	0.222 ± 0.005 b	0.234 ± 0.006 a	0.185 ± 0.006 d
C15:1	0.222 ± 0.066 a	0.228 ± 0.020 a	0.203 ± 0.027 a	0.23 ± 0.05 a	0.205 ± 0.035 a	0.283 ± 0.004 a	0.162 ± 0.003 d	0.246 ± 0.006 b	0.21 ± 0.003 c	0.163 ± 0.007 d	0.243 ± 0.006 b
C16:0	23.39 ± 2.49 a	24.36 ± 3.81 a	25.67 ± 1.98 a	21.97 ± 1.46 b	27.0 ± 1.1 a	21.1 ± 0.2 d	25.6 ± 0.9 b	20.88 ± 0.03 d	27.84 ± 0.07 a	23.8 ± 0.4 c	27.4 ± 0.8 a
C16:1	2.68 ± 0.16 a	2.83 ± 0.35 a	3.009 ± 0.471 a	2.54 ± 0.04 b	3.14 ± 0.27	2.54 ± 0.05 d	2.82 ± 0.1 c	2.51 ± 0.05 d	3.15 ± 0.02 b	2.58 ± 0.03 d	3.4 ± 0.1 a
C17:0	0.97 ± 0.22 a	0.85 ± 0.24 a	1.003 ± 0.365 a	0.69 ± 0.06 b	1.19 ± 0.12 a	0.77 ± 0.04 d	1.17 ± 0.02 b	0.64 ± 0.01 f	1.07 ± 0.02 c	0.67 ± 0.01 e	1.34 ± 0.01 a
C18:0	2.93 ± 0.41 a	3.12 ± 0.65 a	3.24 ± 0.53 a	2.61 ± 0.11 b	3.58 ± 0.21 a	2.55 ± 0.06 d	3.3 ± 0.07 b	2.525 ± 0.001 d	3.71 ± 0.01 a	2.755 ± 0.004 c	3.73 ± 0.05 a
C18:1n9c	1.65 ± 0.33 c	1.865 ± 0.036 b	1.99 ± 0.027 a	1.85 ± 0.15 a	1.82 ± 0.15 a	1.68 ± 0.01 c	1.62 ± 0.03 d	1.86 ± 0.05 b	1.87 ± 0.07 b	2.01 ± 0.02 a	1.97 ± 0.03 a
C18:2n6c	18.56 ± 1.52 b	18.05 ± 1.46 b	20.28 ± 0.80 a	20.11 ± 0.72 a	17.81 ± 1.32 b	19.95 ± 0.02 b	17.2 ± 0.2 d	19.4 ± 0.2 c	16.7 ± 0.1 e	21.1 ± 0.2 a	19.6 ± 0.1 c
C18:3n6	0.350 ± 0.039 a	0.336 ± 0.056 a	0.38 ± 0.04 a	0.338 ± 0.059 a	0.371 ± 0.022 a	0.32 ± 0.01 d	0.386 ± 0.009 b	0.29 ± 0.01 e	0.387 ± 0.002 b	0.41 ± 0.02 a	0.342 ± 0.008 c
C18:3n3	43.90 ± 1.63 a	41.5 ± 4.4 a b	39.01 ± 2.13 b	43.95 ± 2.27 a	39.0 ± 2.6 b	45.4 ± 0.1 a	42.4 ± 0.7 b	45.6 ± 0.2 a	37.5 ± 0.2 d	40.9 ± 0.1 c	37.1 ± 0.9 d
C22:0	0.843 ± 0.087 a	0.93 ± 0.06 a	0.90 ± 0.18 a	0.904 ± 0.133 a	0.88 ± 0.12 a	0.76 ± 0.01 e	0.92 ± 0.02 c	0.881 ± 0.005 d	0.99 ± 0.01 b	1.07 ± 0.01 a	0.73 ± 0.02 f
C22:2	tr	tr	tr	tr	tr	tr	tr	tr	tr	tr	tr
C23:0	0.631 ± 0.138	0.952 ± 0.023 a	0.62 ± 0.04 b	0.78 ± 0.12 a	0.69 ± 0.21 a	0.76 ± 0.01 c	0.51 ± 0.01 f	0.93 ± 0.02 b	0.972 ± 0.003 a	0.66 ± 0.01 d	0.59 ± 0.02 e
C24:0	1.36 ± 0.15 b	1.758 ± 0.052 a	1.089 ± 0.041 b	1.49 ± 0.25 a	1.312 ± 0.456 a	1.23 ± 0.03 d	1.51 ± 0.02 c	1.8 ± 0.1 a	1.72 ± 0.01 b	1.46 ± 0.06 c	0.72 ± 0.03 e
SFA	32.32 ± 3.15 a	34.7 ± 5.4 a	34.86 ± 2.48 a	[illegible] ± 1.52 b	37.31 ± 1.99 a	29.5 ± 0.2 e	35.2 ± 0.8 c	29.75 ± 0.01 e	39.65 ± 0.04 a	32.6 ± 0.3 d	37.1 ± 0.8 b
MUFA	4.878 ± 0.075 b	5.36 ± 0.38 a	5.47 ± 0.48 a	4.994 ± 0.052 b	5.5 ± 0.5 a	4.93 ± 0.07 d	4.82 ± 0.07 e	5.018 ± 0.005 c	5.71 ± 0.07 b	5.03 ± 0.02 c	5.9 ± 0.2 a
PUFA	62.8 ± 3.1 a	59.9 ± 5.8 a	59.67 ± 2.96 a	64.40 ± 1.55 a	57.21 ± 2.37 b	65.6 ± 0.1 a	60.001 ± 0.892 c	65.235 ± 0.004 a	54.64 ± 0.03 e	62.3 ± 0.3 b	57.001 ± 0.969 d

Fatty acids are expressed as a relative percentage of each fatty acid: C11:0—undecanoic acid; C12:0—lauric acid; C13:0—Tridecanoic acid; C14:0—myristic acid; C14:1—tetradecanoic acid; C15:0—pentadecanoic acid; C15:1—pentadecylic acid; C16:0—palmitic acid; C16:1—palmitoleic acid; C17:0—heptadecanoic acid; C18:0—stearic acid; C18:1n9c—oleic acid; C18:2n6c—linoleic acid; C18:3n6—gamma linolenic acid; C18:3n3—α-linolenic acid; C22:0—behenic acid; C23:0—tricosanoic acid; C24:0—lignoceric acid. SFA—saturated fatty acids; MUFA—monounsaturated fatty acids; PUFA—polyunsaturated fatty acids; dw—dry weight; tr—traces. DI: deficit irrigation; FI: full irrigation; CR: crop rotation; NCR no crop rotation. Means in the same row for the same factor (irrigation and crop management) or their interaction followed by different Latin letters are significantly different according to Duncan's Multiple Range test (DMRT) at $p = 0.05$.

3.3. Phenolic Compounds and Bioactive Properties

3.3.1. Phenolic Compounds

Eleven phenolic compounds (Figure 3) were detected in the hydroethanolic extracts of *C. spinosum* leaves (Table 8), including three phenolic acids and six flavonoids, while the profile of phenolic compounds varied among the studied treatments (Table 9). In particular, chicoric acid (peak 2), quercetin-*O*-hexurumoside (peak 3), and luteolin-*O*-hexurumoside (peak 4) were the most abundant compounds, followed by *trans* 5-*O*-Caffeoylquinic acid (peak 1′), isorhamnetin-*O*-hexurunoside (peak 8), apigenin-*O*-hexurunoside (peak 7), and kaempherol-*O*-hexurunoside (peak 6); flavonoids were the richest class of phenolic compounds. The irrigation regime had a significant impact on the phenolic compounds profile with deficit irrigation, resulting in the highest amount for all the detected compounds. On the other hand, crop management did not affect the phenolic compounds profile, except for the content of peaks 9 and 11, which increased when plants were grown under no crop rotation and crop rotation, respectively. Finally, the combined effect of the tested factors was also significant, with seven compounds (including the major ones) having the highest content for the DI × CR treatment, while peaks 7, 9, and 10 benefitted from deficit irrigation and no crop rotation and peak 8 from full irrigation and no crop rotation.

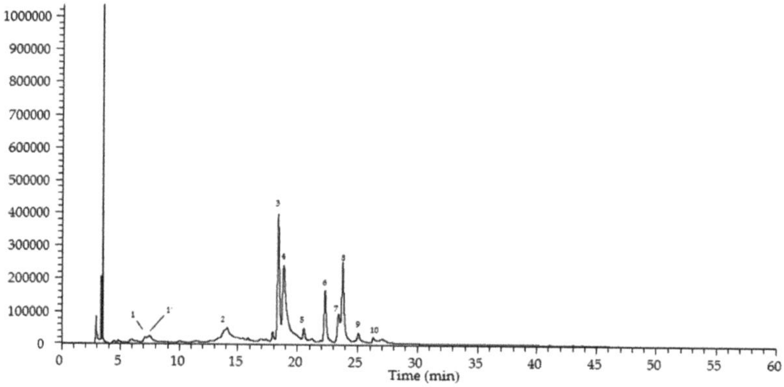

Figure 3. HPLC spectral phenolic profile of hydroethanolic extracts of *C. spinosum* leaves recorded at 370 nm. See Table 8 for tentative identification of detected peaks (peak 1–10).

Table 8. Retention time (Rt), wavelength of the maximum absorption (λmax), deprotonated ion ([M-H]$^-$), main mass fragments (MS2), and tentative identification.

Peak	Rt	λmax	[M-H]$^-$	MS2	Tentative Identification
1	7.07	326	353	191 (100), 179 (23), 135(5)	*cis* 5-*O*-Caffeoylquinic acid
1′	7.41	326	353	191 (100), 179 (12), 135 (5)	*trans* 5-*O*-Caffeoylquinic acid
2	13.98	328	473	391 (39), 311 (100), 293 (28)	Chicoric acid
3	18.22	353	477	301 (100)	Quercetin-*O*-hexurunoside
4	18.73	343	461	285 (100)	Luteolin-*O*-hexurunoside
5	20.4	335	505	463 (11), 301 (100)	Quercetin-*O*-acetylhexoside
6	22.13	346	461	285 (100)	Kaempherol-*O*-hexurunoside
7	23.32	333	445	269 (100)	Apigenin-*O*-hexurunoside
8	23.59	345	491	315 (100)	Isorhamnetin-*O*-hexurunoside
9	24.9	326	489	285 (100)	Kaempherol-*O*-acetylhexoside
10	26.14	327	519	315 (100)	Isorhamnetin-*O*-acetylhexoside

Table 9. Quantification (mg/g extract) of the phenolic compounds found in the hydroethanolic extracts of *Cichorium spinosum* leaves in relation to irrigation and crop management (mean ± SD).

		Irrigation (I)		Crop Management (CM)		I × CM					
Tentative Identification	Control (C)	Deficit Irrigation (DI)	Full Irrigation (FI)	Crop Rotation (CR)	None Crop Rotation (NCR)	C × CR	C × NCR	DI × CR	DI × NCR	FI × CR	FI × NCR
cis 5-*O*-Caffeoylquinic acid	0.54 ± 0.13 b	0.91 ± 0.23 a	0.413 ± 0.122 c	0.613 ± 0.38 a	0.63 ± 0.07 a	0.41 ± 0.01 e	0.66 ± 0.01 c	1.13 ± 0.01 a	0.69 ± 0.03 b	0.29 ± 0.01 f	0.53 ± 0.01 d
trans 5-*O*-Caffeoylquinic acid	0.83 ± 0.25 b	1.92 ± 0.39 a	0.644 ± 0.067 c	1.23 ± 0.78 a	1.03 ± 0.37 a	0.81 ± 0.03 d	0.84 ± 0.02 c	2.301 ± 0.002 a	1.54 ± 0.05 b	0.58 ± 0.03 f	0.71 ± 0.03 e
Chicoric acid	1.095 ± 0.361 b	3.36 ± 0.57 a	0.84 ± 0.18 c	1.90 ± 1.46 a	1.63 ± 0.85 a	1.13 ± 0.01 c	1.07 ± 0.04 d	3.92 ± 0.01 a	2.8 ± 0.05 b	0.67 ± 0.01 f	1.02 ± 0.02 e
Quercetin-*O*-hexurunoside	1.310 ± 0.006 b	3.38 ± 0.03 a	1.51 ± 0.07 c	2.05 ± 0.98 a	2.08 ± 0.92 a	1.31 ± 0.01 e	1.31 ± 0.01 e	3.41 ± 0.03 a	3.354 ± 0.003 b	1.45 ± 0.01 d	1.575 ± 0.001 c
Luteolin-*O*-hexurunoside	1.410 ± 0.098 c	2.55 ± 0.02 a	2.074 ± 0.318 b	1.95 ± 0.46 a	2.07 ± 0.55 a	1.51 ± 0.02 e	1.32 ± 0.03 f	2.569 ± 0.004 a	2.53 ± 0.03 b	1.77 ± 0.04 d	2.38 ± 0.02 c
Quercetin-*O*-acetylhexoside	0.52 ± 0.07 b	0.757 ± 0.027 a	0.512 ± 0.005 b	0.625 ± 0.118 a	0.568 ± 0.122 a	0.585 ± 0.002 c	0.454 ± 0.003 f	0.78 ± 0.02 a	0.73 ± 0.01 b	0.507 ± 0.003 e	0.517 ± 0.004 d
Kaempherol-*O*-hexurunoside	0.811 ± 0.062 c	1.360 ± 0.006 a	0.917 ± 0.122 b	1.009 ± 0.252 a	1.051 ± 0.255 a	0.872 ± 0.004 d	0.75 ± 0.01 f	1.355 ± 0.004 b	1.365 ± 0.006 a	0.799 ± 0.009 e	1.036 ± 0.008 c
Apigenin-*O*-hexurunoside	1.066 ± 0.185 b	1.46 ± 0.021 a	1.57 ± 0.38 a	1.369 ± 0.211 a	1.36 ± 0.44 a	1.25 ± 0.03 d	0.89 ± 0.01 f	1.66 ± 0.03 b	1.26 ± 0.01 c	1.2 ± 0.4 e	1.939 ± 0.001 a
Isorhamnetin-*O*-hexurunoside	1.03 ± 0.03 c	1.47 ± 0.06 a	1.164 ± 0.257 b	1.113 ± 0.223 b	1.336 ± 0.203 a	1.01 ± 0.02 d	1.06 ± 0.01 c	1.42 ± 0.05 b	1.53 ± 0.03 a	0.91 ± 0.01 e	1.41 ± 0.01 b
Kaempherol-*O*-acetylhexoside	0.48 ± 0.04 b	0.616 ± 0.014 a	0.493 ± 0.033 b	0.529 ± 0.059 a	0.53 ± 0.08 a	0.523 ± 0.002 d	0.436 ± 0.003 f	0.603 ± 0.001 b	0.63 ± 0.004 a	0.461 ± 0.001 e	0.526 ± 0.001 c
Isorhamnetin-*O*-acetylhexoside	0.425 ± 0.042 b	0.491 ± 0.002 a	0.421 ± 0.003 b	0.459 ± 0.031 a	0.432 ± 0.044 b	0.466 ± 0.001 c	0.383 ± 0.003 f	0.4927 ± 0.0001 a	0.4887 ± 0.0002 b	0.4179 ± 0.0005 e	0.4234 ± 0.0009 d
Total Phenolic Acids	2.46 ± 0.12 b	6.19 ± 1.19 a	1.90 ± 0.37 c	3.75 ± 0.62 a	3.27 ± 0.26 a	2.34 ± 0.05 d	2.57 ± 0.01 c	7.35 ± 0.01 a	5.03 ± 0.03 b	1.54 ± 0.02 f	2.26 ± 0.02 e
Total Flavonoids	7.06 ± 0.47 c	12.09 ± 0.21 a	8.66 ± 1.18 b	9.1 ± 1.3 a	9.43 ± 1.22 a	7.51 ± 0.03 d	6.602 ± 0.001 e	12.29 ± 0.07 a	11.89 ± 0.05 b	7.51 ± 0.03 d	9.81 ± 0.03 c
Total Phenolic compounds	9.516 ± 0.353 c	18.3 ± 1.4 a	10.56 ± 1.55 b	12.8 ± 1.9 a	12.72 ± 3.26 a	9.9 ± 0.1 d	9.174 ± 0.009 e	19.6 ± 0.1 a	16.92 ± 0.02 b	9.06 ± 0.05 f	12.07 ± 0.01 c

Means in the same row for the same factor (irrigation and crop management) or their interaction followed by different Latin letters are significantly different according to Duncan's Multiple Range test (DMRT) at $p = 0.05$. Calibration curves used for quantification: *p*-Coumaric acid (y = 301,950x + 6966.7, R^2 = 0.9999, LOD (μg/mL) = 0.62 and LOQ (μg/mL) = 1.89, peaks 1 and 1′); caffeic acid (y = 388,345x + 406,369, R^2 = 0.9939, LOD (μg/mL) = 8.57 and LOQ (μg/mL) = 25.97, peak 2); quercetin-3-*O*-glucoside (y = 34,843x − 160,173, R^2 =0.9998, LOD (μg/mL) = 17.01 and LOQ (μg/mL) = 51.54, peaks 3, 4, 5, 6, 8, 9 and 10); apigenin-7-*O*-glucoside (y = 10,683x − 45,794, R^2 = 0.996, LOD (μg/mL) = 136.95 and LOQ (μg/mL) = 414.98, peak 7).

In the previous studies of Petropoulos et al. [56], it was also mentioned that chicoric acid was the predominant phenolic compound; however, the same authors indicated that 5-*O*-Caffeoylquinic acid was the second most abundant compound, whereas they did not detect quercetin-*O*-hexurumoside and luteolin-*O*-hexurumoside which were predominant in our study. Moreover, Liava et al. [59] detected significant amounts of luetolin derivatives and 5-*O*-Caffeoylquinic acid in decoctions and hydroethanolic extracts of *C. spinosum* leaves, whereas chicoric acid content was very low. These contradicting results could be associated with the different growing conditions and agronomic practices implemented, as well as the genetic material tested, since according to the literature a significant variation in phenolic profile should be expected among different *C. spinosum* ecotypes [64].

Moreover, Polyzos et al. [33], who studied the effect of deficit irrigation on *S. hispanicus*, observed that water deficit induced the accumulation of phenolic compounds in either rain-fed conditions or deficit irrigation compared to full irrigation. In pepper (*Capsicum annum* L.) fruit, the greatest TPC values were observed in plants grown under deficit irrigation (50%) [92], while low to moderate (50% field capacity) stress also increased TPC content in *Achillea* species [93]. Moreover, Skrypnik et al. [61] reported that TPC increased with the increasing severity of drought stress in salvia (*S. officinalis* L.) and oregano (*O. vulgare* L.), whereas moderate water stress (50% of soil field capacity) induced the accumulation of phenolic compounds in hyssop (*Hyssopus officinalis* L.) and severe stress (25% of soil field capacity) resulted in lower values of TPC. In contrast, water-deficit conditions (50% field capacity) decreased anthocyanin, quercetin, gallic acid, vanillic acid, chlorogenic acid, and cinnamic acid in sunflower (*H. annuus* L.) leaves [94]. Overall, deficit irrigation regulates the biosynthesis of phytochemical, including phenolic compounds, in a species-specific manner, since, apart from phenolic compounds, other antioxidant compounds could contribute to plant's defense against abiotic stressors [73]. Albergaria et al. [95] also highlighted that the increase in secondary metabolites is not uniform across all plant tissue and organs, and the severity of water stress may significantly impact metabolite synthesis. Therefore, mild water stress could be considered as a eustress and promote the biosynthesis of secondary metabolites, whereas severe stress may result in reduced values when the defense mechanisms of plants cannot cope with the severity of stress [96].

Crop-rotation schemes may also have an impact on phenolic compounds content. For example, Mitchell et al. [97] suggested that flavonoid content in tomato (*S. lycopersicum* L.) fruit increased when organic cultivation combined with various rotation schemes over a ten-year period. Moreover, the same authors recorded the largest increase during the last 4 years of the experiments, regardless of the cropping system (e.g., organic vs. conventional) which explains the lack of effect in our experiment, where a short-term rotation was implemented. Similarly, Buczek et al. [98] observed an increase in phenolic acid content in winter wheat (*T. aestivum* L.) grain when crop rotation was integrated into organic cultivation, whereas Siwek et al. [99] indicated that the integration of cover crops in rotation schemes may increase the phenolic compounds content of the succeeding crop depending on the species both of cover crop and succeeding crop.

3.3.2. Bioactive Properties

The evaluation of the antioxidant activity was performed with cell-based assays, namely TBARS and OxHLIA (Figure 4). In the OxHLIA assay, irrigation had no significant effect, whereas deficit irrigation resulted in the highest antioxidant activity in the TBARS assay. Conversely, crop rotation increased antioxidant activity in the OxHLIA assay, while no significant effect was observed in the TBARS assay. For the combined effect of the tested factors, the highest activity for OxHLIA and TBARS was recorded for DI × CR and C × NCR treatments, respectively. Previously, Polyzos et al. [33] mentioned that the two methods showed a varied response in *S. hispanicus* plants frown under deficit irrigation conditions, with deficit irrigation having the highest activity for TBARS assay and control treatment for OxHLIA assay. These variations may also arise from the involvement of different types of secondary metabolites, further emphasizing the complexity of antioxidant activity assessment and the necessity to evaluate it with more than one method [100].

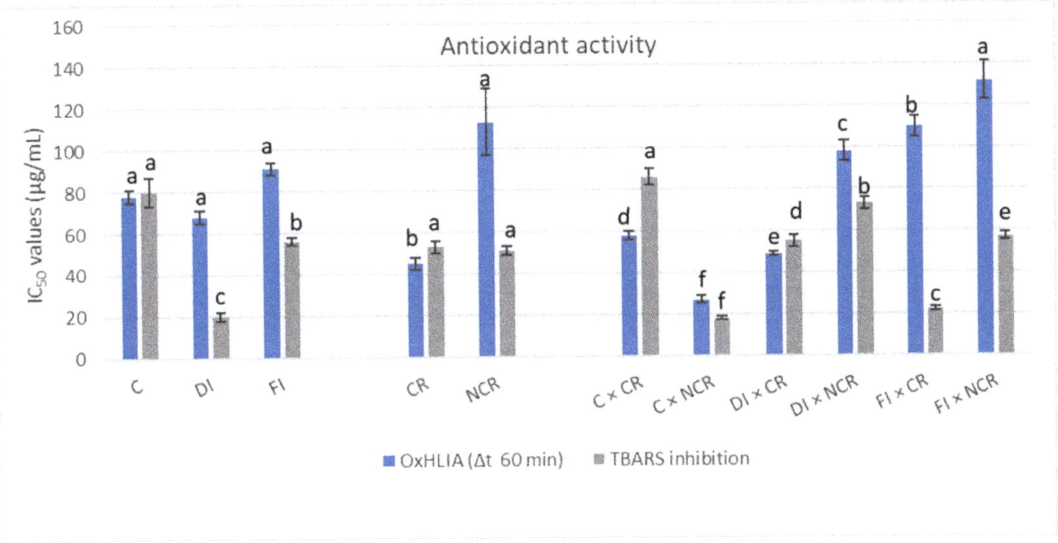

Figure 4. Antioxidant activity of *C. spinosum* L. plants in relation to irrigation and crop management (mean ± SD), determined with OxHLIA and TBARS assays. The results are expressed as IC_{50} values (µg/mL). Different Latin letters above bars of the same color and for the same factor or their interaction are significantly different according to Duncan's Multiple Range test (DMRT) at $p = 0.05$. DI: deficit irrigation; FI: full irrigation; NCR: no crop rotation; CR: crop rotation.

Under drought stress, plants activate both enzymatic and non-enzymatic antioxidants to mitigate the adverse effects [34,101]. For example, Gharibi et al. [93], who assessed various *Achillea* species under different irrigation regimes, observed that each species possessed unique characteristics in its antioxidant systems. Additionally, Bricudes et al. [11] identified chicoric acid as the primary phenolic compound in *C. spinosum* and *C. intybus* leaves and the main contributor to antioxidant activities. In *C. intybus*, quercetin-3,4-O-diglucoside has also been found to be responsible for increased antioxidant activity [102]. Moreover, Petropoulos et al. [103] indicated a negative linear correlation between phenolic compounds content and EC_{50} values of various antioxidant activity assays in several ecotypes of *C. spinosum* which corroborates the importance of phenolic compounds in the antioxidant system of the species. However, besides phenolic compounds and flavonoids, organic acids and tocopherols, particularly α-tocopherol, are also recognized as potent antioxidant molecules [60,66,104].

Means in the same row for the same factor (irrigation and crop management) or their interaction followed by different Latin letters are significantly different according to Duncan's Multiple Range test (DMRT) at $p = 0.05$. The results of the antibacterial properties of hydroethanolic extracts from *C. spinosum* leaves against selected food are presented in Table 10. In terms of food and clinical bacteria, the extracts exhibited moderate inhibitory activity against the studied strains depending on the extract, whereas all of the tested extracts recorded MIC and MBC values higher than the controls. In particular, extracts from plants grown under deficit irrigation conditions showed MIC values against the food bacteria *Enterobacter cloacae*, *Escherichia coli*, *Yersinia enterocolitica*, *Bacillus cereus* and *Listeria monocytogenes*, as well as against the clinical bacteria *Escherichia coli* and Methicillin-resistant *Staphylococcus aureus* (the control treatment was similarly effective, recording the same MIC value). Moreover, the lack of crop rotation resulted in extracts with the highest MIC values against the food bacteria *Escherichia coli Salmonella enterica*, *Yersinia enterocolitica* and *Listeria monocytogenes*, as well as the clinical bacteria *Escherichia coli* and Methicillin-resistant *Staphylococcus aureus*. For the combined effect of the tested factors, C × NCR treatment recorded the highest MIC values against *Listeria monocytogenes* (both food and clinical strains) and *Staphylococcus aureus* (food bacterial strains and Methicillin-resistant clinical strains); DI × CR treatment was the most effective against *Enterobacter cloacae*, while DI × CR treatment had the highest MIC values against *Escherichia coli* (both food and clinical strains), *Salmonella enterica*, *Yersinia enterocolitica*, and Methicillin-resistant *Staphylococcus aureus*. Finally, none of the tested extracts showed any antifungal, NO-production inhibition, antiproliferative, or hepatotoxic effects (Table 10). The only exception was observed in the activity against the AGS cell line, where the hydroethanolic extracts from all the treatments showed the capacity to inhibit cell proliferation, especially extracts from plants grown under deficit irrigation, no crop rotation, or the combination of no crop rotation and full irrigation.

Similarly to our study, Polyzos et al. [41] found no inhibitory effects against *Pseudomonas aeruginosa* for *C. spinosum* leaf extracts. In contrast, Petropoulos et al. [105] reported that *E. cloacae*, *Listeria monocytogenes*, and *Salmonella enterica* were the most sensitive bacterium, while they did not record any inhibitory effects against *E. coli*, which was observed in this study. Additionally, Birsa and Sarbu [102] suggested that *C. intybus* extracts possess antimicrobial and antifungal activities, although no such effects (e.g., antifungal) were recorded in our study. Furthermore, Polyzos et al. [33], who studied the antibacterial properties of *S. hispanicus* leaf extracts, also reported varied efficacy depending on the irrigation regime. In particular, they found that rain-fed plants and plants grown under deficit irrigation exhibited higher efficacy against *Listeria monocytogenes* and *Staphylococcus aureus*, a finding that was not recorded in this study. It is important to note that, apart from the plant species used, the extraction method may also influence the antimicrobial properties, since the choice of solvents affects the extracted bioactive phytochemicals and the antimicrobial properties of the obtained extract [12]. For example, the extraction solvent affected the antimicrobial and anticancer activity of *Mentha piperita* and *Catharanthus. roseus* extracts, while drought stress decreased these activities due to a reduction in phenolic and flavonoid content [106]. Moreover, Petropoulos et al. [105] suggested that apart from phenolic compounds, other bioactive components may also play an important role in the antimicrobial properties of *C. spinosum*. This finding partly explains the higher antibacterial effects recorded in our study for the extracts obtained from plants grown under deficit irrigation. However, this is not consistent with the effects of the cropping system, where no crop rotation showed better results than crop rotation without significant differences in total phenolic compounds between these two treatments. Therefore, it should be noted that the growing conditions, plant species, and extraction methods may lead to changes in antimicrobial and antifungal properties due to differences in the phytochemical composition of the obtained extracts.

Table 10. The antioxidant, antimicrobial, anti-inflammatory, cytotoxic, and hepatotoxic activities of the hydroethanolic extracts of *Cichorium spinosum* leaves in relation to irrigation and crop management (mean ± SD).

Treatments	Irrigation (I)						Crop Management				I × CM											
	Control (C)		Deficit Irrigation (DI)		Full Irrigation (FI)		Crop Rotation (CR)		None Crop Rotation (NCR)		C × CR		C × NCR		DI × CR		DI × NCR		FI × CR		FI × NCR	
Antioxidant activity IC_{50} values (μg/mL) [A]																						
OxHLIA (Δt 60 min)	78 ± 3 a		68 ± 3 a		91 ± 3 a		45 ± 3 b		113 ± 16 a		58 ± 2 d		27 ± 2 f		49 ± 1 e		98 ± 5 c		110 ± 5 b		132 ± 9 a	
TBARS inhibition	80 ± 7 a		20 ± 2 c		56 ± 2 b		53 ± 3 a		51 ± 2 a		86 ± 4 a		18 ± 1 f		55 ± 3 d		73 ± 3 b		22 ± 1 e		57 ± 2 c	
Antibacterial activity (mg/mL) [B]																						
	MIC	MBC	MIC	MBC	MIC	MBC	MIC	MBC	MIC	MBC	MIC	MBC	MIC	MBC	MIC	MBC	MIC	MBC	MIC	MBC	MIC	MBC
Food bacteria																						
Gram-negative bacteria																						
Enterobacter cloacae	10	>10	7.5	>10	>10	>10	>10	>10	>10	>10	10	>10	10	>10	5	>10	10	>10	>10	>10	10	>10
Escherichia coli	5	>10	3.75	>10	10	>10	6.7	>10	5.8	>10	5	>10	5	>10	5	>10	2.5	>10	10	>10	10	>10
Pseudomonas aeruginosa	10	>10	10	>10	>10	>10	>10	>10	10	>10	10	>10	10	>10	10	>10	10	>10	>10	>10	10	>10
Salmonella enterica	>10	>10	>10	>10	10	>10	>10	>10	5.58	>10	>10	>10	5	>10	>10	>10	2.5	>10	10	>10	10	>10
Yersinia enterocolitica	10	>10	7.5	>10	10	>10	10	>10	8.3	>10	10	>10	10	>10	10	>10	5	>10	10	>10	10	>10
Gram-positive bacteria																						
Bacillus cereus	10	>10	7.5	>10	>10	>10	>10	>10	10	>10	>10	>10	10	>10	>10	>10	10	>10	10	>10	>10	>10
Listeria monocytogenes	5	>10	3.75	>10	10	>10	6.7	>10	5.8	>10	>10	>10	2.5	>10	10	>10	5	>10	10	>10	10	>10
Staphylococcus aureus	10	>10	10	>10	>10	>10	>10	>10	10	>10	10	>10	2.5	>10	10	>10	5	>10	10	>10	10	>10
Clinical bacteria																						
Gram-negative bacteria																						
Escherichia coli	5	>10	3.75	>10	10	>10	6.7	>10	7.5	>10	5	>10	5	>10	5	>10	2.5	>10	10	>10	10	>10
Klebsiella pneumoniae	>10	>10	10	>10	>10	>10	>10	>10	>10	>10	>10	>10	>10	>10	10	>10	10	>10	>10	>10	>10	>10
Morganella morganii	>10	>10	>10	>10	>10	>10	>10	>10	>10	>10	>10	>10	>10	>10	>10	>10	>10	>10	>10	>10	>10	>10
Proteus mirabilis	>10	>10	>10	>10	>10	>10	>10	>10	>10	>10	>10	>10	>10	>10	>10	>10	>10	>10	>10	>10	>10	>10
Pseudomonas aeruginosa	10	>10	>10	>10	>10	>10	>10	>10	>10	>10	10	>10	10	>10	>10	>10	10	>10	>10	>10	10	>10

Table 10. *Cont.*

Treatments	Irrigation (I)						Crop Management				I × CM											
	Control (C)		Deficit Irrigation (DI)		Full Irrigation (FI)		Crop Rotation (CR)		None Crop Rotation (NCR)		C × CR		C × NCR		DI × CR		DI × NCR		FI × CR		FI × NCR	
Gram-positive bacteria																						
Enterococcus faecalis	>10	>10	>10	>10	>10	>10	>10	>10	>10	>10	>10	>10	10	>10	>10	>10	10	>10	10	>10	>10	>10
Listeria monocytogenes	>10	>10	>10	>10	>10	>10	>10	>10	>10	>10	>10	>10	2.5	>10	10	>10	5	>10	10	>10	10	>10
MRSA	6.25	>10	6.25	>10	10	>10	10	>10	5	>10	10	>10	2.5	>10	10	>10	2.5	>10	10	>10	10	>10
Antifungal activity (mg/mL) [B]																						
	MIC	MBC	MIC	MBC	MIC	MBC	MIC	MBC	MIC	MBC	MIC	MBC	MIC	MBC	MIC	MBC	MIC	MBC	MIC	MBC	MIC	MBC
Aspergillus brasiliensis	>10	>10	>10	>10	>10	>10	>10	>10	>10	>10	>10	>10	>10	>10	>10	>10	>10	>10	>10	>10	>10	>10
Aspergillus fumigatus	10	>10	10	>10	10	>10	10	>10	10	>10	10	>10	10	>10	10	>10	10	>10	10	>10	10	>10
NO-production inhibition (IC_{50} values µg/mL) [C]																						
RAW 264,7	>400		>400		>400		>400		>400		>400		>400		>400		>400		>400		>400	
Antiproliferative Activity (GI_{50} values µg/mL) [D]																						
AGS	291 ± 16.04 a		221 ± 21 c		240 ± 46 b		264 ± 32 a		237 ± 42 b		289 ± 28 ab		225 ± 23 c		279 ± 26 b		293 ± 22 a		218 ± 14 c		202 ± 13 d	
Caco-2	>400		>400		>400		>400		>400		>400		>400		>400		>400		>400		>400	
MCF7	>400		>400		>400		>400		>400		>400		>400		>400		>400		>400		>400	
NCI-H460	>400		>400		>400		>400		>400		>400		>400		>400		>400		>400		>400	
Hepatotoxicity (GI_{50} values µg/mL) [D]																						
PLP2	>400		>400		>400		>400		>400		>400		>400		>400		>400		>400		>400	

For antioxidant activity results, means in the same row for the same factor (irrigation and crop management) or their interaction followed by different Latin letters are significantly different according to Duncan's Multiple Range test (DMRT) at $p = 0.05$; [A] Trolox IC_{50} values: 5.8 ± 0.6 µg/mL (TBARS), 21.5 ± 0.2 µg/mL (OxHLIA 60 min); [B] Food bacteria—Streptomycin, 1 mg/mL (*E. cloacae*, *S. entericolitis*, *Y. enterocolitica*, *B. cereus*, *L. monocytogenes* and *S. aureus*—MIC/MBC 0.007; *E. coli*—MIC/MBC 0.01; *P. aeruginosa*—MIC/MBC 0.06), Methicillin, 1 mg/mL (*S. aureus*—MIC/MIB 0.007), Ampicillin, 10 mg/mL (*E. cloacae*, *E. coli*, *S. enterica*, *Y. enterocolitica*, *L. monocytogenes* and *S. aureus*—MIC/MBC 0.15; —MIC/MBC 0.01; *P. aeruginosa*—MIC/MBC 0.63); clinical bacteria—Ampicillin, 10 mg/mL (*E. coli*, *P. mirabilis*, *E. faecalis*, *L. monocytogenes*, MRSA MIC/MBC—<0.15; *K. pneumoniae*, *M. morganii*, *P. aeruginosa* MIC/MBC—>10); Imipenem, 1 mg/mL (*E. coli*, *K. pneumoniae*, *M. morganii*, *P. mirabilis*, *L. monocytogenes* MIC/MBC—<0.0078; *P. aeruginosa* MIC/MBC—0.5/1); Vancomycin, 1 mg/mL (*E. faecalis* MIC/MBC—<0.0078 and MRSA MIC/MBC—0.25/0.5; antifungal activity—Ketoconazole (*A. brasiliensis* MIC/MFC 0.06/0.125, *A. fumigatus* MIC/MFC 0.5/1); MIC—Minimum Inhibitory Concentration, MBC—Minimum Bactericidal Concentration, and MFC—Minimum Fungicidal Concentration; [C] Dexamethasone IC_{50} value: 6.3 ± 0.4 µg/mL; [D] Ellipticine GI_{50} values: 1.23 ± 0.03 µg/mL (AGS), 1.21 ± 0.02 µg/mL (Caco-2), 1.01 ± 0.01 µg/mL (NCI-H460), 1.02 ± 0.02 µg/mL (MCF-7), and 1.4 ± 0.1 µg/mL (PLP2).

4. Conclusions

The results of this study indicate that *C. spinosum* can be grown under deficit irrigation conditions while being incorporated into a crop-rotation system, without compromising the biomass yield and the quality of the final product. Specifically, the fatty-acid profile was optimized, showing an increase in PUFA and a decrease in SFA, while total phenolic compounds, flavonoids, and phenolic acid content in the leaf extracts increased, thus enhancing their health-promoting properties. Moreover, leaf extracts exhibited inhibitory activity against certain bacteria, including *E. cloacae*, *E. coli*, *L. monocytogenes*, *Y. enterocolitica*, *B. cereus*, and *MRSA*, while all the extracts were moderately effective against AGS cell lines. In conclusion, based on our results, *C. spinosum* is a promising alternative/complementary crop that could be valorized in small-scale farms of the Mediterranean, thus facilitating its commercial exploitation through its integration into sustainable cropping systems. Moreover, agronomic practices such as deficit irrigation and crop rotation may positively affect the growth and quality of *C. spinosum* plants contributing towards the mitigation of climate change adverse effects on crop production, while contributing to saving water and the diversification of crops. However, further research is required to refine crop management through the diversification of crops integrated into rotation schemes and to better study the long-term effects of crop rotation both on plant growth and chemical composition, as well as on soil physicochemical properties and functional microbiome.

Author Contributions: Conceptualization, I.C.F.R.F., L.B. and S.A.P.; methodology, N.P. and V.L.; B.H.P., T.C.S.P.P., M.A.-O., F.M. and M.I.D.; validation, B.H.P., F.M., T.C.S.P.P. and M.I.D.; formal analysis, N.P., T.C.S.P.P., F.M., M.A.-O. and B.H.P.; investigation, N.P., T.C.S.P.P., F.M., M.A.-O. and B.H.P.; resources, L.B. and S.A.P.; data curation, N.P. and S.A.P.; writing—original draft preparation, N.P. and B.H.P.; writing—review and editing, T.C.S.P.P., I.C.F.R.F., L.B. and S.A.P.; visualization, S.A.P.; supervision, S.A.P.; project administration, S.A.P. and L.B.; funding acquisition, S.A.P. and L.B. All authors have read and agreed to the published version of the manuscript.

Funding: This work was funded by the General Secretariat for Research and Technology of Greece (project VALUEFARM PRIMA2019-11) and PRIMA foundation in FCT Portugal, under the project VALUEFARM (PRIMA/0009/2019).

Data Availability Statement: The original contributions presented in this study are included in the article. Further inquiries can be directed to the corresponding author.

Acknowledgments: This work was supported by national funds through FCT/MCTES (PIDDAC): CIMO, UIDB/00690/2020 (DOI: 10.54499/UIDB/00690/2020) and UIDP/00690/2020 (DOI: 10.54499/UIDP/00690/2020); and SusTEC, LA/P/0007/2020 (DOI: 10.54499/LA/P/0007/2020). L. Bar-ros (DOI: 10.54499/CEECINST/00107/2021/CP2793/CT0002) and M.I. Dias (DOI: 10.54499/CEE CINST/00016/2018/CP1505/CT0004) thank the national funding by FCT through the institutional scientific employment programs for their contracts. The doctoral scholarships from B.H. Paschoalinotto (2023.02731.BD) and M. Añibarro-Ortega (2020.06297.BD).

Conflicts of Interest: The authors declare no conflicts of interest.

References

1. Ceccanti, C.; Landi, M.; Benvenuti, S.; Pardossi, A.; Guidi, L. Mediterranean wild edible plants: Weeds or "new functional crops"? *Molecules* **2018**, *23*, 2299. [CrossRef] [PubMed]
2. Corrêa, R.C.G.; Di Gioia, F.; Ferreira, I.C.F.R.; Petropoulos, S.A. Wild greens used in the Mediterranean diet. In *The Mediterranean Diet: An Evidence-based Approach*; Preedy, V., Watson, R., Eds.; Academic Press: London, UK, 2020; pp. 209–228, ISBN 9788578110796.
3. Sánchez-Mata, M.d.C.; Tardío, J. *Mediterranean Wild Edible Plants*; Springer: New York, NY, USA, 2016; ISBN 978-1-4939-3327-3.
4. Chatzopoulou, E.; Carocho, M.; Di Gioia, F.; Petropoulos, S.A. The beneficial health effects of vegetables and wild edible greens: The case of the mediterranean diet and its sustainability. *Appl. Sci.* **2020**, *10*, 9144. [CrossRef]

5. Borelli, T.; Hunter, D.; Powell, B.; Ulian, T.; Mattana, E.; Termote, C.; Pawera, L.; Beltrame, D.; Penafiel, D.; Tan, A.; et al. Born to eat wild: An integrated conservation approach to secure wild food plants for food security and nutrition. *Plants* **2020**, *9*, 1299. [CrossRef] [PubMed]
6. Psaroudaki, A.; Nikoloudakis, N.; Skaracis, G.; Katsiotis, A. Genetic structure and population diversity of eleven edible herbs of Eastern Crete. *J. Biol. Res.* **2015**, *22*, 7. [CrossRef]
7. Klados, E.; Tzortzakis, N. Effects of substrate and salinity in hydroponically grown *Cichorium spinosum*. *J. Soil. Sci. Plant Nutr.* **2014**, *14*, 211–222. [CrossRef]
8. Petropoulos, S.A.; Karkanis, A.; Martins, N.; Ferreira, I.C.F.R. Edible halophytes of the Mediterranean basin: Potential candidates for novel food products. *Trends Food Sci. Technol.* **2018**, *74*, 69–84. [CrossRef]
9. Vardavas, C.I.; Majchrzak, D.; Wagner, K.H.; Elmadfa, I.; Kafatos, A. The antioxidant and phylloquinone content of wildly grown greens in Crete. *Food Chem.* **2006**, *99*, 813–821. [CrossRef]
10. Zegichi, S.; Kalithraka, S.; Simopoulos, A. Nutritional composition of molokhia (*Corchorus olitorius*) and stamnagathi (*Cichorium spinosum*). *World Rev. Nutr. Diet.* **2003**, *91*, 1–21.
11. Brieudes, V.; Angelis, A.; Vougogiannopoulos, K.; Pratsinis, H.; Kletsas, D.; Mitakou, S.; Halabalaki, M.; Skaltsounis, L.A. Phytochemical analysis and antioxidant potential of the phytonutrient-rich decoction of *Cichorium spinosum* and *C. intybus*. *Planta Med.* **2016**, *82*, 1070–1078. [CrossRef]
12. Petropoulos, S.A.; Fernandes, Â.; Tzortzakis, N.; Sokovic, M.; Ciric, A.; Barros, L.; Ferreira, I.C.F.R. Bioactive compounds content and antimicrobial activities of wild edible Asteraceae species of the Mediterranean flora under commercial cultivation conditions. *Food Res. Int.* **2019**, *119*, 859–868. [CrossRef]
13. Chatzigianni, M.; Alkhaled, B.; Livieratos, I.; Stamatakis, A.; Ntatsi, G.; Savvas, D. Impact of nitrogen source and supply level on growth, yield and nutritional value of two contrasting ecotypes of *Cichorium spinosum* L. grown hydroponically. *J. Sci. Food Agric.* **2018**, *98*, 1615–1624. [CrossRef] [PubMed]
14. Mancosu, N.; Snyder, R.L.; Kyriakakis, G.; Spano, D. Water scarcity and future challenges for food production. *Water* **2015**, *7*, 975–992. [CrossRef]
15. Lovelli, S.; Perniola, M.; Scalcione, E.; Troccoli, A.; Ziska, L.H. Future climate change in the Mediterranean area: Implications for water use and weed management. *Ital. J. Agron.* **2012**, *7*, 44–49. [CrossRef]
16. Adu, M.O.; Yawson, D.O.; Armah, F.A.; Asare, P.A.; Frimpong, K.A. Meta-analysis of crop yields of full, deficit, and partial root-zone drying irrigation. *Agric. Water Manag.* **2018**, *197*, 79–90. [CrossRef]
17. Claro, A.M.; Fonseca, A.; Fraga, H.; Santos, J.A. Future Agricultural Water Availability in Mediterranean Countries under Climate Change: A Systematic Review. *Water* **2024**, *16*, 2484. [CrossRef]
18. Fischer, G.; Tubiello, F.N.; van Velthuizen, H.; Wiberg, D.A. Climate change impacts on irrigation water requirements: Effects of mitigation, 1990–2080. *Technol. Forecast. Soc. Change* **2007**, *74*, 1083–1107. [CrossRef]
19. Nikolaou, G.; Neocleous, D.; Christou, A.; Kitta, E.; Katsoulas, N. Implementing sustainable irrigation in water-scarce regions under the impact of climate change. *Agronomy* **2020**, *10*, 1120. [CrossRef]
20. Neupane, J.; Guo, W. Agronomic basis and strategies for precision water management: A review. *Agronomy* **2019**, *9*, 87. [CrossRef]
21. Savé, R.; de Herralde, F.; Aranda, X.; Pla, E.; Pascual, D.; Funes, I.; Biel, C. Potential changes in irrigation requirements and phenology of maize, apple trees and alfalfa under global change conditions in Fluvià watershed during XXIst century: Results from a modeling approximation to watershed-level water balance. *Agric. Water Manag.* **2012**, *114*, 78–87. [CrossRef]
22. Galindo, A.; Collado-González, J.; Griñán, I.; Corell, M.; Centeno, A.; Martín-Palomo, M.J.; Girón, I.F.; Rodríguez, P.; Cruz, Z.N.; Memmi, H.; et al. Deficit irrigation and emerging fruit crops as a strategy to save water in Mediterranean semiarid agrosystems. *Agric. Water Manag.* **2018**, *202*, 311–324. [CrossRef]
23. Ünlü, M.; Kanber, R.; Şenyigit, U.; Onaran, H.; Diker, K. Trickle and sprinkler irrigation of potato (*Solanum tuberosum* L.) in the Middle Anatolian Region in Turkey. *Agric. Water Manag.* **2006**, *79*, 43–71. [CrossRef]
24. De Boni, A.; D'Amico, A.; Acciani, C.; Roma, R. Crop Diversification and Resilience of Drought-Resistant Species in Semi-Arid Areas: An Economic and Environmental Analysis. *Sustainability* **2022**, *14*, 9552. [CrossRef]
25. Wang, X.; Li, X.; Zhao, W.; Hou, X.; Dong, S. Current views of drought research: Experimental methods, adaptation mechanisms and regulatory strategies. *Front. Plant Sci.* **2024**, *15*, 1371895. [CrossRef] [PubMed]
26. Fita, A.; Rodríguez-Burruezo, A.; Boscaiu, M.; Prohens, J.; Vicente, O. Breeding and domesticating crops adapted to drought and salinity: A new paradigm for increasing food production. *Front. Plant Sci.* **2015**, *6*, 978. [CrossRef] [PubMed]
27. Jia, P.Y.; Zhang, L.X.; Huang, Z.; Tian, F.P.; Hu, Y.; Wu, G.L. Physiological characteristics of three wild *Sonchus* species to prolonged drought tolerance in arid regions. *Pakistan J. Bot.* **2018**, *50*, 9–17.
28. Guarise, M.; Borgonovo, G.; Bassoli, A.; Ferrante, A. The Effect of Drought on *Sisymbrium officinale* (L.) Wild Species for Potential Cultivation as a Leafy Vegetable. *Horticulturae* **2023**, *9*, 111. [CrossRef]
29. Christoforidi, I.; Kollaros, D.; Manios, T.; Daliakopoulos, I.N. Drought- and Salt-Tolerant Plants of the Mediterranean and Their Diverse Applications: The Case of Crete. *Land* **2022**, *11*, 2038. [CrossRef]

30. Litskas, V.D.; Chrysargyris, A.; Tzortzakis, N.; Stavrinides, M.C.; Petropoulos, S.A. Can the commercial cultivation of wild edible species contribute to sustainable food production? A case study of golden thistle (*Scolymus hispanicus* L.). *Int. J. Life Cycle Assess.* **2025**, *30*, 446–461. [CrossRef]
31. Borelli, T.; Hunter, D.; Padulosi, S.; Amaya, N.; Meldrum, G.; de Oliveira Beltrame, D.M.; Samarasinghe, G.; Wasike, V.W.; Güner, B.; Tan, A.; et al. Local solutions for sustainable food systems: The contribution of orphan crops and wild edible species. *Agronomy* **2020**, *10*, 231. [CrossRef]
32. Ray, A.; Ray, R.; Sreevidya, E.A. How Many Wild Edible Plants Do We Eat—Their Diversity, Use, and Implications for Sustainable Food System: An Exploratory Analysis in India. *Front. Sustain. Food Syst.* **2020**, *4*, 56. [CrossRef]
33. Polyzos, N.; Paschoalinotto, B.H.; Pires, T.C.S.P.; Añibarro-Ortega, M.; Calhelha, R.; Ferreira, I.C.F.R.; Dias, M.I.; Barros, L.; Petropoulos, S.A. The Impact of Deficit Irrigation on the Agronomic Performance and Chemical Composition of *Scolymus hispanicus* L. *Horticulturae* **2024**, *10*, 479. [CrossRef]
34. Calone, R.; Mircea, D.M.; González-Orenga, S.; Boscaiu, M.; Lambertini, C.; Barbanti, L.; Vicente, O. Recovery from Salinity and Drought Stress in the Perennial *Sarcocornia fruticosa* vs. the Annual *Salicornia europaea* and *S. veneta*. *Plants* **2022**, *11*, 1058. [CrossRef]
35. Bowles, T.M.; Mooshammer, M.; Socolar, Y.; Calderón, F.; Cavigelli, M.A.; Culman, S.W.; Deen, W.; Drury, C.F.; Garcia y Garcia, A.; Gaudin, A.C.M.; et al. Long-Term Evidence Shows that Crop-Rotation Diversification Increases Agricultural Resilience to Adverse Growing Conditions in North America. *One Earth* **2020**, *2*, 284–293. [CrossRef]
36. Li, J.; Huang, L.; Zhang, J.; Coulter, J.A.; Li, L.; Gan, Y. Diversifying crop rotation improves system robustness. *Agron. Sustain. Dev.* **2019**, *39*, 38. [CrossRef]
37. Yu, T.; Mahe, L.; Li, Y.; Wei, X.; Deng, X.; Zhang, D. Benefits of Crop Rotation on Climate Resilience and Its Prospects in China. *Agronomy* **2022**, *12*, 436. [CrossRef]
38. Preissel, S.; Reckling, M.; Schläfke, N.; Zander, P. Magnitude and farm-economic value of grain legume pre-crop benefits in Europe: A review. *F. Crop. Res.* **2015**, *175*, 64–79. [CrossRef]
39. Lötjönen, S.; Ollikainen, M. Does crop rotation with legumes provide an efficient means to reduce nutrient loads and GHG emissions? *Rev. Agric. Food Environ. Stud.* **2017**, *98*, 283–312. [CrossRef]
40. Reckling, M.; Hecker, J.M.; Bergkvist, G.; Watson, C.A.; Zander, P.; Schläfke, N.; Stoddard, F.L.; Eory, V.; Topp, C.F.E.; Maire, J.; et al. A cropping system assessment framework—Evaluating effects of introducing legumes into crop rotations. *Eur. J. Agron.* **2016**, *76*, 186–197. [CrossRef]
41. Polyzos, N.; Paschoalinotto, B.H.; Compocholi, M.; Pinela, J.; Heleno, S.A.; Calhelha, R.C.; Dias, M.I.; Barros, L.; Petropoulos, S.A. Fertilization of pot-grown *Cichorium spinosum* L.: How it can affect plant growth, chemical profile, and bioactivities of edible parts? *Horticulturae* **2022**, *8*, 890. [CrossRef]
42. AOAC. *Official Methods of Analysis of Association of Official Analytical Chemists*; Horwitz, W., Latimer, G., Eds.; AOAC Inter.: Gaithersburg, MD, USA, 2019; ISBN 0935584773.
43. Mohamed, M.H.M.; Ali, M.M.E.; Zewail, R.M.Y.; Liava, V.; Petropoulos, S.A. The Mitigating Effects of Biostimulant Amendments on the Response of Purslane Plants Grown under Drought Stress Conditions. *Horticulturae* **2024**, *10*, 858. [CrossRef]
44. Saheri, F.; Barzin, G.; Pishkar, L.; Boojar, M.M.A.; Babaeekhou, L. Correction to: Foliar spray of salicylic acid induces physiological and biochemical changes in purslane (*Portulaca oleracea* L.) under drought stress. *Biologia* **2020**, *75*, 2189–2200. [CrossRef]
45. Delfine, S.; Fratianni, A.; D'Agostino, A.; Panfili, G. Influence of Drought Stress on Physiological Responses and Bioactive Compounds in Chicory (*Cichorium intybus* L.): Opportunity for a Sustainable Agriculture. *Foods* **2022**, *11*, 3725. [CrossRef] [PubMed]
46. Nemecek, T.; von Richthofen, J.S.; Dubois, G.; Casta, P.; Charles, R.; Pahl, H. Environmental impacts of introducing grain legumes into European crop rotations. *Eur. J. Agron.* **2008**, *28*, 380–393. [CrossRef]
47. Venter, Z.S.; Jacobs, K.; Hawkins, H.J. The impact of crop rotation on soil microbial diversity: A meta-analysis. *Pedobiologia* **2016**, *59*, 215–223. [CrossRef]
48. Zhao, J.; Chen, J.; Beillouin, D.; Lambers, H.; Yang, Y.; Smith, P.; Zeng, Z.; Olesen, J.E.; Zang, H. Global systematic review with meta-analysis reveals yield advantage of legume-based rotations and its drivers. *Nat. Commun.* **2022**, *13*, 4926. [CrossRef]
49. Jensen, C.R.; Joernsgaard, B.; Andersen, M.N.; Christiansen, J.L.; Mogensen, V.O.; Friis, P.; Petersen, C.T. The effect of lupins as compared with peas and oats on the yield of the subsequent winter barley crop. *Eur. J. Agron.* **2004**, *20*, 405–418. [CrossRef]
50. Sánchez-Navarro, V.; Zornoza, R.; Faz, Á.; Fernández, J.A. Does the use of cowpea in rotation with a vegetable crop improve soil quality and crop yield and quality? A field study in SE Spain. *Eur. J. Agron.* **2019**, *107*, 10–17. [CrossRef]
51. Cui, Z.; Yan, B.; Gao, Y.; Wu, B.; Wang, Y.; Xie, Y.; Xu, P.; Wang, H.; Wen, M.; Wang, Y.; et al. Crop yield and water use efficiency in response to long-term diversified crop rotations. *Front. Plant Sci.* **2022**, *13*, 1024898. [CrossRef]
52. Ghadirnezhad Shiade, S.R.; Fathi, A.; Taghavi Ghasemkheili, F.; Amiri, E.; Pessarakli, M. Plants' responses under drought stress conditions: Effects of strategic management approaches—A review. *J. Plant Nutr.* **2023**, *46*, 2198–2230. [CrossRef]

53. Yousefvand, P.; Sohrabi, Y.; Mastinu, A.; Heidari, G.; Weisany, W. Salicylic acid altered the fatty acids compositions and nutrient status of shallot (*Allium hirtifolium*) grown under drought stress. *J. Agric. Food Res.* **2024**, *18*, 101502. [CrossRef]
54. Allamine, H.M.; Buyuktas, D.; Karaca, C.; Aydinsakir, K.; Erdurmus, C. Effect of regulated deficit irrigation on productivity, evapotranspiration and quality of grain sorghum. *Irrig. Sci.* **2023**, *41*, 277–293. [CrossRef]
55. Abdou, N.M.; Roby, M.H.H.; AL-Huqail, A.A.; Elkelish, A.; Sayed, A.A.S.; Alharbi, B.M.; Mahdy, H.A.A.; Abou-Sreea, A.I.B. Compost Improving Morphophysiological and Biochemical Traits, Seed Yield, and Oil Quality of *Nigella sativa* under Drought Stress. *Agronomy* **2023**, *13*, 1147. [CrossRef]
56. Petropoulos, S.A.; Levizou, E.; Ntatsi, G.; Fernandes, Â.; Petrotos, K.; Akoumianakis, K.; Barros, L.; Ferreira, I.C.F.R. Salinity effect on nutritional value, chemical composition and bioactive compounds content of *Cichorium spinosum* L. *Food Chem.* **2017**, *214*, 129–136. [CrossRef]
57. Petropoulos, S.A.; Fernandes, Â.; Dias, M.I.; Pereira, C.; Calhelha, R.C.; Chrysargyris, A.; Tzortzakis, N.; Ivanov, M.; Sokovic, M.D.; Barros, L.; et al. Chemical composition and plant growth of *Centaurea raphanina* subsp. *mixta plants cultivated under saline conditions. Molecules* **2020**, *25*, 2204. [CrossRef]
58. Petropoulos, S.; Fernandes, Â.; Karkanis, A.; Antoniadis, V.; Barros, L.; Ferreira, I.C.F.R. Nutrient solution composition and growing season affect yield and chemical composition of *Cichorium spinosum* plants. *Sci. Hortic.* **2018**, *231*, 97–107. [CrossRef]
59. Liava, V.; Fernandes, Â.; Reis, F.; Finimundy, T.; Mandim, F.; Pinela, J.; Stojković, D.; Ferreira, I.C.F.R.; Barros, L.; Petropoulos, S.A. How Does Domestic Cooking Affect the Biochemical Properties of Wild Edible Greens of the Asteraceae Family? *Foods* **2024**, *13*, 2677. [CrossRef] [PubMed]
60. Al Muhairi, M.A.; Cheruth, A.J.; Kurup, S.S.; Rabert, G.A.; Al-Yafei, M.S. Effect of abscisic acid on biochemical constituents, enzymatic and non enzymatic antioxidant status of lettuce (*Lactuca sativa* L.) under varied irrigation regimes. *Cogent Food Agric.* **2015**, *1*, 1080888. [CrossRef]
61. Skrypnik, L.; Maslennikov, P.; Antipina, M.; Katserov, D.; Feduraev, P. Comparative Study on the Response of Hyssop (*Hyssopus officinalis* L.), Salvia (*Salvia officinalis* L.), and Oregano (*Origanum vulgare* L.) to Drought Stress Under Foliar Application of Selenium. *Plants* **2024**, *13*, 2986. [CrossRef] [PubMed]
62. Zhao, W.; Wu, K.; Wu, Y.; Yu, H.; Cao, W.; Ma, H. Effects of biochar amendment on greenhouse tomato quality, nutrient uptake and use efficiency under various irrigation and fertilization regimes. *Sci. Hortic.* **2024**, *337*, 113441. [CrossRef]
63. Petropoulos, S.A.; Fernandes, Â.; Calhelha, R.C.; Di Gioia, F.; Kolovou, P.; Barros, L.; Ferreira, I.C.F.R. Chemical composition and bioactive properties of *Cichorium spinosum* L. in relation to nitrate/ammonium nitrogen ratio. *J. Sci. Food Agric.* **2019**, *99*, 6741–6750. [CrossRef]
64. Petropoulos, S.A.; Fernandes, Â.; Ntatsi, G.; Levizou, E.; Barros, L.; Ferreira, I.C.F.R. Nutritional profile and chemical composition of *Cichorium spinosum* ecotypes. *LWT-Food Sci. Technol.* **2016**, *73*, 95–101. [CrossRef]
65. Petropoulos, S.; Fernandes, Â.; Karkanis, A.; Ntatsi, G.; Barros, L.; Ferreira, I. Successive harvesting affects yield, chemical composition and antioxidant activity of *Cichorium spinosum* L. *Food Chem.* **2017**, *237*, 83–90. [CrossRef]
66. Faizan, M.; Alam, P.; Rajput, V.D.; Shareen; Kaur, K.; Faraz, A.; Minkina, T.; Maqbool Ahmed, S.; Rajpal, V.R.; Hayat, S. Potential role of tocopherol in protecting crop plants against abiotic stresses. *Physiol. Mol. Biol. Plants* **2023**, *29*, 1563–1575. [CrossRef] [PubMed]
67. Ali, E.; Hussain, S.; Hussain, N.; Kakar, K.U.; Shah, J.M.; Zaidi, S.H.R.; Jan, M.; Zhang, K.; Khan, M.A.; Imtiaz, M. Tocopherol as plant protector: An overview of Tocopherol biosynthesis enzymes and their role as antioxidant and signaling molecules. *Acta Physiol. Plant.* **2022**, *44*, 20. [CrossRef]
68. Munné-Bosch, S.; Falk, J. New insights into the function of tocopherols in plants. *Planta* **2004**, *218*, 323–326. [CrossRef] [PubMed]
69. Sadiq, M.; Aisha, N.; Muhammad, A.; Al, F.; Parvaiz, Q. Alpha-Tocopherol-Induced Regulation of Growth and Metabolism in Plants Under Non-stress and Stress Conditions. *J. Plant Growth Regul.* **2019**, *38*, 1325–1340. [CrossRef]
70. Kapur, B.; Karaca, C.; Sarıdaş, M.A.; Ağçam, E.; Çeliktopuz, E.; Kargı, S.P. Enhancing secondary compounds in strawberry fruit through optimized irrigation and seaweed application. *Sci. Hortic.* **2024**, *324*, 112609. [CrossRef]
71. Ibrahim, E.A.; Ebrahim, N.E.S.; Mohamed, G.Z. Mitigation of water stress in broccoli by soil application of humic acid. *Sci. Rep.* **2024**, *14*, 2765. [CrossRef]
72. Thomas, A.; Beena, R.; Laksmi, G.; Soni, K.B.; Swapna, A.; Viji, M.M. Changes in sucrose metabolic enzymes to water stress in contrasting rice genotypes. *Plant Stress* **2022**, *5*, 100088. [CrossRef]
73. Xu, J.; Li, X.; Wan, W.; Zhu, X.; Li, C.; Zhao, X.; Zhao, Y.; Pang, S.; Diao, M. Impact of regulated deficit irrigation on the dynamics of quality changes in processing tomato fruits during ripening. *Agric. Water Manag.* **2024**, *304*, 109068. [CrossRef]
74. Götze, P.; Rücknagel, J.; Wensch-Dorendorf, M.; Märländer, B.; Christen, O. Crop rotation effects on yield, technological quality and yield stability of sugar beet after 45 trial years. *Eur. J. Agron.* **2017**, *82*, 50–59. [CrossRef]
75. Papafilippaki, A.; Nikolaidis, N.P. Comparative study of wild and cultivated populations of *Cichorium spinosum*: The influence of soil and organic matter addition. *Sci. Hortic.* **2020**, *261*, 108942. [CrossRef]

76. Gómez-Bellot, M.J.; Guerrero, L.; Yuste, J.E.; Vallejo, F.; Sánchez-Blanco, M.J. Identifying Bioactive Compounds in Common Bean (*Phaseolus vulgaris* L.) Plants under Water Deficit Conditions. *Horticulturae* **2024**, *10*, 663. [CrossRef]
77. Obadi, A.; Alharbi, A.; Alomran, A.; Alghamdi, A.G.; Louki, I.; Alkhasha, A.; Alqardaeai, T. Enhancement in Tomato Yield and Quality Using Biochar Amendments in Greenhouse under Salinity and Drought Stress. *Plants* **2024**, *13*, 1634. [CrossRef] [PubMed]
78. Zhang, R.; Zhang, H.; Yang, C.; Li, H.; Wu, J. Effects of water stress on nutrients and enzyme activity in rhizosphere soils of greenhouse grape. *Front. Microbiol.* **2024**, *15*, 1376849. [CrossRef] [PubMed]
79. Alomari-Mheidat, M.; Corell, M.; Martín-Palomo, M.J.; Castro-Valdecantos, P.; Medina-Zurita, N.; de Sosa, L.L.; Moriana, A. Moderate Water Stress Impact on Yield Components of Greenhouse Tomatoes in Relation to Plant Water Status. *Plants* **2024**, *13*, 128. [CrossRef]
80. Thompson, R.B.; Gallardo, M.; Valdez, L.C.; Fernández, M.D. Using plant water status to define threshold values for irrigation management of vegetable crops using soil moisture sensors. *Agric. Water Manag.* **2007**, *88*, 147–158. [CrossRef]
81. Yfantopoulos, D.; Ntatsi, G.; Karkanis, A.; Savvas, D. Evaluation of the Role of Legumes in Crop Rotation Schemes of Organic or Conventionally Cultivated Cabbage. *Agronomy* **2024**, *14*, 297. [CrossRef]
82. Haruna, S.I.; Nkongolo, N.V. Influence of cover crop, tillage, and crop rotation management on soil nutrients. *Agriculture* **2020**, *10*, 225. [CrossRef]
83. Adesanya, T.; Zvomuya, F.; Fernandez, M.R.; Luce, M. Crop rotation diversity and tillage effects on soil and wheat grain nutrient concentration in an organically-managed system. *J. Agric. Food Res.* **2024**, *18*, 101411. [CrossRef]
84. Dwivedi, S.L.; Nigam, S.N.; Nageswara Rao, R.C.; Singh, U.; Rao, K.V.S. Effect of drought on oil, fatty acids and protein contents of groundnut (*Arachis hypogaea* L.) seeds. *F. Crop. Res.* **1996**, *48*, 125–133. [CrossRef]
85. Ghaffari, M.; Gholizadeh, A.; Rauf, S.; Shariati, F. Drought-stress induced changes of fatty acid composition affecting sunflower grain yield and oil quality. *Food Sci. Nutr.* **2023**, *11*, 7718–7731. [CrossRef] [PubMed]
86. Harisha, C.B.; Rane, J.; Halagunde Gowda, G.R.; Chavan, S.B.; Chaudhary, A.; Verma, A.K.; Ravi, Y.; Asangi, H.; Halli, H.M.; Boraiah, K.M.; et al. Effect of Deficit Irrigation and Intercrop Competition on Productivity, Water Use Efficiency and Oil Quality of Chia in Semi-Arid Regions. *Horticulturae* **2024**, *10*, 101. [CrossRef]
87. Upchurch, R.G. Fatty acid unsaturation, mobilization, and regulation in the response of plants to stress. *Biotechnol. Lett.* **2008**, *30*, 967–977. [CrossRef]
88. Xu, L.; Han, L.; Huang, B. Membrane fatty acid composition and saturation levels associated with leaf dehydration tolerance and post-drought rehydration in Kentucky bluegrass. *Crop Sci.* **2011**, *51*, 273–281. [CrossRef]
89. Stepien, A.; Wojtkowiak, K.; Pietrzak-Fiecko, R. Nutrient content, fat yield and fatty acid profile of winter rapeseed (*Brassica napus* L.) grown under different agricultural production systems. *Chil. J. Agric. Res.* **2017**, *77*, 266–272. [CrossRef]
90. Mohammadi, K.; Ghalavand, A.; Aghaalikhani, M.; Heidari, G.; Shahmoradi, B.; Sohrabi, Y. Effect of different methods of crop rotation and fertilization on canola traits and soil microbial activity. *Aust. J. Crop Sci.* **2011**, *5*, 1261–1268.
91. Wacal, C.; Ogata, N.; Basalirwa, D.; Sasagawa, D.; Kato, M.; Handa, T.; Masunaga, T.; Yamamoto, S.; Nishihara, E. Fatty Acid Composition of Sesame (*Sesamum indicum* L.) Seeds in Relation to Yield and Soil Chemical Properties on Continuously Monocropped Upland Fields Converted from Paddy Fields. *Agronomy* **2019**, *9*, 801. [CrossRef]
92. Gokkus, M.K. Effects of urea fertilizer and nettle extract on the biochemical and morphological characteristics of ornamental peppers (*Capsicum frutescens* L.) under deficit irrigation conditions. *Irrig. Drain.* 2024; *Early View*. [CrossRef]
93. Gharibi, S.; Tabatabaei, B.E.S.; Saeidi, G.; Goli, S.A.H. Effect of Drought Stress on Total Phenolic, Lipid Peroxidation, and Antioxidant Activity of *Achillea* Species. *Appl. Biochem. Biotechnol.* **2016**, *178*, 796–809. [CrossRef]
94. Mahmood, S.; Afzal, B.; Bashir, R.; Shakoor, M.B.; Nisa, Z.U.; Rizwan, M.; Awais, M.; Azeem, M.; Wahid, A.; Yong, J.W.H. Melatonin priming could modulate primary and secondary metabolism of sunflower with better nutraceutical value and tolerance against water deficit environment. *Plant Stress* **2024**, *13*, 100533. [CrossRef]
95. Albergaria, E.T.; Oliveira, A.F.M.; Albuquerque, U.P. The effect of water deficit stress on the composition of phenolic compounds in medicinal plants. *South African J. Bot.* **2020**, *131*, 12–17. [CrossRef]
96. Muszyńska, E.; Dziurka, K.; Labudda, M. What Makes the Life of Stressed Plants a Little Easier? Defense Mechanisms against Adverse Conditions. *Plants* **2023**, *12*, 1040. [CrossRef]
97. Mitchell, A.E.; Hong, Y.J.; Koh, E.; Barrett, D.M.; Bryant, D.E.; Denison, R.F.; Kaffka, S. Ten-year comparison of the influence of organic and conventional crop management practices on the content of flavonoids in tomatoes. *J. Agric. Food Chem.* **2007**, *55*, 6154–6159. [CrossRef] [PubMed]
98. Buczek, J.; Jańczak-Pieniążek, M.; Harasim, E.; Kwiatkowski, C.A.; Kapusta, I. Effect of Cropping Systems and Environment on Phenolic Acid Profiles and Yielding of Hybrid Winter Wheat Genotypes. *Agriculture* **2023**, *13*, 834. [CrossRef]
99. Siwek, P.; Bucki, P.; Domagała-Świątkiewicz, I.; Lalewicz, P. Effect of cover crops integration in crop rotation on the yield and chemical composition of edible parts of vegetables grown in an organic system in high tunnel. *Sci. Hortic.* **2024**, *332*, 4–10. [CrossRef]

100. Gonçalves, S.; Moreira, E.; Andrade, P.B.; Valentão, P.; Romano, A. Effect of in vitro gastrointestinal digestion on the total phenolic contents and antioxidant activity of wild Mediterranean edible plant extracts. *Eur. Food Res. Technol.* **2019**, *245*, 753–762. [CrossRef]
101. Sarker, U.; Oba, S. Drought Stress Effects on Growth, ROS Markers, Compatible Solutes, Phenolics, Flavonoids, and Antioxidant Activity in *Amaranthus tricolor*. *Appl. Biochem. Biotechnol.* **2018**, *186*, 999–1016. [CrossRef]
102. Birsa, M.L.; Sarbu, L.G. Health Benefits of Key Constituents in *Cichorium intybus* L. *Nutrients* **2023**, *15*, 1322. [CrossRef]
103. Petropoulos, S.; Fernandes, A.; Barros, L.; Ferreira, I. A comparison of the phenolic profile and antioxidant activity of different *Cichorium spinosum* L. ecotypes. *J. Sci. Food Agric.* **2017**, *98*, 183–189. [CrossRef]
104. Sishu, N.K.; Selvaraj, C.I. *Phytochemistry, Pharmacological Applications, and Therapeutic Effects of Green Synthesized Nanomaterials Using Cichorium Species—A Comprehensive Review*; Springer: Berlin/Heidelberg, Germany, 2024; Volume 397, ISBN 0123456789.
105. Petropoulos, S.; Fernandes, Â.; Stojković, D.; Pereira, C.; Taofiq, O.; Di Gioia, F.; Tzortzakis, N.; Soković, M.; Barros, L.; Ferreira, I. Cotton and cardoon byproducts as potential growing media components for *Cichorium spinosum* L. commercial cultivation. *J. Clean. Prod.* **2019**, *240*, 118254. [CrossRef]
106. Alhaithloul, H.A.; Soliman, M.H.; Ameta, K.L.; El-Esawi, M.A.; Elkelish, A. Changes in ecophysiology, osmolytes, and secondary metabolites of the medicinal plants of *Mentha piperita* and *Catharanthus roseus* subjected to drought and heat stress. *Biomolecules* **2020**, *10*, 43. [CrossRef] [PubMed]

horticulturae

MDPI

Article

Biochemical and Morphological Traits of Wild Myrtle Populations for Horticultural Use

Donya Shahbazian [1], Akbar Karami [1,*] and Nazim S. Gruda [2]

1 Department of Horticultural Sciences, School of Agriculture, Shiraz University, Shiraz 71441, Iran; shahbaziandonya@gmail.com
2 Department of Horticultural Sciences, INRES—Institute of Crop Science and Resource Conservation, University of Bonn, 53113 Bonn, Germany; ngruda@uni-bonn.de
* Correspondence: akarami2004@gmail.com or akbarkarami@shirazu.ac.ir; Tel.: +98-713-2286133

Abstract: Myrtle (*Myrtus communis* L.), an evergreen shrub belonging to the Myrtaceae family, is widely valued for its applications in the cosmetic, hygienic, and medicinal industries. This plant produces berries in two colors—white and black—with black berries receiving increasing attention due to their superior antioxidant properties. This study investigates black myrtle berries' morphological and biochemical characteristics from eight prominent natural habitats in Fars Province, southwestern Iran. The results reveal significant variations in the morphological (such as fruit length and diameter, fruit length-to-diameter ratio, seed length and diameter, fruit weight, pulp weight, seed weight, pulp-to-seed ratio, and number of seeds) and biochemical attributes among the studied populations. The longest fruit was from the Kherqeh (KH) (8.29 mm) population, while the shortest was from the Baghnari (BN) (5.85 mm) population. The largest fruit diameter was also from KH (6.83 mm), which did not show a significant difference compared to the Zanjiran (ZF) population, while the smallest fruit diameter was from BN (5.12 mm), which did not differ significantly from the Kavar (KA), Simakan (SM), Kouhmareh Road (JK), or Atashkadeh (AT) populations. Notably, all populations exhibited high levels of phenolic compounds, ranging from 660 to 1846 mg per 100 g of fresh weight, and potent antioxidant activity, as indicated by low half-maximal inhibitory concentrations ranging from 0.018 to 0.187 mg per gram. Correlation analyses further demonstrated that altitude and specific soil properties influenced the biochemical traits of the berries to varying degrees. These findings offer valuable scientific insights for selecting and utilizing specific myrtle populations in horticulture, particularly for breeding programs to maximize antioxidant properties and phenolic content.

Keywords: *Myrtus communis*; antioxidant activity; phenolic compounds; wild berries

1. Introduction

Myrtle (*Myrtus communis* L.), commonly known as common myrtle, belongs to the Myrtaceae family. This plant is native to Mediterranean regions, including Portugal, southern France, Spain, southern Turkey, southern Italy, Greece, Israel, Lebanon, Iran, Afghanistan, Libya, Morocco, Algeria, and Tunisia [1,2]. It is a shrub or small tree with dense foliage, reaching heights of over 2.5 m. The leaves are small, leathery, pointed or oval-shaped, dark green, and morphologically classified into three forms—decussate, spiral, and tricussate—with the decussate arrangement being the most common [3]. The fruit appears in black and white varieties, containing numerous seeds (Figure 1). Insects and wind pollinate these plants, while birds generally disperse their seeds [4].

(a) (b)

Figure 1. The fruits of myrtle are (**a**) black or (**b**) white. The black fruit is more common and has recently gained attention due to its higher antioxidant properties, while the white fruit is rare. The ripening of the fruits occurs in autumn, between October and November (photos: Donya Shahbazian).

The nutritional and medicinal value of myrtle fruit has been extensively documented. This fruit is used in alcoholic beverages and the well-known Italian soft drink "Chinotto" in Sardinia, Italy, due to its high concentration of oenothein B [5]. Myrtle has been associated with various medicinal benefits, including wound healing in diabetic patients, potential support in managing respiratory symptoms observed during COVID-19, enhancing thyroid function, healing burn wounds, lowering high blood pressure, alleviating respiratory and pulmonary disorders, reducing stress and insomnia, treating oral ulcers, and protecting the liver [5,6]. Research by Cruciani et al. [7] indicates that myrtle may slow the aging process in cells exposed to oxidative stress and can help prevent age-related diseases. Additionally, myrtle extract has been found to increase the post-harvest shelf life of fruits and reduce microbial loads in salmon and red meat. When added to milk, myrtle fruit extract enhances nutritional value and lowers blood urea levels without affecting milk coagulation or shelf life. Moreover, the extract derived from the industrial waste of myrtle fruit can protect cells from oxidative stress damage, making it valuable in food formulations [5]. Myrtle fruit is rich in anthocyanins (31 types), flavonoids (11 types), antioxidants, tannins, vitamins (A, E, B), and essential elements such as selenium, zinc, iron, potassium, phosphorus, fiber, protein, fat, oleic acid, linolenic acid, palmitic acid, and linoleic acid [5,7–10]. The European Scientific Associations have recognized myrtle fruit as a significant source of anthocyanins, flavonoids, and antioxidants [11].

The biochemical composition of the fruit, particularly its phenolic and antioxidant compounds, is primarily influenced by genotype, growing conditions, climatic factors, fruit maturity stage, and soil properties, which are typically enhanced under environmental stress [2,12–14]. Despite the myrtle plant's widespread distribution in Iran, research on its fruit's morphological and biochemical characteristics remains limited. This study evaluates the biochemical and morphological characteristics of black myrtle fruit collected from eight key Fars Province, Iran, distribution areas. Additionally, it examines the effects of altitude and soil characteristics on the fruit's biochemical properties. The results provide a scientific foundation for future research in breeding programs and food industries.

2. Materials and Methods

2.1. Plant Material and Fruit Collection

Black myrtle fruit samples were collected from mid-November to early December 2024 from eight natural habitats in Fars Province (Zanjiran (ZF), Kouhmareh Road (JK), Tang Kherqeh (KH), Kavar (KA), Simakan (SM), Atashkadeh (AT), Sarvegarm (SF), and Baghnari (BN)) (Table 1) at full maturity. Three replications per population were conducted, with data collected from nine individuals per population. The fruit samples were transferred to the Medicinal Plants Laboratory of the School of Agriculture at Shiraz University. After removing external contaminants, 60 fruits from each replication were set aside for morphological measurements, while 300 g per replication was stored at −80 °C for biochemical analysis. Soil samples were also collected from 0–30 cm depth and analyzed at the Soil Science Laboratory of Shiraz University (Table 1).

Table 1. Collection site, soil, and geographical characteristics of myrtle populations from southern Iran.

NO.	Accession Name	Collection Site	Latitude	Longitude	Altitude (m)	Clay	Silt	Sand	Organic Matter (%)	pH	EC
1	ZF	Zanjyran, Firozabad, Fars, Iran	29°03′38″ N	52°38′33″ E	1697	5.4	64.6	30	2.18	7.43	1.059
2	JF	Juddehkuhmareh, Firozabad, Fars, Iran	29°06′37″ N	52°33′29″ E	1570	7.4	38	54.6	2.31	7.11	0.71
3	KH	Khergheh, Firozabad, Fars, Iran	28°54′22″ N	52°22′22″ E	1497	11.96	8	90.4	1.15	7.62	1.1
4	KA	Kavar, Fars, Iran	29°07′42″ N	53°35′31″ E	1525	7.95	28	64.04	2.62	7.35	1.14
5	SM	Simakan, Jahrom, Fars, Iran	28°41′32″ N	52°57′42″ E	1344	13.4	2.6	64.06	2.45	7.12	1.05
6	AT	Atashkadeh, Fars, Iran	28°53′23″ N	52°32′31″ E	1478	1.96	10	88.04	1.14	7.42	1.5
7	SF	Sarvegarm, Fasa, Fars, Iran	29°18′05″ N	53°23′09″ E	1743	5.96	30	64.04	1.67	7.28	1.5
8	BN	Baghnari, Noorabad mamasani, Fars, Iran	30°11′10″ N	51°47′58″ E	1293	7.96	20	72.04	2.05	7.34	0.97

The populations used, with their abbreviations, were Zanjiran (ZF), Kouhmareh Road (JK), Tang Kherqeh (KH), Kavar (KA), Simkan (SM), Atashkadeh (AT), Sarvegarm (SF), and Baghnari (BN).

2.2. Chemicals and Reagents

The chemicals and reagents were purchased from Sigma-Aldrich (St. Louis, MO, USA) and Fluka (Heidelberg, Germany). All other chemicals and reagents used were of the highest purity commercially available.

2.3. Measurement of Morphological Characteristics

2.3.1. Weight of Flesh, Seeds, and Fruit

To determine the weight of flesh, seeds, and fruit, 20 fruits from each individual were selected and measured using a digital scale with an accuracy of 0.001 g. Two groups of fruits, consisting of 20 from each individual, were used to determine the weight accurately. One group was measured for pulp and seed weight, while the other was measured separately for total fruit weight.

2.3.2. Length and Diameter of Fruit and Seeds

To determine the length and diameter of the fruit and seeds, 20 fruits from each individual were randomly selected and measured using a digital caliper with an accuracy of 0.01 mm. For the fruit length-to-diameter ratio, the length data were divided by the diameter of the fruits.

2.3.3. Number of Seeds

To determine the number of seeds, 20 fruits from each individual were randomly selected and counted.

2.4. Measurement of Biochemical Characteristics

2.4.1. Total Phenol Content (TPC)

A methanol solvent (1:3 ratio) was used to extract phenolic compounds. Initially, 1 g of fruit flesh was mixed with 3 mL of ethanol and stored in a cold room for 12 to 14 h. The sample was then thoroughly mixed with methanol using a mortar and pestle and transferred to a sealed tube. The samples were centrifuged for 15 min at 6000 rpm, and the supernatant was used to measure phenol content. The Folin–Ciocalteu method was employed, where 700 µL of the methanolic extract was combined with 140 µL of 50% Folin reagent and kept at room temperature for 3 min. Then, 700 µL of 2% sodium carbonate was added, and the mixture was left at room temperature for 30 min before being centrifuged for 5 min at 13,400 rpm. The absorbance was read at 750 nm using an Epoch device (BioTek, Charlotte, VT, USA, with Gen5™ Data Analysis Software). The absorbance data were converted to various concentrations of gallic acid (prepared from the Sigma brand in Germany) by plotting the standard curve of gallic acid using Excel software. The recorded absorbances from the samples were placed on the y-axis, and the x-axis, or concentration, was obtained. The results were expressed as milligrams of gallic acid equivalent per 100 g of fresh weight (mg GAE/100 gFW).

2.4.2. Diphenyl-1-picrylhydrazyl (DPPH) Radical Scavenging Capacity Assay

The percentage of free radical inhibition was determined using the DPPH assay method by Burits et al. [15]. A volume of 100 µL of the methanolic extract at different concentrations (0–250 mg/mL) was added to 5 mL of DPPH solution. In the control (blank) solution, 100 µL of methanol was added instead of the extract. The solutions were vigorously shaken for 10 s and kept for 30 min at room temperature in the dark. Absorbance was measured at 517 nm using a BioTek Epoch device (USA). The percentage of free radical inhibition (I%) was calculated using the following equation:

$$I\% = [(A517\ \text{blank} - A517\ \text{sample})/A517\ \text{blank}] \times 100$$

A517 blank and A517 sample refer to the absorbance of the blank and the sample at 517 nm, respectively. The IC50 value, representing the concentration required to neutralize 50% free radicals, was used to compare antioxidant activity.

2.4.3. Total Anthocyanin Content

A 10 mg sample of frozen fruit flesh was placed in a centrifuge tube, and 1 mL of acidic methanol (1% hydrochloric acid) was added. The extract was stored overnight at 4 °C. Then, it was centrifuged for 10 min at 10,500 rpm and read at two wavelengths of 530 and 657 nm [16].

2.4.4. Total Soluble Solids (TSS)

A quantity of 0.1 gr of dried fruit powder was placed in a Falcon tube, and 10 mL of 80% ethanol was added and mixed. The tubes were centrifuged for 10 min at 5000 rpm, then the supernatant was separated, and another 10 mL of 80% ethanol was added to the sediment in the tube and centrifuged again. The supernatant was added to the previous solution. A 5% phenol solution was prepared. A total of 1 mL of the target solution and 1 mL of Folin's solution were placed in a test tube, and 5 mL of concentrated sulfuric acid was immediately added. The resulting mixture was stirred for 15 s. Afterward, the tubes were placed in ice water to cool. Finally, their absorbance was measured at a wavelength of 490 nm using a visible spectrophotometer. Pure D-glucose was used to draw the standard curve. For this purpose, solutions containing glucose with concentrations ranging from 10 to 200 μg mL^{-1} were prepared, and their absorbance was read at a wavelength of 490 nm using a BioTek Epoch device (USA) [17].

2.4.5. Titratable Acidity (TA)

To extract total acid, 2.5 gr of fruit was thoroughly crushed using a mortar and pestle, and then 10 mL of distilled water was added. The extract was then centrifuged, and titration was performed with 0.1 standard NaOH to reach a pH of 2.8 in the presence of phenolphthalein. The results were expressed in terms of malic acid [18]. The percentage of acid was calculated using the following equation:

$$\% \text{ Acid} = N \times V1 \times \text{Eq.wt}/V2 \times 10$$

V1 = normalized NaOH, V2 = sample size, Eq.wt = molecular weight of the predominant acid (malic acid $C_4H_6O_5$ = 134.09 g/mol).

2.4.6. Total Protein Content

To measure protein concentration using the Bradford method, 100 μL of protein extract was added to test tubes containing 5 mL of Bradford reagent and was quickly vortexed. After 2 min, the absorbance was read using an Epoch device (USA) at a wavelength of 595 nm. Bovine serum albumin was used to construct the standard curve. For this purpose, solutions containing bovine serum albumin at various concentrations were prepared, and their absorbance was read at a wavelength of 595 nm. The total protein content in the samples was calculated and reported based on the absorbance values read from the standard curve [19].

2.5. Statistical Analysis

A completely randomized design with three replications was used for the experiment. All data were analyzed using SAS software (version 4.2), and means were compared using the LSD test at 1% and 5% significance levels and correlation analysis between traits.

3. Results and Discussion

3.1. Evaluation of Morphological Characteristics of Myrtle Fruit from Different Populations

The results of the evaluation of morphological characteristics are presented in Table 2. The longest fruit was from KH (8.29 mm), while the shortest was from BN (5.85 mm). The largest fruit diameter was also from KH (6.83 mm), which did not show a significant difference compared to the ZF population, while the smallest fruit diameter was from BN (5.12 mm), which did not differ significantly from the KA, SM, JF, and AT populations. The highest length-to-diameter ratio was found in the SF region (1.25), while the lowest ratio was in ZF (1.06). The BN population had the longest seed (3.22 mm), while the shortest seed was from SM (2.21 mm), which did not show a significant difference compared to KH. BN had the largest seed diameter (1.11 mm) among the different populations, while the smallest seed diameter was from SM (0.81 mm), which did not differ significantly from KA. The heaviest fruit was from KH (2.35 g), and the lightest was from BN (1.72 g). The heaviest seed was from KH (1.19 g), while BN had the lowest seed weight (0.81 g) among the studied populations. The highest flesh weight of the fruit was from SF (1.82 g), while the lowest flesh weight was from BN (1.51 g), which did not show a significant difference compared to SM. The highest flesh-to-seed ratio was found in KH (0.78), while the lowest was in SM (0.50), which did not differ significantly from BN. The highest number of seeds was recorded in ZF (5.1), which did not show a significant difference compared to the KH population. The lowest number of seeds was found in BN (1.67). A high number of seeds in the fruit and a low flesh-to-seed ratio negatively affect fresh consumption and processing industries, such as in beverage production, for myrtle. This is because the amount of tannin in the fruit flesh reaches a minimum at ripening; however, the seeds retain tannins until the fruit ripens and enter beverages during processing, causing an astringent taste [2]. However, the high tannin content in the pharmaceutical industry, especially ellagitannins found in seeds, can be advantageous. Ellagitannins are a complex class of bioactive compounds or hydrolyzable tannins, identified by one or more units of hexa-hydroxy diphenyl, found in limited quantities in the seeds of dicotyledonous angiosperm myrtle [20]. These compounds have been shown to improve overall health, reduce inflammation, and lower the risk of cardiovascular diseases, neurological disorders, and cancer. There have been few studies on the morphological characteristics of myrtle fruit. Wahid et al. [21] reported significant differences in myrtle fruit morphology among populations in Morocco. Their findings on the morphological characteristics of the fruit from three growth sites are consistent with our results. They also stated that the differences in fruit morphology are primarily of genetic origin. This study examined the morphological characteristics of two varieties, "Barbara" and "Daniela". "Barbara" produced fewer seeds (1.8) and had a higher flesh-to-seed ratio (3.19) than "Daniela", making it more suitable for commercial fresh consumption. Meanwhile, "Daniela", due to its significant vegetative growth potential, can be used as a dual-purpose variety, ideal for both leaf and fruit production in the pharmaceutical industry and beverage production [22]. An examination of the morphological characteristics of myrtle fruit across seven growth sites in Italy showed significant differences in fruit morphology across regions. The fruit length varied between 9.03 mm and 18.7 mm, the fruit diameter ranged from 5.74 mm to 8.22 mm, and the length-to-diameter ratio varied from 1.10 to 1.29 [23]. The results of this study indicated a high degree of morphological variation in ecologically similar sites, possibly due to genetic diversity or an adaptive mechanism in response to climatic conditions [23,24].

Table 2. Morphological characteristics of myrtle fruits in Fars myrtle populations. The table presents the average fruit characteristics derived from the 20 sampled fruits from each individual.

Region	Fruit Length (mm)	Fruit Diameter (mm)	Length/ Diameter Ratio	Seed Length (mm)	Seed Diameter (mm)	Fruit Weight (g)	Pulp Weight (g)	Seed Weight (g)	NO. Seeds
BN	5.85 ^{g}	5.12 ^{d}	1.13 ^{e}	3.22 ^{a}	1.11 ^{a}	1.72 ^{g}	1.51 ^{f}	0.81 ^{h}	1.67 ^{f}
KA	7.47 ^{c}	5.80 ^{bcd}	1.28 ^{a}	2.47 ^{e}	0.82 ^{e}	2.17 ^{b}	1.66 ^{c}	1.07 ^{b}	3.66 ^{d}
SM	6.68 ^{e}	5.76 ^{bcd}	1.16 ^{c}	2.21 ^{f}	0.81 ^{e}	1.93 ^{e}	1.54 ^{f}	1.05 ^{c}	3.32 ^{e}
JF	6.49 ^{f}	5.72 ^{bcd}	1.13 ^{e}	2.53 ^{d}	0.91 ^{bc}	2.08 ^{c}	1.77 ^{b}	0.93 ^{f}	3.61 ^{d}
ZF	6.79 ^{b}	6.34 ^{ab}	1.06 ^{f}	2.71 ^{c}	0.85 ^{d}	1.78 ^{f}	1.61 ^{d}	0.95 ^{e}	5.1 ^{a}
KH	8.29 ^{a}	6.83 ^{a}	1.23 ^{b}	2.23 ^{f}	0.90 ^{c}	2.35 ^{a}	1.58 ^{e}	1.19 ^{a}	5.06 ^{a}
AT	6.66 ^{e}	5.52 ^{cd}	1.21 ^{b}	2.45 ^{e}	0.92 ^{bc}	1.93 ^{d}	1.42 ^{g}	1.01 ^{d}	4.64 ^{b}
SF	7.67 ^{b}	6.05 ^{bc}	1.28 ^{a}	2.93 ^{b}	0.93 ^{b}	2.08 ^{c}	1.82 ^{a}	0.91 ^{g}	4.32 ^{c}

Mean values with the different letters within a column are significantly different ($p < 0.01$); LSD test. The populations used, with their abbreviations, were Zanjiran (ZF), Kouhmareh Road (JK), Tang Kherqeh (KH), Kavar (KA), Simkan (SM), Atashkadeh (AT), Sarvegarm (SF), and Baghnari (BN).

3.2. *Evaluation of Biochemical Characteristics of Myrtle Fruit from Different Populations in Fars Province*

The results of the biochemical characteristics evaluation of the fruit are presented in Table 3. The IC50 values ranged from 0.018 mg of dry matter per milliliter to 0.882 mg per milliliter across different regions. The IC50 values in various areas were compared with the IC50 of the synthetic antioxidant butylated hydroxytoluene (BHT)

(IC50 = 0.0407 mg/mL DW) as a positive control. The lowest IC50 was recorded for SF (IC50 = 0.018), which was lower than that of BHT, indicating that the antioxidant activity of SF fruit was higher than that of the synthetic antioxidant BHT. The highest IC50 was associated with AT, which did not significantly differ from ZF, KH, or JF. Numerous studies have reported the high antioxidant activity of myrtle fruit [2,11,13,25,26]. An investigation into the differences in antioxidant activity among various parts of the fruit (whole fruit, seeds, and pericarp) of *Myrtus communis* var. *italica* in Tunisia showed that all parts exhibited high antioxidant activity. Among these, the seeds had the highest antioxidant activity (IC50 = 8 μg/mL), even greater than the synthetic antioxidant BHT (IC50 = 25 μg/mL). The antioxidant activities of the other parts of the fruit were close to that of the synthetic antioxidant [26], which was consistent with our results. The phenol content ranged from 660 to 1846 mg per 100 g of fresh material. BN had the highest phenol content among the studied areas (1846 mg per 100 g of fresh material), while KH had the lowest (660 mg per 100 g of fresh material), which was not significantly different from KA and JF. Melito et al. [23] reported the phenol content of myrtle fruit in seven different habitats in Sicily, Italy, ranging from 2466 to 3800 mg/100 g DW. Additionally, a study examining the phenol content in different parts of *Myrtus communis* var. *italica* fruit found that the phenol content varied among different parts of the fruit. The phenol content in the seeds was 23.87 mg/g DW, in the pericarp (fruit flesh), 2.76 mg/g DW, and in the whole fruit, 13.73 mg/g DW [27]. Since this study measured only the phenol content in the pericarp, the pericarp of the myrtle fruits analyzed here had a higher phenol content (825–1824 mg/100 g DW) than those in the previous study. The phenol content of the two cultivars, "Barbara" and "Daniela", grown in Alghero, Italy, over two consecutive years and at different stages of fruit development, showed significant variation each year and at different stages of growth. The phenol content of the fruit decreased as it approached the ripening stage [27].

Table 3. Biochemical characteristics of myrtle fruits in Fars myrtle populations.

Region	IC50 (mg dw/mL)	TPC (mg/100 gfw)	Anthocyanin (mg/100 gfw)	TSS (%)	TA (%)	TSS/TA Ratio	Protein (mg/g FW)
JF	0.87 [a]	820 [de]	102.2 [d]	2.4 [a]	0.1 [cd]	24 [a]	1.33 [ab]
ZF	0.84 [a]	1002 [cd]	218.3 [c]	1.87 [bc]	0.13 [bc]	15 [bc]	1.70 [a]
AT	0.88 [a]	1212 [c]	200.8 [c]	2.15 [ab]	0.17 [a]	13 [cd]	0.60 [de]
KH	0.87 [a]	706 [d]	121.3 [d]	1.033 [d]	0.16 [ab]	10 [f]	0.74 [cd]
KA	0.13 [c]	851 [de]	122 [d]	1.83 [c]	0.12 [c]	16 [b]	1.08 [bc]
SM	0.25 [b]	1218 [c]	53.1 [e]	0.8 [d]	0.07 [d]	12 [de]	1.022 [bc]
SF	0.018 [d]	1600 [b]	311 [a]	1.04 [d]	0.12 [c]	9 [ef]	0.32 [ef]
BN	0.11 [c]	1832 [a]	271.5 [b]	1.84 [c]	0.18 [a]	10 [de]	0.027 [f]

Mean values with different letters within a column are significantly different ($p < 0.01$); LSD test. The abbreviations stand for total phenol content (TPC), total soluble solids (TSS), and titratable acidity (TA).

A study on four genotypes of myrtle fruit collected from different regions of Antalya, Turkey (two black genotypes and two white genotypes), revealed that black fruits contained higher levels of phenolic compounds than white fruits. Additionally, the phenol content varied even among genotypes with the same fruit color, possibly due to genetic and ecological factors. The predominant phenolic compounds in black myrtle fruit included chlorogenic acid, syringic acid, naringin, gallic acid, caffeic acid, and p-hydroxybenzoic acid. In contrast, in white fruit, naringin, gallic acid, and chlorogenic acid were present [28].

The anthocyanin content ranged from 1.53 to 311 mg per 100 g of fresh weight. The highest anthocyanin content (311 mg per 100 g of fresh weight) was found in the SF genotype, while the lowest was in the SM region (53 mg per 100 g of fresh weight). Fadda and Mulas [18] reported that anthocyanin content in myrtle fruit varied over consecutive years and at different stages of growth. In 2005, the anthocyanin content was 117 mg/100 g FW; in 2006, it decreased to 70 mg/100 g FW. With the onset of fruit ripening, the level of anthocyanins increased. The variation in anthocyanin levels over consecutive years may be due to climatic conditions. Temperature is a significant factor influencing anthocyanin production, as lower nighttime temperatures enhance the activity of the enzymes involved in the biosynthetic pathway of anthocyanins [29]. A study on the anthocyanin content in different parts of the fruit (pericarp, whole fruit, seeds) of *M. communis* var. *italica* in Tunisia showed that the highest anthocyanin content (4.64 mg/g FW) was found in the entire fruit. The anthocyanin content in the fruit's pericarp was 3.47 mg/g DW, while the seeds contained no anthocyanins. The soluble solid content varied between 0.8% and 2.4%. JF had the highest soluble solid content (2.5%), which was not significantly different from the AT region. The lowest soluble solid content was measured in SM (0.81%), which did not differ substantially from the KH and SF regions. The acid content varied between 0.07% and 0.18%. The highest acid content was associated with BN (0.18%), which did not significantly differ from KH and AT. The lowest acid content was found in SM (0.07%), although it was not significantly different from JF. The ratio of soluble solids to acid varied between 6.31 and 24. The JF population had the highest ratio of soluble solids to acid (22.42) among the different regions. The lowest ratio of soluble solids to acid was found in KH, which did not differ significantly from SF. The protein content varied between 0.27 and 1.7 mg/g DW. The highest protein content was found in ZF (1.7 mg/g DW), which did not differ significantly from JF. The lowest protein content was associated with the BN region (0.27 mg/g DW), which did not differ substantially from the SF region. The fruit sugar content of the two cultivars, Barbara and Daniela, from Italy, was determined. The sugar content values at full ripeness in the two varieties, "Barbara" and "Daniela", were 8.28% and 7.56%, respectively. The authors also stated that the sugar content increases as the fruit ripens, and there was no significant difference in sugar content over two consecutive years. In another study, the crude protein content of myrtle fruit was reported to be 17.41% [30].

Çakmak and colleagues [10] investigated the effects of freezing, sun drying, and microwave drying on the nutritional value of wild and cultivated fruit populations in Osmaniye, Turkey. The results showed that both types of fruit were rich in vitamins A, B, and E and essential elements such as selenium, zinc, iron, and copper. The levels of vitamins A and E, cyanocobalamin, thiamine, nicotinamide, and folic acid were high in wild and cultivated fruits. In contrast, the level of riboflavin was higher in cultivated fruits compared to wild ones. No significant differences were observed between fresh and frozen fruits regarding biochemical parameters. Thus, freezing was identified as the most suitable method for preserving myrtle fruits.

AlJuhaimi and colleagues [8] reported potassium, phosphorus, and iron levels in fruits collected from Mersin, Turkey, as 5212.77, 949.08, and 3.76 mg/kg, respectively. A study comparing the antioxidant activity, phenolic compounds, and fatty acids of myrtle fruit, tree bark, and seeds revealed that all three exhibited high antioxidant activity and significant amounts of phenolic compounds. Therefore, these products can serve as sources of natural antioxidants and potential substitutes for synthetic antioxidants. The highest levels of fatty acids in the bark and seeds were those of linoleic acid, while the lowest were those of γ-linolenic and α-linolenic acids in both parts [31].

3.3. Correlation Analysis Between Traits

The correlation analysis (Table 4) showed a strong positive relationship (0.435) between phenol content and altitude above sea level and a strong negative relationship (−0.425) with soil EC. Additionally, there was a strong negative relationship (−0.544) between clay content and IC50 and a strong positive relationship (0.482) with anthocyanins. A strong positive relationship was also found between TSS/TA and silt (0.507), sand (0.501), soil pH (0.527), and organic matter (0.541). There was a robust negative relationship (−0.708) between TA and clay, a strong positive relationship (0.511) with sand, and a strong negative relationship (−0.451) with organic matter. Different results have been reported regarding the effect of altitude on phenol and antioxidant content. Some studies indicated an increase, others indicated a decrease, and some showed no effect [29,32,33]. Medda et al. [2] reported that altitude has a varying impact on the phenolic content and antioxidant levels in different parts of myrtle. As altitude increased, the levels of phenols and antioxidants in the leaves increased, whereas in the fruit, they decreased. Some studies suggest that genetic factors have a significant influence, more than environmental factors, on the biochemical characteristics of plants [23,24,29]. According to the correlation results (Table 4), the levels of phenols and proteins increased with altitude, which may be one of the mechanisms by which the plant adapts to adverse environmental factors such as high light intensity and ultraviolet radiation [26]. Depending on whether leaves or fruits are the primary targets, selecting specific altitudinal ranges for cultivation may optimize the desired biochemical traits. It has been established that environmental factors can influence the biosynthesis, accumulation, and qualitative composition of phenolic compounds in various plant organs. The increase in the concentration of phenolic compounds under biotic and abiotic stress is due to the PAL gene's enhanced expression and the PAL enzyme's activity, which provides a substrate for other reactions [9]. An assessment of the impact of environmental factors on phenolic compounds, tannins, and anthocyanins in 22 cultivars in Italy showed that high temperatures negatively affect phenolic compounds. Anthocyanin content increases with elevation but decreases with rising temperatures. This reduction may be due to lower expression of the genes responsible for anthocyanin production or the degradation of anthocyanins. Additionally, the levels of anthocyanins and tannins positively correlated with rainfall, while total phenols remained unaffected. Water and light availability had a more significant impact on the fruit's tannin content than temperature [9]. Overall, the concentration of secondary metabolites may serve as a genotypic marker for identifying the origin of cultivars or as an indicator of quality. For example, in kiwi, the level of phenolic compounds can indicate the specific growth site among five locations in China. Besides climate, other abiotic factors may also affect the content of secondary metabolites [34]. Soil is a complex physical, chemical, and biological environment, and various factors influence the relationship between soil, water, and plants [35]. Since roots are the primary agents for absorbing external materials into the plant system, soil edaphic conditions are crucial in producing biochemical compounds and metabolites [30,36]. Ogundola et al. [37] stated that soil texture significantly impacts the biochemical and phytochemical composition of *Solanum nigrum*, with the highest levels of tannins, flavonoids, anthocyanins, and antioxidant activity measured in plants grown in silty clay loam soils. They also noted that the increase in biochemical and phytochemical compounds in silty clay loam soils may be attributed to adequate moisture retention, which is essential for biochemical and phytochemical functions. Myrtle naturally grows in sandy loam soil with high organic matter, as shown in Table 1 and our previous study [38].

Table 4. Correlation of biochemical characteristics of myrtles and some environmental factors.

Altitude (m)	1													
Clay (Soil)	0.378	1												
Silt (Soil)	0.641 **	0.19	1											
Sand (Soil)	−0.545	−0.393	−0.97 **	1										
Organic Matter % (Soil)	0.026	0.45 *	0.398	0.248	1									
pH	−0.105	−0.103	−0.174	0.185	−0.171	1								
EC (Soil)	0.273	−0.375	0.236	0.261	−0.058	0.212	1							
TPC	0.435 *	−0.036	−0.104	0.106	0.023	0.05	−0.435 *	1						
TSS	0.396	−0.429	0.331	−0.206	0.382	−0.348	−0.132	0.136	1					
Protein	0.751 **	0.161	0.7 **	−0.724 **	0.247	−0.259	0.073	0.441 *	0.106	1				
IC50	−0.227	−0.544 *	0.118	−0.02	−0.042	−0.09	0.335	−0.367	0.275	0.369	1			
TA	−0.069	−0.708 **	−0.353	0.511 *	−0.451 *	0.219	0.194	0.243	0.26	−0.486 *	0.175	1		
TSS/TA	0.431	−0.071	0.507 *	0.501 *	0.541 *	0.527 *	−0.329	−0.087	0.732 **	0.445 *	0.152	−0.418	1	
Anthocyanin	−0.024	0.482 *	0.15	0.019	−0.092	0.291	−0.02	0.297	0.412	−0.339	0.002	0.69 **	−0.173	1

** and * are significantly different ($p < 0.01$ and $p < 0.05$ respectively). The abbreviations stand for total phenol content (TPC), total soluble solids (TSS), and titratable acidity (TA).

4. Conclusions

The present research highlights the considerable diversity in the morphology and biochemistry of myrtle fruit populations in southwestern Iran. It also emphasizes the influence of genotypes and environmental factors, such as altitude and soil conditions, on fruit biochemical composition, offering valuable insights for selecting suitable cultivation areas. The studied populations demonstrated high levels of phenolic compounds and antioxidant activity, with the SF population showing more significant antioxidant levels. Myrtle fruits hold great promise as a horticultural crop and a natural alternative to synthetic antioxidants, offering potential applications in medicine and fresh consumption. This research also provides a basis for further breeding research in these areas.

Author Contributions: Conceptualization, A.K. and N.S.G.; methodology, D.S.; investigation, D.S.; writing—original draft preparation, D.S.; writing—review and editing, A.K. and N.S.G. All authors have read and agreed to the published version of the manuscript.

Funding: This research was funded by Shiraz University (Fund NO: 96GCU1M154198).

Data Availability Statement: The raw data supporting the conclusions of this article will be made available by the authors on request.

Conflicts of Interest: The authors declare no conflicts of interest.

References

1. Aslan, D.; Alan, B.; Yay, N.Ö.; Karaoğlu, S.Y.; Ertaş, B.; Şen, A.; Ceylan, C.; Akbay, T.T.; Ercan, F.; Atasoy, B.M. Neuroprotective effect of *Myrtus communis* against ionizing radiation-induced brain injury: Insights from histopathological and biochemical analysis in rats. *J. Radiat. Res. Appl. Sci.* **2024**, *17*, 101082. [CrossRef]
2. Medda, S.; Fadda, A.; Dessena, L.; Mulas, M. Quantifying total phenols, tannins, anthocyanins content in *Myrtus communis* L. and antioxidant activity evaluation in function of plant development stages and altitude of origin site. *Agronomy* **2021**, *11*, 1059. [CrossRef]
3. Zilkah, S.; Goldschdmidt, E.E. Myrtle (*Myrtus communis* L.)—A native Mediterranean and cultured crop species. In *Medicinal and Aromatic Plants of the Middle-East*; Springer: Dordrecht, The Netherlands, 2014; pp. 253–267.
4. Giampieri, F.; Cianciosi, D.; Forbes-Hernández, T.Y. Myrtle (*Myrtus communis* L.) berries, seeds, leaves, and essential oils: New undiscovered sources of natural compounds with promising health benefits. *Food Front.* **2020**, *1*, 276–295. [CrossRef]
5. Gorjian, H.; Khaligh, N.G. Myrtle: A versatile medicinal plant. *Nutrire* **2023**, *48*, 10. [CrossRef] [PubMed]
6. Sumbul, S.; Ahmad, M.A.; Asif, M.; Akhtar, M. *Myrtus communis* Linn.—A review. *Indian J. Nat. Prod. Resour.* **2011**, *2*, 395–402.
7. Cruciani, S.; Garroni, G.; Ginesu, G.C.; Fadda, A.; Ventura, C.; Maioli, M. Unravelling cellular mechanisms of stem cell senescence: An aid from natural bioactive molecules. *Biology* **2020**, *9*, 57. [CrossRef]
8. AlJuhaimi, F.; Kulluk, D.A.; Ahmed IA, M.; Özcan, M.M.; Karrar, E. Investigation of accumulation of element contents in some wild and cultivated dried fruits. *Biol. Trace Elem. Res.* **2024**, *203*, 544–548. [CrossRef]
9. Medda, S.; Fadda, A.; Mulas, M. Climate variables of the sites of origin and genotype influence on phenolic compounds accumulation in cultivars of *Myrtus communis* L. *Horticulturae* **2022**, *8*, 928. [CrossRef]
10. Çakmak, M.; Bakar, B.; Özer, D.; Geckil, H.; Karatas, F.; Saydam, S. Investigation of some biochemical parameters of wild and cultured *Myrtus communis* L. fruits subjected to different conservation methods. *J. Food Meas. Charact.* **2021**, *15*, 983–993. [CrossRef]
11. Tuberoso CI, G.; Rosa, A.; Bifulco, E.; Melis, M.P.; Atzeri, A.; Pirisi, F.M.; Dessì, M.A. Chemical composition and antioxidant activities of *Myrtus communis* L. berries extracts. *Food Chem.* **2010**, *123*, 1242–1251. [CrossRef]
12. Wannes, W.A.; Marzouk, B. Characterization of myrtle seed (*Myrtus communis* var. *baetica*) as a source of lipids, phenolics, and antioxidant activities. *J. Food Drug Anal.* **2016**, *24*, 316–323. [PubMed]
13. Messaoud, C.; Boussaid, M. *Myrtus communis* berry color morphs: A comparative analysis of essential oils, fatty acids, phenolic compounds, and antioxidant activities. *Chem. Biodivers.* **2011**, *8*, 300–310. [CrossRef] [PubMed]
14. Shahbazian, D.; Karami, A.; Raouf Fard, F.; Eshghi, S.; Maggi, F. Essential Oil Variability of Superior Myrtle (*Myrtus communis* L.) Accessions Grown Under the Same Conditions. *Plants* **2022**, *11*, 3156. [CrossRef] [PubMed]
15. Burits, M.; Bucar, F. Antioxidant activity of *Nigella sativa* essential oil. *Phytother. Res.* **2000**, *14*, 323–328. [CrossRef] [PubMed]
16. Hashemi, M.B.; Niakousari, M.; Saharkhiz, M.J. Antioxidant activity of *Satureja bachtiarica* Bunge essential oil in rapeseed oil irradiated with UV rays. *Eur. J. Lipid Sci. Technol.* **2011**, *113*, 1132–1137. [CrossRef]

17. Maatallah, S.; Guizani, M.; Lahbib, K.; Montevecchi, G.; Santunione, G.; Hessini, K.; Dabbou, S. Physiological traits, fruit morphology and biochemical performance of six old fig genotypes grown in warm climates "Gafsa oasis" in Tunisia. *J. Agric. Food Res.* **2024**, *17*, 101253. [CrossRef]
18. Fadda, A.; Mulas, M. Chemical changes during myrtle (*Myrtus communis* L.) fruit development and ripening. *Sci. Hortic.* **2010**, *125*, 477–485. [CrossRef]
19. Bradford, M.M. A rapid and sensitive method for the quantitation of microgram quantities of protein utilizing the principle of protein-dye binding. *Anal. Biochem.* **1976**, *72*, 248–254. [CrossRef]
20. Franco, A.M.; Tocci, N.; Guella, G.; Dell'Agli, M.; Sangiovanni, E.; Perenzoni, D.; Vrhovsek, U.; Mattivi, F.; Manca, G. Myrtle seeds (*Myrtus communis* L.) as a rich source of the bioactive ellagitannins oenothein B and eugeniflorin D_2. *ACS Omega* **2019**, *4*, 15966–15974. [CrossRef]
21. Wahid, N.; Chkichekh, A.; Bakry, M. Morphological traits and essential oil yield variation of three *Myrus communis* L. populations: Implication for domestication. *Direct Res. J. Agric. Food Sci.* **2016**, *4*, 199–207.
22. Mulas, M.; Francesconi AH, D.; Perinu, B.; Fadda, A. 'Barbara' and 'Daniela': Two cultivars for myrtle berries production. *Acta Hortic.* **2001**, *576*, 169–175. [CrossRef]
23. Melito, S.; La Bella, S.; Martinelli, F.; Cammalleri, I.; Tuttolomondo, T.; Leto, C.; Mulas, M. Morphological, chemical, and genetic diversity of wild myrtle (*Myrtus communis* L.) populations in Sicily. *Turk. J. Agric. For.* **2016**, *40*, 249–261. [CrossRef]
24. Gao, S.; Ren, Y.; Masabni, J.; Zou, F.; Xiong, H.; Zhu, J. Influence of geographical and climatic factors on *Quercus variabilis* Blume fruit phenotypic diversity. *Diversity* **2021**, *13*, 329. [CrossRef]
25. Anwar, S.; Ahmed, N.; Al Awwad, N.; Ansari, S.Y.; Wagih, M.E. Myrtle (*Myrtus communis* L.) oils. In *Essential Oils in Food Preservation, Flavor and Safety*; Academic Press: Cambridge, MA, USA, 2016; pp. 581–592.
26. Wannes, W.A.; Mhamdi, B.; Kchouk, M.E.; Marzouk, B. Composition of fruit volatiles and annual changes in the leaf volatiles of *Myrtus communis* var. *baetica* in Tunisia. *Riv. Ital. Sostanze Grasse* **2011**, *88*, 128–134.
27. Aidi Wannes, W.; Saidani Tounsi, M.; Marzouk, B. Morphological and chemical characterization of two wild Tunisian myrtle (*Myrtus communis* L.) populations. *Trends Phytochem. Res.* **2019**, *3*, 231–242.
28. Şan, B.; Yildirim, A.N.; Polat, M.; Yildirim, F. Chemical Compositions of Myrtle (*Myrtus communis* L.) Genotypes Having Bluish-Black and Yellowish-White Fruits. *Erwerbsobstbau* **2015**, *57*, 203–210. [CrossRef]
29. Zeng, Q.; Dong, G.; Tian, L.; Wu, H.; Ren, Y.; Tamir, G.; Yu, H. High altitude is beneficial for antioxidant components and sweetness accumulation of rabbiteye blueberry. *Front. Plant Sci.* **2020**, *11*, 573531. [CrossRef]
30. Aydın, C.; Özcan, M.M. Determination of nutritional and physical properties of myrtle (*Myrtus communis* L.) fruits growing wild in Turkey. *J. Food Eng.* **2007**, *79*, 453–458. [CrossRef]
31. Guo, J.; Yuan, Y.; Dou, P.; Yue, T. Multivariate statistical analysis of the polyphenolic constituents in kiwifruit juices to trace fruit varieties and geographical origins. *Food Chem.* **2017**, *232*, 552–559. [CrossRef]
32. Crespo, P.; Bordonaba, J.G.; Terry, L.A.; Carlen, C. Characterisation of major taste and health-related compounds of four strawberry genotypes grown at different Swiss production sites. *Food Chem.* **2010**, *122*, 16–24. [CrossRef]
33. Guerrero-Chavez, G.; Scampicchio, M.; Andreotti, C. Influence of the site altitude on strawberry phenolic composition and quality. *Sci. Hortic.* **2015**, *192*, 21–28. [CrossRef]
34. Akyüz, M.; Güzel, A.; Elmastas, M. Fatty acid composition and antioxidant capacity of Myrtus (*Myrtus communis* L.). *Malays. Appl. Biol.* **2019**, *48*, 101–112.
35. Pizano, G.S.; Pacheco, C.H.; Salmerón, J.H.; León, R.H. The role of abiotic factors modulating the plant-microbe-soil interactions: Toward sustainable agriculture. A review. *Span. J. Agric. Res.* **2017**, *15*, 13.
36. Nirala, V.K.; Srivastava, A.; Chaudhary, M.K.; Srivastava, S. Effect of edaphic factors on phenolic contents in the natural population of *Ichnocarpus frutescens* (L.) R. Br. collected from Central India and Eastern Ghats. *Biochem. Syst. Ecol.* **2024**, *112*, 104759. [CrossRef]
37. Ogundola, A.F.; Bvenura, C.; Ehigie, A.F.; Afolayan, A.J. Effects of soil types on phytochemical constituents and antioxidant properties of *Solanum nigrum*. *S. Afr. J. Bot.* **2022**, *151*, 325–333. [CrossRef]
38. Shahbazian, D.; Karami, A.; Eshghi, S.; Maggi, F. Variation in the essential oil (EO) contents and compositions of Myrtle (*Myrtus communis* L.) populations collected from natural habitats in the different geographical regions of southern Iran. *J. Essent. Oil Res.* **2018**, *30*, 369–378. [CrossRef]

horticulturae

MDPI

Communication

Variations in Physical and Chemical Characteristics of *Terminalia catappa* Nuts

Shahla Hosseini Bai [1,*], Bruce Randall [1], Repson Gama [2], Basil Gua [3], Doni Keli [4], Kim Jones [5], Brittany Elliott [6] and Helen M. Wallace [6]

1 School of Environment and Science, Griffith University, Nathan, Brisbane, QLD 4111, Australia
2 School of Science, Technology and Engineering, University of the Sunshine Coast, Maroochydore, QLD 4558, Australia
3 Beg's Sol Seed Technology and Supplies, Honiara, Solomon Islands
4 Solfarm Fresh, Honiara, Solomon Islands
5 Cropwatch Independent Laboratories, Wardell, NSW 2477, Australia
6 School of Biology and Environmental Science, Faculty of Science, Queensland University of Technology, Brisbane, QLD 4000, Australia
* Correspondence: s.hosseini-bai@griffith.edu.au

Abstract: Indigenous forest foods have great potential to diversify cropping systems and increase food security and the resilience of food systems to climate change. Underutilised indigenous tree nuts in particular can provide health benefits to local communities and improve livelihoods when commercialised. However, for many tree nut species, there is little knowledge of important kernel characteristics. Kernel size and oil composition are important factors for commercialisation and health benefits, respectively. We assessed kernel attributes of *Terminalia catappa* L. (Combretaceae), a traditional forest food in the Pacific, in the Solomon Islands, Vanuatu and Fiji. We assessed kernel mass and kernel-to-fruit mass ratio, explored the fatty acid profile of oil, and oil stability against oxidation using accelerated ageing at 45 °C for 21 days. The largest kernels were found in the Solomon Islands with an average mass of 1.66 g. Similarly, kernel-to-fruit mass ratios were higher in the Solomon Islands and Vanuatu than in Fiji. *Terminalia catappa* contained higher concentrations of unsaturated fatty acids than saturated fatty acids. Among the unsaturated fatty acids, oleic acid and linoleic acid were the two most abundant. Kernels incubated at 45 °C exhibited significantly higher hexanal concentrations on day 7 compared to those on day 0 of incubation. This rapid oil oxidation may be associated with high unsaturated fatty acid concentrations in kernels. These findings may have implications for oil shelf life. Our study suggests *T. catappa* trees from the Solomon Islands exhibit desirable kernel characteristics that make them suitable for selection and commercialization. The commercialization of an underutilised forest food tree like *T. catappa* will enhance food and nutrition security for local communities.

Keywords: *Terminalia*; tropical almond; commercialisation; tree selection; post-harvest

1. Introduction

Forests are one of the major sources of food and income for many communities worldwide [1–3]. Forests in the tropics also provide livelihood, environmental and nutritional benefits in rural populations, although tropical trees are often poorly integrated into food systems [4]. There is growing interest in scaling up forest-sourced foods such as wild edible nuts to increase food security [1,3,5]. However, very few culinary nut species have been commercialised, and over 95% of global trade in culinary nuts consists of just five species:

 https://doi.org/10.3390/horticulturae11050540

almond (*Prunus amygdalus* Batch.), walnut (*Juglans regia* L.), cashew (*Anacardium occidentale* L.), hazelnut (*Corylus avellana* L.) and pistachio (*Pistacia vera* L.) [6]. The world market for processed nuts has almost doubled in the last decade driven by consumers' growing preference for healthy snacks [6]. New indigenous tree nut species offer the potential to improve livelihoods when traded commercially and also to increase food security in rural populations.

Canarium (*Canarium indicum* L.) and Brazil nuts (*Bertholletia excelsa* Bonpl.) are two indigenous tree nuts that have been successfully commercialised [7–9]. Nuts from *Canarium indicum* are the basis of an emerging industry that has increased income for small-scale farmers, especially women in Pacific countries [10]. Brazil nut provides approximately 40% of household income in the Amazon basin [11]. Both canarium and Brazil nuts are sourced from forests and hence help to maintain forest cover and biodiversity while producing income for smallholders [8,11]. Thus, commercialising indigenous tree nut species from forests has the potential to empower small-scale farmers in developing countries and maintain the biodiversity of forests [12,13].

Key considerations for commercialising new nuts species are kernel size and kernel recovery [14–16]. Kernel recovery is calculated based on the kernel recovered from the weight of the intact nut including the shell (nut in shell) [17]. Kernel-to-fruit ratio can also be used as an alternative for kernel recovery in tree selection if the kernel is extracted from the fresh fruit, rather than from the nut in shell. High kernel mass and high kernel recovery or kernel-to-fruit ratio lead to high financial returns for processors, growers and smallholder farmers [15]. Premium prices are also usually paid for large kernels, which leads to additional income for both growers and processors [15]. Kernel mass can be influenced by various factors including soil nutrient availability, management practices and environmental factors [15,17,18]. In addition, kernels are often larger close to villages, compared with those in the forests, as indigenous people have selected and domesticated nut species over centuries [19]. As a result, there can be large differences in kernel mass and kernel-to-fruit ratio throughout the geographic range of a species [14,15]. In general, trees with large kernel mass or kernel-to-fruit ratio need to be identified and propagated to commercialise a new nut crop [14,20].

Tree nuts are particularly rich in essential minerals, high in protein and contain beneficial fatty acids, antioxidants and some vitamins (i.e., A and E) [5,9,21]. Both saturated and unsaturated fatty acids are beneficial [22]. Unsaturated fatty acids, however, are more susceptible to oxidation than saturated fatty acids [5,22,23]. High unsaturated lipid concentrations may result in decreased oil stability against oxidation [23]. Lipid oxidation leads to reduced oil quality for human health and shortens the shelf life of nuts [22]. Lipid oxidation should be minimised while processing and packing nuts. The fatty acid profile can also be associated with different health benefits [21], but there is no information on the fatty acid profile for many indigenous tree nuts. The fatty acid profile is influenced by both environment and genetics [24]. There is a growing interest in including chemical composition as a selection trait in nut breeding programmes [25]. Hence, it is important to explore the fatty acid composition of tree nuts and understand the oil health benefits and oil stability to enable commercialisation.

This study examines the kernel attributes in *Terminalia catappa* L. (Combretaceae) collected from three Pacific countries. *Terminalia catappa* is a monoecious, large tropical tree that inhabits Australia, southeast Asia, Malesia, Polynesia and the Pacific Islands [26]. The tree can grow up to 40 m tall and is characterised by its drupaceous fruits, which can vary between 2.5 and 10 cm in length and approximately 2 and 4 cm in width [27]. The transition of fruit to maturity is distinguished by a change in the exterior fruit flesh colour from green to yellow and then, finally, red when fully mature or ripe [27]. This species produces flattened-obovoid seeds that are edible (hereafter referred to as 'kernels') and encased by an outer protective layer known as testa [27,28]. *Terminalia catappa*, commonly known as tropical almond, or Alite in the Solomon Islands, Natapoa in Vanuatu and Tarvola in Fiji, has been traditionally consumed throughout the Pacific and has recently been flagged for potential commercialisation [14]. In the Pacific, *T. catappa* usually grows within coastal areas where the soil is poor (Figure 1) [27]. This tree can play an important role in stabilising soil, as it establishes in coastal areas where the soil is highly prone to erosion [27]. This tree species commences bearing fruit when it reaches three years of age and it is also a timber tree [27].

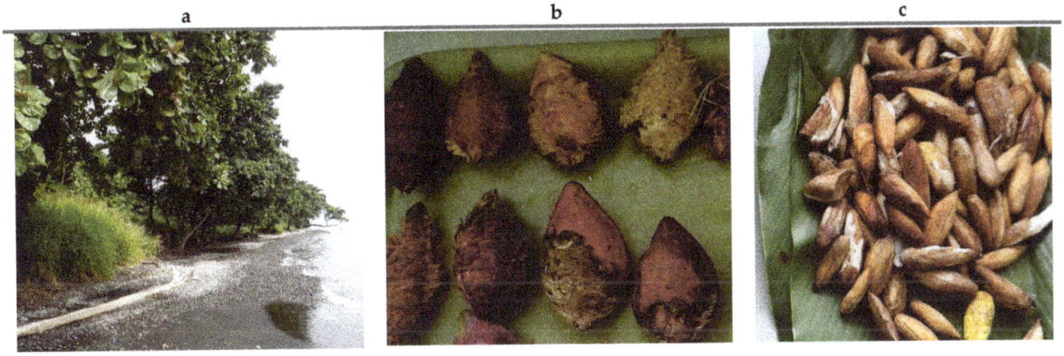

Figure 1. (**a**) *Terminalia catappa* trees occurring in the coastal areas of Solomon Islands, (**b**) ripe fruit and (**c**) kernels of *T. catappa* (photo credit: (**a**,**b**) Bruce Randall and (**c**) Brittany Elliott).

In this study, we aimed to (1) investigate differences in fruit and kernel sizes in trees between countries; (2) determine total oil and fatty acid composition of oil; and (3) examine oil stability and potential for long shelf life using an accelerated ageing trial. It was hypothesised that there would be variations in fruit and kernel size among countries and it was also expected that the fatty acid profile of *T. catappa* would affect the shelf life of kernels.

2. Materials and Methods

2.1. Sample Collection and Preparation

We had 30 fruit replicates randomly collected from the canopy of each tree and 69 tree replicates within countries. In brief, the fruits of *T. catappa* were collected from villages located in the Solomon Islands (18 trees), Vanuatu (21 trees) and Fiji (30 trees) over 24 months between March 2017 and 2019 (Figure 2). The fruiting season of *T. catappa* varies across the Pacific Islands [27]. In lower latitudes, fruiting can be intermittent throughout the year [27]. Fruit production in Fiji and Vanuatu is typically highest between March and June [27]. In this study, the fruit collection aligned with the timing of fruiting; however, there were some occasions where fruit were collected at different times due to differences in fruiting patterns. The weight of individual fresh fruit was recorded using digital scales (PA4101 Pioneer Analytical Balance, OHAUS, Parsippany, NJ, USA). Traditional methods were used to crack fruit samples, where applying the force of stone hammers on either the edge or apex of the fruit revealed the kernels inside [27]. Kernels were then dried for 9 h at 40 °C to approximately 10% moisture content and weighed.

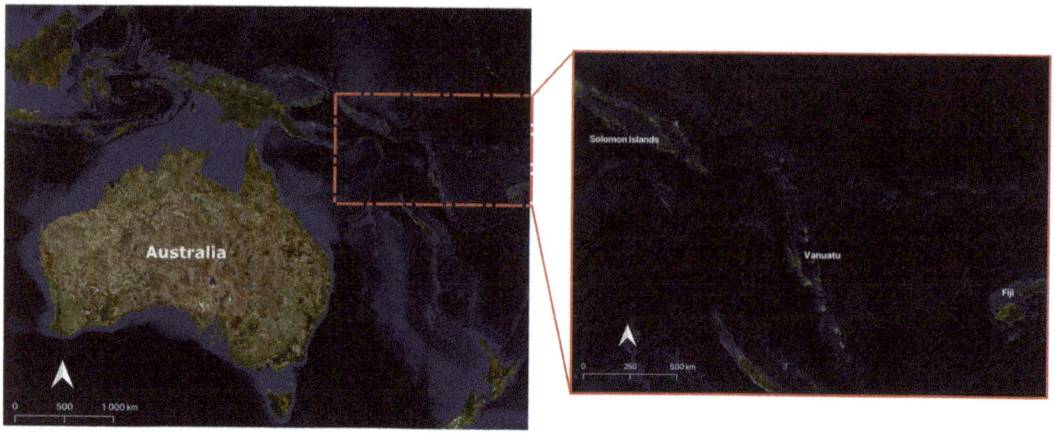

Figure 2. *Terminalia catappa* trees occurring in the coastal areas of Solomon Islands, Vanuatu and Fiji. This map was produced using the free and open-source software, QGIS v 3.36.2, and ESRI base map layer (World Imagery (Clarity) Beta).

2.2. Oil Extraction and Chemical Analyses

The 30 kernels collected from each tree were randomly assigned into 5 replicates per tree, with each replicate containing 6 kernels. Kernels of each replicate were pooled and crushed using a mortar and pestle. The crushed kernels were added to Pentane (45–50 mL) and stirred for 45 min using a magnetic stirrer followed by centrifuging at 2600 rpm for 6 min. The procedure was repeated to ensure kernel oil was fully extracted. The Pentane was then evaporated a Bucci Rotovac (BÜCHI Labortechnik AG, Flawil, Switzerland). The weight of oil was recorded and kept in sealed vials at 4 °C for fatty acid composition analysis. A PerkinElmer Clarus 580 GC coupled to a SQ 8S MS (PerkinElmer, Waltham, MA, USA) was used to measure the fatty acid composition of the oil samples, following the same conditions as described in Bai et al. [5]. In summary, an Elite-5MS (30 m × 0.25 mm × 0.25 μm) column was used, with helium carrier gas at a constant flow of 1.0 mL/min set to a temperature programme of 50 °C for 0.5 min, ramping at 10 °C/min until 300 °C and holding for 1.0 min [5]. Masses ranging from 40 to 400 (m/z) and from 3.1 to 26.5 min at 70 eV were analysed.

2.3. Oil Stability Using Accelerated Ageing

Terminalia catappa fruit were sourced from Vanuatu for this part of our experiment. Approximately 60 fruit were collected from three trees. Kernels were manually extracted from fruits as described above and then dried for 17 h at 45 °C to a moisture content of 1.5%. All kernels were pooled and mixed and then randomly divided into three replicates. Each replicate contained approximately 20 g of the kernel and was sealed in an aluminium foil pouch of 14 cm × 8 cm and incubated at 45 °C for 21 days. Hexanal concentration was measured at day 0 (initial), after 7, 14 and 21 days following the incubation commencement. Approximately 40 mL gas of headspace at each day of sample collection was extracted using a syringe and analysed for hexanal concentration using the e-nose, *OdourScan®*, developed by Next Instruments [29].

2.4. Calculations

The kernel-to-fruit ratio (%) was calculated as (weight of individual kernel/individual fruit) × 100. Oil content was calculated as (mass of oil × 100)/mass of kernel. Total unsaturated fatty acids (TUS) were calculated as the sum of C16:1 cis (palmitoleic acid),

C18:2 (linoleic acid), C18:1 cis (oleic acid), C18:1 trans (elaidic acid) and C20:1 (eicosenoic acid). Total saturated fatty acids (TS) were calculated as the sum of C14:0 (myristic acid), C16:0 (palmitic acid), C18:0 (stearic acid), C20:0 (arachidic acid) and C22:0 (behenic acid).

2.5. Statistical Analyses

All data met the assumptions of normality. We compared differences across countries in kernel attributes including fruit mass, kernel mass, kernel-to-fruit, ratio (%) and total oil and fatty acid composition using a one-way ANOVA followed by Tukey's test at $p < 0.05$. The differences in hexanal concentrations among sampling days were examined using a one-way ANOVA followed by Tukey's test at $p < 0.05$. Finally, two principal component analyses (PCA) were implemented to (a) visualise how fatty acid compositions, including C16:1, C18:2, C18:1 cis, C20:1, C14:0, C16:0, C18:0, C20:0 and C22:0, were distributed based on the countries and (b) show the extent to which kernel characteristics of fruit mass, kernel mass, kernel-to-fruit weight ratio, total oil content, total saturated fatty acids and total unsaturated fatty acids would vary among three countries. We calculated eigenvalues, percentages of variance and cumulative percentages of variance of PCA scores for fatty acid compositions and kernel characteristics. We explored relationships between different fatty acids using Pearson's correlation. We also explored correlations between different kernel characteristics using Pearson's correlation. We explored relationships between fatty acids and kernel characteristics using Pearson's correlation, with a 2-tailed significance test. The SPSS version 24 (IBM Corp, Chicago, IL, USA) was used to perform all data analyses.

3. Results

The average fruit mass was significantly higher in Vanuatu (33.72 g) than in the Solomon Islands (29.21 g) and Fiji (7.97 g) (Table 1). The smallest fruit mass was found in Fiji. Kernel mass was on average 1.66 g in the Solomon Islands, significantly higher than samples from Vanuatu and Fiji (Table 1). The average kernel mass in Vanuatu was also significantly higher than in Fiji (Table 1). The average kernel-to-fruit ratio (%) was greater in both the Solomon Islands and Vanuatu compared with that of Fiji (Table 1).

Table 1. Differences in fruit mass, kernel mass, kernel-to-fruit weight (%), total oil (%), total saturated fatty acid (TS) and total unsaturated fatty acid (TUS) of *Terminalia catappa* at the country level. Different lower-case letters indicate significant differences among countries at $p < 0.05$ (Tukey's HSD).

	Solomon Islands	Vanuatu	Fiji
Fruit mass (g)			
Mean ± SE	29.21 ± 0.52 b	33.72 ± 0.90 a	7.97 ± 0.17 c
N	418	486	746
Minimum	10.17	5.90	2.60
Maximum	67.49	142.19	35.00
Kernel mass (g)			
Mean ± SE	1.66 ± 0.04 a	1.58 ± 0.03 b	0.33 ± 0.01 c
N	416	486	746
Minimum	0.10	0.05	0.01
Maximum	3.93	3.94	0.82
Kernel-to-fruit weight (%)			
Mean ± SE	5.64 ± 0.10 a	5.61 ± 0.13 a	4.75 ± 0.09 b
N	416	486	746
Minimum	0.28	0.35	0.12
Maximum	15.53	15.28	11.13

Table 1. *Cont.*

	Solomon Islands	Vanuatu	Fiji
Total oil (%)			
Mean ± SE	59.37 ± 1.08 a	55.34 ± 1.21 b	56.86 ± 0.30 ab
N	85	105	150
Minimum	46.27	28.21	47.32
Maximum	77.55	84.19	65.56
TS (%)			
Mean ± SE	46.04 ± 0.55 a	44.30 ± 0.37 b	38.60 ± 0.28 c
N	68.00	90.00	150.00
Minimum	36.92	32.86	31.96
Maximum	64.65	52.11	45.99
TUS (%)			
Mean ± SE	54.44 ± 0.52 c	55.70 ± 0.37 b	61.40 ± 0.27 a
N	68.00	90.00	150.00
Minimum	49.33	47.89	54.01
Maximum	68.55	67.14	68.04

Kernels from the Solomon Islands exhibited significantly higher total oil content compared to those from Vanuatu (Table 1). Kernels from Fiji contained significantly higher unsaturated fatty acid concentrations than those from the Solomon Islands and Vanuatu (Table 1). Across all countries, the predominant fatty acids identified were palmitic acid (C16:0), linoleic acid (C18:2) and oleic acid (C18:1Cis) (Table 2).

Table 2. Descriptive statistics (mean ± standard error) of *Terminalia catappa* for fatty acid compositions collected from Solomon Islands (n = 68), Vanuatu (n = 90) and Fiji (n = 150).

	Solomon Islands	Vanuatu	Fiji
C14:0 (myristic)	0.04 ± 0.001	0.03 ± 0.002	0.02 ± 0.001
C16:0 (palmitic)	40.37 ± 0.570	39.86 ± 0.354	34.16 ± 0.255
C18:0 (stearic)	5.22 ± 0.068	4.28 ± 0.079	4.29 ± 0.061
C20:0 (arachidic)	0.35 ± 0.014	0.11 ± 0.003	0.11 ± 0.002
C22:0 (behenic)	0.06 ± 0.004	0.03 ± 0.002	0.03 ± 0.001
C16:1 (palmitoleic)	0.26 ± 0.012	0.08 ± 0.003	0.06 ± 0.001
C18:2 (linoleic)	23.81 ± 0.411	20.31 ± 0.388	20.83 ± 0.337
C18:1 cis (oleic)	29.73 ± 0.485	34.9 ± 0.343	40.15 ± 0.326
C18:1 trans (elaidic)	0.61 ± 0.015	0.37 ± 0.034	0.33 ± 0.006
C20:1 (eicosenoic)	0.04 ± 0.003	0.03 ± 0.002	0.03 ± 0.001

We identified a strong negative correlation between the two predominant fatty acid concentrations found in *T. catappa* kernels: palmitic acid (C16:0) and linoleic acid (C18:2) ($r = -0.76$) (Supplementary Table S1). Positive correlations were found between fruit mass and kernel mass ($r = 0.70$) and between total saturated fatty acids (TS%) and total unsaturated fatty acids (TUS%) ($r = 0.50$) (Supplementary Table S2). Notably, palmitic acid (C16:0) exhibited significant positive correlations with kernel characteristics including kernel mass ($r = 0.28$) and kernel-to-fruit weight ($r = 0.28$) (Supplementary Table S3). In contrast, linoleic acid (C18:2) was negatively correlated with kernel mass ($r = -0.43$) and kernel-to-fruit weight ($r = -0.44$).

We found no distinct clustering in the fatty acid compositions of the countries, with PCA1 explaining 35.4% of the total variance and PCA2 explaining 57.7% of the total variance (Figure 3a). Two principal components (i.e., PC_1 and PC_2) with eigenvalues of 4.85 and 1.52, respectively, were extracted (a in Table 3). Palmitoleic acid (C16:1), arachidic acid

(C20:0), myristic acid (C14:0) and behenic acid (C22:0) had strong positive loadings of 0.93, 0.90 and 0.81, respectively, on PCA1 (Table 3). The predominant fatty acids in *T. catappa* kernels—oleic acid (C18:1 *cis*), palmitic acid (C16:0) and linoleic acid (C18:2)—had strong negative loadings of −0.66, −0.83 and −0.75 in PCA 1, PCA2 and PCA3, respectively (a in Table 3). Kernel characteristics between the Solomon Islands and Fiji did not overlap and the two principal components of PCA1 and PCA2 explained 45.4% and 21.8% of total variance, respectively (Figure 3b). Two principal components (i.e., PCA1 and PCA2) with eigenvalues of 2.85 and 1.16, respectively, were extracted (b in Table 3). Kernel mass, total saturated fatty acids and fruit mass had strong positive loadings of 0.82, 0.56 and 0.72 on PCA1. Kernel-to-fruit weight had a positive loading of 0.91 on PCA2, and total oil had a positive loading of 0.86 on PCA3 (b in Table 3).

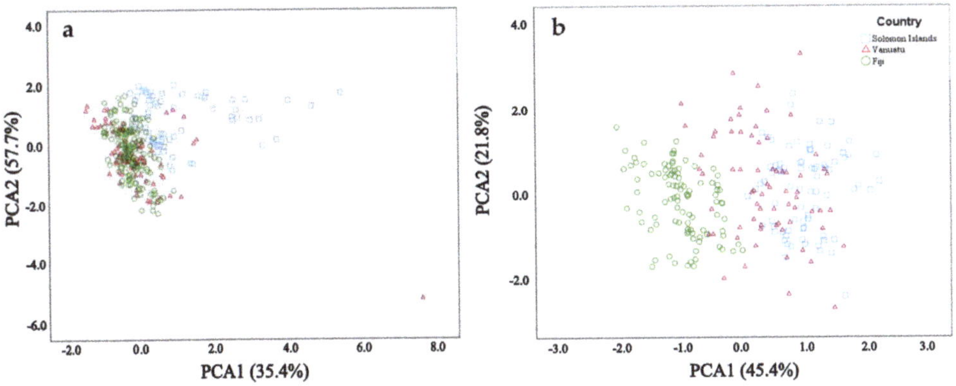

Figure 3. Principal component analyses (PCA) showing the clustering of *Terminalia catappa* based on (**a**) fatty acid composition and (**b**) kernel characteristics per country of origin: Solomon Islands (open blue rectangular), Vanuatu (open red triangular) and Fiji (open green circle).

Table 3. Principal component analyses (PCA) values of *Terminalia catappa*: (**a**) fatty acid composition and (**b**) kernel characteristics.

	(a) Fatty Acid Compositions			(b) Kernel Characteristics *		
	PCA1	PCA2	PCA3	PCA1	PCA2	PCA3
Eigenvalues	4.85	1.52	1.18	2.85	1.16	1.07
Percentages of variance	35.44	22.32	17.87	45.49	21.49	18.10
Cumulative percentages of variance	35.44	57.77	75.64	45.49	66.98	85.09
C16:1 (palmitoleic)	0.90	0.03	0.01			
C20:0 (arachidic)	0.90	0.11	−0.04			
C14:0 (myristic)	0.81	−0.11	0.18			
C22:0 (behenic)	0.81	0.34	0.15			
C18:1 cis (oleic)	−0.67	0.63	0.27			
C18:1 trans (elaidic)	0.65	0.15	0.41			
C20:1 (eicosenoic)	0.60	0.50	0.21			
C18:0 (stearic)	0.58	−0.14	−0.34			
C16:0 (palmitic)	0.31	−0.83	0.40			
C18:2 (linoleic)	0.51	0.11	−0.75			
Fruit mass (g)				0.72	−0.19	0.39
Kernel mass (g)				0.82	0.395	0.19
Kernel-to-fruit weight (%)				0.26	0.91	−0.27
Total oil (%)				−0.02	0.18	0.86
TS (%)				0.89	−0.24	−0.20
TUS (%)				−0.89	0.23	0.20

* Kernel characteristics included fruit mass, kernel mass, kernel-to-fruit weight ratio, total oil content, total saturated fatty acids and total unsaturated fatty acids.

Hexanal concentrations initially measured at 30.2 ppm increased to 142.9 ppm after 21 days of incubation (Table 4). Moreover, hexanal concentrations on day 7 were significantly higher than on day 0 (Table 4).

Table 4. Hexanal concentrations (mean ± standard error) of *Terminalia catappa* over 21 days of incubation at 45 °C. Different lower-case letters indicate significant differences between incubation days at $p < 0.05$ (Tukey's HSD).

	Incubation Period (Days)			
	Day 0	**Day 7**	**Day 14**	**Day 21**
Hexanal Concentration (ppm)	30.20 ± 6.67 c	97.59 ± 5.67 b	132.27 ± 9.17 ab	142.99 ± 12.51 a

4. Discussion

There were significant differences in kernel mass among the three countries. Kernel mass is considered one of the more important factors in tree selection [13,30,31]. Nuts with larger kernels increase financial returns to farmers and processors when commercialised [13,32]. We found an 80% difference between the weight of the largest (Solomon Islands) and smallest (Fiji) kernels. Therefore, farmers from Fiji would need to process five nuts more for every nut collected in the Solomon Islands to obtain the same kernel return. *Canarium indicum* has been recently commercialised in the Pacific [10,33]. The recommended kernel mass for *C. indicum* is at least 3 g to make this nut economically viable [15]. The average kernel mass of *T. catappa* was less than 3 g, although some were very close to 3 g. Hence, the commercialisation of nuts should focus on selecting trees with large kernels, as processing fruits with small kernels is not likely to be economically viable [10,15].

Kernel-to-fruit mass of *T. catappa* also differed significantly among the three countries. In general, high kernel recovery (kernel mass to nut-in-shell mass) or kernel-to-fruit mass is sought for tree selection [31,34]. These characteristics may be weighted for tree selection but other traits are also important for nut selection. For example, some macadamia nuts with high kernel recovery have thin shells and, hence, are susceptible to pests [31,34]. Soil nutrient availability is another factor in kernel recovery and yield [35]. In macadamia, boron spray has increased kernel recovery and yield [35]. Therefore, it is important to understand pest tolerances and nutrient needs of trees that have high kernel recovery values before tree selection is undertaken.

We observed rapid oil oxidation in *T. catappa*, perhaps because of the higher total unsaturated fatty acid than saturated fatty acid concentrations [5,22,23]. The concentration of oleic acid and linoleic acid is another driving factor affecting oil stability. Oleic acid is a monounsaturated fatty acid, but linoleic acid is a polyunsaturated fatty acid. Polyunsaturated fatty acids possess higher double-bond carbons in the molecular structure than those of monosaturated fatty acids and hence are less stable against oxidation [22,36]. Therefore, higher oleic acid concentrations than linoleic acid concentrations can increase oil stability [22,23]. Oleic acid concentrations were higher than linoleic acid concentrations in *T. catappa*. We had sourced kernels for the accelerated ageing from Vanuatu where oleic acid concentrations varied between 27% and 41%. However, we also observed the highest oleic acid concentrations in Fiji (at 50%) and the smallest oleic acid concentrations in the Solomon Islands (at 22%). Hence, our observation of rapid oil oxidation might imply a shortened shelf life of this nut under elevated temperatures. In the current study, the fast oxidation occurred under high temperatures; hence, low storage temperatures may prolong the shelf life of the oil.

The fatty acid composition of the *T. catappa* was very close to other tropical nuts including *Barringtonia procera* (Miers) R. Knuth and *C. indicum* [5,9,37]. Our results imply that *T. catappa* contains oil with beneficial fatty acids when consumed as food [5]. *Terminalia catappa* is rich in oleic acid and linoleic acid, and both of these unsaturated fatty acids are known to confer health benefits in other culinary nuts [5,38]. These fatty acids are known to be protective against several cardiovascular diseases [39], which are an increasingly significant public health concern in Pacific Island nations [40]. It should also be noted that *T. catappa* had higher palmitic acid concentrations than the majority of culinary nuts [5]. Palmitic acid has been found to be a dominant saturated fatty acid in another tropical nut, *C. indicum* [5]. Culinary nuts, which have beneficial fatty acids and antioxidants, may offer potential benefits in mitigating the risk of cardiovascular disease when consumed instead of processed alternatives [41].

Variations in the physical and chemical properties of *T. catappa* nuts were found. Fruit mass showed the highest variation in Vanuatu. Interestingly, the large variation in fruit mass was not translated into the kernel mass variation because there were no significant differences in average kernel mass between the Solomon Islands and Vanuatu where their stand errors were overlapping. Many factors drive interpopulation variability in plants [42]. Generally, there are common factors that drive the kernel variations in interpopulation such as the origin of the nuts, environmental conditions, pollination, soil properties, nutrient availability and genetics [14,17,43–45]. In a related study on this species, there was significant variability between trees within each country in kernel mass [14]. Variation in kernel properties is common in tree nuts and is often reported between trees of different genotypes, and even within a single genotype [14,15]. For example, self-pollinated macadamia nuts are smaller than cross-pollinated nuts from the same maternal genotype, and self- and cross-pollinated nuts have different chemical compositions [46]. All three countries were also in locations that did not share similar tropical climatic conditions. As a result, a lack of seasonal alignment on some occasions existed in our dataset, which might partly play a part in increasing variation among countries. It should also be noted that we found that some characteristics, such as fatty acid compositions, did not significantly vary from one country to another. It is very common in the Pacific countries that a plant like *T. catappa* has been intensively selected for domestication by villagers and seeds may have been transferred from one country to another over centuries [27]. For example, this species has even been naturalised in countries such as Brazil and those in the Caribbean and East Africa [27]. Similarities found between countries in terms of fatty acid composition could be the result of selection.

5. Conclusions

This study identified trees from the Solomon Islands have the largest kernels and kernel-to-fruit ratio. As a result, trees from the Solomon Islands could be selected for tree breeding and commercialisation. In all locations, *T. catappa* had higher total unsaturated fatty acid concentrations than saturated fatty acid concentrations, which might facilitate rapid oil oxidation. However, high total unsaturated fatty acid concentrations indicate potential health benefits for this tree nut. Our study suggests *T. catappa* could improve food security and provide health benefits to local communities, and select populations with large kernels and high kernel-to-fruit ratios could be targeted for commercialisation.

Supplementary Materials: The following supporting information can be downloaded at: https://www.mdpi.com/article/10.3390/horticulturae11050540/s1, Table S1: Correlation matrix of fatty acids present in *Terminalia catappa* kernels; Table S2: Correlation matrix for the variables of fruit mass, kernel mass, kernel-to-fruit weight (%), total oil (%), total saturated fatty acid (TS) and total unsaturated fatty acid (TUS) of *Terminalia catappa*; Table S3: Correlation matrix between variables of fruit mass, kernel mass, kernel-to-fruit weight (%) and fatty acids present in *Terminalia catappa* kernels (n = 275); Tables S4 and S5: Supplementary Data for Tables.

Author Contributions: Conceptualisation, S.H.B. and H.M.W.; methodology, S.H.B.; validation, S.H.B., B.R., R.G., B.G., D.K., K.J., B.E. and H.M.W.; formal analysis, S.H.B.; data curation, S.H.B., B.R., R.G., B.G., D.K., K.J. and H.M.W.; writing—original draft preparation, S.H.B.; writing—review and editing, S.H.B., B.R., R.G., B.G., D.K., K.J., B.E. and H.M.W.; visualisation, S.H.B. and B.E.; supervision, S.H.B.; funding acquisition, S.H.B. and H.M.W. All authors have read and agreed to the published version of the manuscript.

Funding: This research was funded by the Australian Centre for International Agricultural Research, project numbers: FST-2014-067, FST-2017-038 and FST-2022-124.

Data Availability Statement: The original contributions presented in this study are included in the article/Supplementary Materials. Further inquiries can be directed to the corresponding author(s).

Acknowledgments: This work was only possible with the support provided in-country to collect samples by the Secretariat of the South Pacific (Fiji), Lapita Café and Department of Forestry (Vanuatu), SolFresh and Beg's Sol Seed Technology and Supplies (Solomon Islands). We thank Peter Brooks for his help in the analysis of the oil samples for FAME. We acknowledge John Oakeshott and Paitia Negalevu of the Secretariat of Pacific Community, Suva, Fiji.

Conflicts of Interest: The authors declare no conflicts of interest. The funders had no role in the design of the study; in the collection, analyses or interpretation of data; in the writing of the manuscript; or in the decision to publish the results.

References

1. Dao, T.H.H.; Hölscher, D. Impact of Non-Timber Forest Product Use on the Tree Community in North-Western Vietnam. *Forests* **2018**, *9*, 431. [CrossRef]
2. Epanda, M.A.; Tsafack Donkeng, R.; Ngo Nonga, F.; Frynta, D.; Adi, N.N.; Willie, J.; Speelman, S. Contribution of Non-Timber Forest Product Valorisation to the Livelihood Assets of Local People in the Northern Periphery of the Dja Faunal Reserve, East Cameroon. *Forests* **2020**, *11*, 1019. [CrossRef]
3. Sardeshpande, M.; Shackleton, C. Wild Edible Fruits: A Systematic Review of an Under-Researched Multifunctional NTFP (Non-Timber Forest Product). *Forests* **2019**, *10*, 467. [CrossRef]
4. Jansen, M.; Guariguata, M.R.; Raneri, J.E.; Ickowitz, A.; Chiriboga-Arroyo, F.; Quaedvlieg, J.; Kettle, C.J. Food for thought: The underutilized potential of tropical tree-sourced foods for 21st century sustainable food systems. *People Nat.* **2020**, *2*, 1006–1020. [CrossRef]
5. Bai, S.H.; Brooks, P.; Gama, R.; Nevenimo, T.; Hannet, G.; Hannet, D.; Randall, B.; Walton, D.; Grant, E.; Wallace, H.M. Nutritional quality of almond, canarium, cashew and pistachio and their oil photooxidative stability. *J. Food Sci. Technol.* **2019**, *56*, 792–798. [CrossRef]
6. International Nut & Dried Fruit Council. World Nut and Dried Fruit Trade Map. Available online: https://inc.nutfruit.org/inc-publishes-trade-maps-series/ (accessed on 21 October 2024).
7. Wallace, H.; Randall, B.; Grant, E.; Jones, K.; Walton, D.; Poienou, M.; Nevenimo, T.; Moxon, J.; Pauku, R.L. Processing methods for Canarium nuts in the Pacific. *Acta Hortic.* **2016**, *1128*, 145–149. [CrossRef]
8. Wallace, H.; Komolong, B.; Nevenimo, T.; Waaii, C.; Hannett, D.; Hannett, G.; Kapi Ling, S.; Grant, E.; Hodges, B.; Kill, E.; et al. *Enhancing Private Sector-Led Development of the Canarium Industry in PNG: ACIAR Project FST/2014/099 Final Report*; Australian Centre for International Agricultural Research (ACIAR): Canberra, Australia, 2019; pp. 1–135. Available online: https://www.aciar.gov.au/publication/technical-publications/enhancing-private-sector-led-development-canarium-nut-industry-papua-new-guinea-final (accessed on 21 October 2024).
9. Hosseini Bai, S.; Darby, I.; Nevenimo, T.; Hannet, G.; Hannet, D.; Poienou, M.; Grant, E.; Brooks, P.; Walton, D.; Randall, B.; et al. Effects of roasting on kernel peroxide value, free fatty acid, fatty acid composition and crude protein content. *PLoS ONE* **2017**, *12*, e0184279. [CrossRef]
10. Wallace, H.M.; Hannet, D.; Hannet, G.; Hosseini-Bai, S.; Jones, K.; Komolong, B. Commercialising an indigenous agroforestry tree: Overview of commercial processing methods for *Canarium indicum* (galip) nuts in Papua New Guinea. *Acta Hortic.* **2022**, *1355*, 345–350. [CrossRef]
11. Soriano, M.; Mohren, F.; Ascarrunz, N.; Dressler, W.; Peña-Claros, M. Socio-ecological costs of Amazon nut and timber production at community household forests in the Bolivian Amazon. *PLoS ONE* **2017**, *12*, e0170594. [CrossRef]

12. Leakey, R.R. Converting 'trade-offs' to 'trade-ons' for greatly enhanced food security in Africa: Multiple environmental, economic and social benefits from 'socially modified crops'. *Food Secur.* **2018**, *10*, 505–524. [CrossRef]
13. Randall, B.W.; Walton, D.A.; Grant, E.L.; Zekele, P.; Gua, B.; Pauku, R.; Wallace, H.M. Selection of the tropical nut *Canarium indicum* for early fruiting, nut-in-shell size and kernel size. *Acta Hortic.* **2016**, *1109*, 169–174. [CrossRef]
14. Bai, S.H.; Randall, B.; Grant, E.; Gama, R.; Gua, B.; Keli, D.; Negalevu, P.; Oakeshott, J.; Wallace, H.M. Tree-to-tree variation of kernel size in two underutilized tree nuts in Pacific. *Acta Hortic.* **2022**, *1340*, 141–144. [CrossRef]
15. Leakey, R.; Fuller, S.; Treloar, T.; Stevenson, L.; Hunter, D.; Nevenimo, T.; Binifa, J.; Moxon, J. Characterization of tree-to-tree variation in morphological, nutritional and medicinal properties of *Canarium indicum* nuts. *Agrofor. Syst.* **2008**, *73*, 77–87. [CrossRef]
16. Pauku, R.; Lowe, A.; Rrb, L. Domestication of indigenous fruit and nut trees for agroforestry in the Solomon Islands. *For. Trees Livelihoods* **2010**, *19*, 269–287. [CrossRef]
17. Richards, T.E.; Kämper, W.; Trueman, S.J.; Wallace, H.M.; Ogbourne, S.M.; Brooks, P.R.; Nichols, J.; Hosseini Bai, S. Relationships between nut size, kernel quality, nutritional composition and levels of outcrossing in three macadamia cultivars. *Plants* **2020**, *9*, 228. [CrossRef]
18. Leakey, R.; Shackleton, S.; Plessis, P.D. Domestication potential of Marula (*Sclerocarya birrea* subsp *caffra*) in South Africa and Namibia: 1. Phenotypic variation in fruit traits. *Agrofor. Syst.* **2005**, *64*, 25–35. [CrossRef]
19. Levis, C.; Flores, B.M.; Moreira, P.A.; Luize, B.G.; Alves, R.P.; Franco-Moraes, J.; Lins, J.; Konings, E.; Peña-Claros, M.; Bongers, F.; et al. How People Domesticated Amazonian Forests. *Front. Ecol. Evol.* **2018**, *5*, 171. [CrossRef]
20. Leakey, R.R.B.; Page, T. The 'ideotype concept' and its application to the selection of cultivars of trees providing agroforestry tree products. *For. Trees Livelihoods* **2006**, *16*, 5–16. [CrossRef]
21. Ros, E. Health benefits of nut consumption. *Nutrients* **2010**, *2*, 652–682. [CrossRef]
22. Gama, T.; Wallace, H.; Trueman, S.; Hosseini Bai, S. Quality and shelf life of tree nuts: A review. *Sci. Hortic.* **2018**, *242*, 116–126. [CrossRef]
23. Miraliakbari, H.; Shahidi, F. Oxidative stability of tree nut oils. *J. Agric. Food Chem.* **2008**, *56*, 4751–4759. [CrossRef]
24. Gonçalves, B.; Pinto, T.; Aires, A.; Morais, M.C.; Bacelar, E.; Anjos, R.; Ferreira-Cardoso, J.; Oliveira, I.; Vilela, A.; Cosme, F. Composition of Nuts and Their Potential Health Benefits-An Overview. *Foods* **2023**, *12*, 942. [CrossRef]
25. Kodad, O.; Socias i Company, R.; Prats, M.S.; LÓpez Ortiz, M.C. Variability in tocopherol concentrations in almond oil and its use as a selection criterion in almond breeding. *J. Hortic. Sci. Biotech.* **2006**, *81*, 501–507. [CrossRef]
26. Pedley, L.; Kodela, P.G. *Terminalia catappa*. Available online: https://profiles.ala.org.au/opus/foa/profile/Terminalia%20catappa (accessed on 9 April 2025).
27. Thomson, L.A.; Evans, B. *Terminalia catappa* (tropical almond). *Species Profiles Pac. Isl. Agrofor.* **2006**, *2*, 1–20.
28. Smith, A.C. Studies of Pacific Island Plants, XXIV. The genus *Terminalia* (Combretaceae) in Fiji, Samoa, and Tonga. *Brittonia* **1971**, *23*, 394–412. [CrossRef]
29. Hosseini Bai, S.; Trueman, S.J.; Gama, T.; Jones, K.; Walton, D.; Randall, B.; Wallace, H.M. Shelf life of macadamia kernels of different origin. *Acta Hortic.* **2019**, *1256*, 375–378. [CrossRef]
30. Blaikie, S.J.; O'Farrell, P.J.; Chacko, E.K.; Müller, W.J.; Wei, X.; Steele Scott, N.; Sykes, S.R. Assessment and selection of new hybrids from the Australian cashew breeding program. *Aust. J. Exp. Agric.* **2002**, *42*, 615–623. [CrossRef]
31. Walton, D.A.; Wallace, H. Genetic and postharvest factors affecting macadamia kernel quality. *Afri J. Agric. Res.* **2012**, *7*, 2490–2495. [CrossRef]
32. Ogunsina, B.S.; Bamgboye, A.I. Pre-shelling parameters and conditions that influence the whole kernel out-turn of steam-boiled cashew nuts. *J. Saudi Soc. Agric. Sci.* **2014**, *13*, 29–34. [CrossRef]
33. Hosseini Bai, S.; Gama, R.; Jones, K.; Hannet, D.; Hannet, G.; Komolong, B.; Brooks, P.; Grant, E.; Elliott, B.; Wallace, H.M. Presence of Testa and Shell Maintains Oil Stability in Almond and Canarium Nuts. *Horticulturae* **2023**, *9*, 1003. [CrossRef]
34. O'Connor, K.; Hayes, B.; Topp, B. Prospects for increasing yield in macadamia using component traits and genomics. *Tree Genet. Genomes* **2018**, *14*, 7. [CrossRef]
35. Stephenson, R. Macadamia: Domestication and commercialization. *Chr. Hortic.* **2005**, *45*, 11–15.
36. Amaral, J.S.; Casal, S.; Seabra, R.M.; Oliveira, B.P. Effects of roasting on hazelnut lipids. *J. Agric. Food Chem.* **2006**, *54*, 1315–1321. [CrossRef]
37. Hosseini Bai, S.; Nevenimo, T.; Hannet, G.; Hannet, D.; Jones, K.; Trueman, S.; Grant, E.; Walton, D.A.; Randall, B.; Wallace, H.M. Freezing, roasting and salt dipping impacts on peroxide value, free fatty acid and fatty acid concentrations of nut kernels. *Acta Hortic.* **2019**, *1256*, 71–76. [CrossRef]
38. Griffin, L.; Dean, L. Nutrient Composition of Raw, Dry-Roasted, and Skin-On Cashew Nuts. *J. Food Res.* **2024**, *6*, 13. [CrossRef]
39. Alasalvar, C.; Salvadó, J.-S.; Ros, E. Bioactives and health benefits of nuts and dried fruits. *Food Chem.* **2020**, *314*, 126192. [CrossRef]
40. Winter-Smith, J.; Selak, V.; Harwood, M.; Ameratunga, S.; Grey, C. Cardiovascular disease and its management among Pacific people: A systematic review by ethnicity and place of birth. *BMC Cardiov. Disord.* **2021**, *21*, 515. [CrossRef]

41. Glenn, A.J.; Aune, D.; Freisling, H.; Mohammadifard, N.; Kendall, C.W.C.; Salas-Salvadó, J.; Jenkins, D.J.A.; Hu, F.B.; Sievenpiper, J.L. Nuts and Cardiovascular Disease Outcomes: A Review of the Evidence and Future Directions. *Nutrients* **2023**, *15*, 911. [CrossRef]
42. Vidaković, A.; Radunić, M.; Poljak, I. Variation in chemical composition and fruit morphometric traits of almond-leaved pear (*Pyrus spinosa* Forssk.) natural populations. *Genet. Resour. Crop Evol.* **2025**, *72*, 1495–1510. [CrossRef]
43. Gama, T.; Wallace, H.M.; Trueman, S.; Hosseini Bai, S. Variability in crude protein and mineral nutrient concentrations of almonds. *Acta Hortic.* **2018**, *1219*, 213–218. [CrossRef]
44. Kucukyumuk, Z.; Erdal, I. Rootstock and cultivar effect on mineral nutrition, seasonal nutrient variation and correlations among leaf, flower and fruit nutrient concentrations in apple trees. *Bulg. J. Agric. Sci.* **2011**, *17*, 633–641.
45. Sathe, S.K.; Seeram, N.P.; Kshirsagar, H.H.; Heber, D.; Lapsley, K.A. Fatty Acid Composition of California Grown Almonds. *J. Food Sci.* **2008**, *73*, C607–C614. [CrossRef] [PubMed]
46. Trueman, S.J.; Penter, M.G.; Malagodi-Braga, K.S.; Nichols, J.; De Silva, A.L.; Ramos, A.T.M.; Moriya, L.M.; Ogbourne, S.M.; Hawkes, D.; Peters, T.; et al. High Outcrossing Levels among Global Macadamia Cultivars: Implications for Nut Quality, Orchard Designs and Pollinator Management. *Horticulturae* **2024**, *10*, 203. [CrossRef]

MDPI AG
Grosspeteranlage 5
4052 Basel
Switzerland
Tel.: +41 61 683 77 34

Horticulturae Editorial Office
E-mail: horticulturae@mdpi.com
www.mdpi.com/journal/horticulturae

MDPI

www.ingramcontent.com/pod-product-compliance
Lightning Source LLC
LaVergne TN
LVHW070454061126
851858LV00011B/238
9783725881932